VOLUME ONE HUNDRED AND SEVENTY THREE

PROGRESS IN MOLECULAR BIOLOGY AND TRANSLATIONAL SCIENCE

Autism

VOLUME ONE HUNDRED AND SEVENTY THREE

PROGRESS IN MOLECULAR BIOLOGY AND TRANSLATIONAL SCIENCE

Autism

Edited by

MIROLYUBA ILIEVA

Psychiatry in the Region of Southern Denmark, Odense University Hospital; Department of Psychiatry, Department of Clinical Research; BRIDGE—Brain Research—Inter-Disciplinary Guided Excellence, Department of Clinical Research, University of Southern Denmark, Odense, Denmark

WAY KWOK-WAI LAU

Department of Special Education and Counselling, The Education University of Hong Kong, Hong Kong, China

ELSEVIER

ACADEMIC PRESS

An imprint of Elsevier

Academic Press is an imprint of Elsevier
50 Hampshire Street, 5th Floor, Cambridge, MA 02139, United States
525 B Street, Suite 1650, San Diego, CA 92101, United States
The Boulevard, Langford Lane, Kidlington, Oxford OX5 1GB, United Kingdom
125 London Wall, London EC2Y 5AS, United Kingdom

First edition 2020

Notices
Knowledge and best practice in this field are constantly changing. As new research and experience broaden our understanding, changes in research methods, professional practices, or medical treatment may become necessary.

Practitioners and researchers must always rely on their own experience and knowledge in evaluating and using any information, methods, compounds, or experiments described herein. In using such information or methods they should be mindful of their own safety and the safety of others, including parties for whom they have a professional responsibility.

To the fullest extent of the law, neither the Publisher nor the authors, contributors, or editors, assume any liability for any injury and/or damage to persons or property as a matter of products liability, negligence or otherwise, or from any use or operation of any methods, products, instructions, or ideas contained in the material herein.

ISBN: 978-0-12-821242-4
ISSN: 1877-1173

For information on all Academic Press publications
visit our website at https://www.elsevier.com/books-and-journals

Publisher: Zoe Kruze
Acquisitions Editor: Ashlie M. Jackman
Editorial Project Manager: Leticia Lima
Production Project Manager: James Selvam
Cover Designer: Matt Limbert

Typeset by SPi Global, India

Contents

Contributors

Megumi Andoh
Laboratory of Chemical Pharmacology, Graduate School of Pharmaceutical Sciences, The University of Tokyo, Bunkyo-ku, Tokyo, Japan

Melika Babadi
School of Interdisciplinary Science, McMaster University, Hamilton, ON, Canada

Elizabeth Baker
University of California, Graduate School of Education, Riverside, CA, United States

Luigi Bellocchio
CNRS, INCIA, UMR 5287, Bat B2, Pessac Cedex; INSERM, U1215 NeuroCentre Magendie, Bordeaux Cedex, France

Inés Bouzón-Arnáiz
Institute for Bioengineering of Catalonia (IBEC), The Barcelona Institute of Science and Technology; Barcelona Institute for Global Health (ISGlobal, Hospital Clínic—Universitat de Barcelona), Barcelona, Spain

Oswald Chen
Division of Translational Neuroscience, Department of Clinical Neurosciences, Hotchkiss Brain Institute, Alberta Children's Hospital Research Institute, Cumming School of Medicine, University of Calgary, Calgary, AB, Canada

Phoebe Pui Pui Cheung
Department of Rehabilitation Sciences, Hong Kong Polytechnic University, Hung Hom, Hong Kong

Taylor Chomiak
Division of Translational Neuroscience, Department of Clinical Neurosciences, Hotchkiss Brain Institute, Alberta Children's Hospital Research Institute, Cumming School of Medicine, University of Calgary, Calgary, AB, Canada

Hannah J. Ferrara
Division of Translational Neuroscience, Department of Clinical Neurosciences, Hotchkiss Brain Institute, Alberta Children's Hospital Research Institute, Cumming School of Medicine, University of Calgary, Calgary, AB, Canada

Xiujuan Geng
Brain and Mind Institute, The Chinese University of Hong Kong, Shatin, Hong Kong

Sonja Simone Hohmann
Psychiatry in the Region of Southern Denmark, Odense University Hospital, Odense, Denmark

Bin Hu
Division of Translational Neuroscience, Department of Clinical Neurosciences, Hotchkiss Brain Institute, Alberta Children's Hospital Research Institute, Cumming School of Medicine, University of Calgary, Calgary, AB, Canada

Yuji Ikegaya
Laboratory of Chemical Pharmacology, Graduate School of Pharmaceutical Sciences, The University of Tokyo, Bunkyo-ku, Tokyo, Japan

Mirolyuba Ilieva
Psychiatry in the Region of Southern Denmark, Odense University Hospital; Department of Psychiatry; BRIDGE—Brain Research—Inter-Disciplinary Guided Excellence, Department of Clinical Research, University of Southern Denmark, Odense, Denmark

Xin Kang
Brain and Mind Institute; Department of Linguistics and Modern Languages, The Chinese University of Hong Kong, Shatin, Hong Kong

Ryuta Koyama
Laboratory of Chemical Pharmacology, Graduate School of Pharmaceutical Sciences, The University of Tokyo, Bunkyo-ku, Tokyo, Japan

Benson Wui Man Lau
Department of Rehabilitation Sciences, Hong Kong Polytechnic University, Hung Hom, Hong Kong

Way Kwok-Wai Lau
Department of Special Education and Counselling, The Education University of Hong Kong, Hong Kong, China

Mei-Kei Leung
Department of Special Education and Counselling, The Education University of Hong Kong, Hong Kong, China

Tanja Maria Michel
Psychiatry in the Region of Southern Denmark, Odense University Hospital; Department of Psychiatry; BRIDGE—Brain Research—Inter-Disciplinary Guided Excellence, Department of Clinical Research, University of Southern Denmark, Odense, Denmark

Hasti Namvarhaghighi
School of Interdisciplinary Science, McMaster University, Hamilton, ON, Canada

Dong Ik Park
Danish Research Institute of Translational Neuroscience (DANDRITE)-Nordic EMBL Partnership for Molecular Medicine; The Danish National Research Foundation Center, PROMEMO, Department of Biomedicine, Aarhus University, Aarhus, Denmark

Susanna Pietropaolo
University of Bordeaux, Bordeaux Cedex; CNRS, INCIA, UMR 5287, Bat B2, Pessac Cedex, France

Florence I. Roullet
Department of Psychiatry and Behavioural Neurosciences, McMaster University, Hamilton, ON, Canada

Roheeni Saxena
Department of Environmental Health Sciences, Mailman School of Public Health, Columbia University, New York, NY, United States

Caterina Scuderi
Department of Physiology and Pharmacology "Vittorio Erspamer", SAPIENZA University of Rome, Rome, Italy

Katherine K.M. Stavropoulos
University of California, Graduate School of Education, Riverside, CA, United States

Gary Steinman
Visiting Researcher, Department of Obstetrics & Gynecology, Hadassah Hospital—Hebrew University, Ein Kerem, Jerusalem, Israel

Isabelle Tahmazian
Division of Translational Neuroscience, Department of Clinical Neurosciences, Hotchkiss Brain Institute, Alberta Children's Hospital Research Institute, Cumming School of Medicine, University of Calgary, Calgary, AB, Canada

Morten Thorsen
Department of Child and Adolescent Psychiatry, Aalborg, Denmark

Alexei Verkhratsky
Faculty of Biology, Medicine and Health, The University of Manchester, Manchester, United Kingdom; Achucarro Center for Neuroscience, IKERBASQUE, Basque Foundation for Science, Bilbao, Spain

Patrick C.M. Wong
Brain and Mind Institute; Department of Linguistics and Modern Languages; Department of Otorhinolaryngology, Head and Neck Surgery, The Chinese University of Hong Kong, Shatin, Hong Kong

Benjamin K. Yee
Department of Rehabilitation Sciences, Faculty of Health & Social Sciences, The Hong Kong Polytechnic University, Hong Kong, China

Preface

Autism deserves a high level of attention considering the trend to increase worldwide. According to WHO, 1 in 160 children has an autism spectrum disorder (ASD). It is life-long complex neurodevelopmental disability related to impaired social behavior, language and communication, and a narrow range of interests and activities. It is called "spectrum" because affects people on a different way. All of them develop on their own track. While some people with ASD can live independently, others experience mental health issues, learning difficulties and have severe disabilities requiring life-long care and support.

As other psychiatric patients, people with ASD are often subject to stigma, discrimination, and human rights violations. Frequently, the access to social services and support for people with ASD is insufficient.

The etiology of ASD is heterogenous. To date, there is no known cause of this complex disorder or neither an effective treatment. It is a fact that the number of diagnosed people has increased significantly in the last 2 decades. Early and accurate diagnosis as well as effective intervention or treatments that build on scientific findings are essential for tackling this problem. Therefore, understanding the etiology and pathophysiology of ASD is currently the hot topic in the field. Existing scientific data reveals various pathways or mechanisms of ASD.

This book summarizes the current knowledge and provides an in depth review on the progress in research of autism spectrum disorders. A broad range of topics are covered in this issue including genetics (whole genome sequencing, gene expression studies, RNA editing), epigenetics, in vitro models of ASD (patients derived stem cells and cerebral organoids as tools to understand ASD), in vivo models of ASD, neurodevelopmental deficits of ASD (neuron–glia interactions, brain architecture and neuronal network, neuroimaging), biomarkers, epidemiology, diagnosis, and treatment.

"The molecular biology of autism" will be of interest of broad range of specialists working in the field and interested in autism—psychiatrists, neuroscientists, molecular and cell biologists, psychologists, nurses, and medical students.

CHAPTER ONE

The putative etiology and prevention of autism

Gary Steinman*
Visiting Researcher, Department of Obstetrics & Gynecology, Hadassah Hospital—Hebrew University, Ein Kerem, Jerusalem, Israel
*Corresponding author: e-mail address: dav4601@aol.com

Contents

Abstract

Since the initial psychological report by Leo Kanner in 1943, relatively little formal biochemical/neurological research on the cause of autism, other than peripheral searches for genomic mutations, had been carried until the end of the 20th century. As a result of studies on twin sets and the conclusion that autism was largely a hereditary defect, numerous investigations have sought various genetic faults in particular. However, such studies were able to reveal a plausible etiology for this malady in only a small percentage of instances.

Key bio-molecular characteristics of this syndrome have been uncovered when the potential roles of the glia were studied in depth. Findings related to biochemical deficiencies appearing early in the newborn, such as depressed IGF-1 (insulin-like growth factor #1) in neurogenesis/myelination, are becoming emphasized in many laboratories. Progress leading to timely diagnoses and subsequent prevention of central nervous system dysconnectivity now seems plausible. The tendency for an infant to develop autism may currently be determinable and preventable before irreversible psychosocial disturbances become established.

Progress in Molecular Biology and Translational Science, Volume 173
ISSN 1877-1173
https://doi.org/10.1016/bs.pmbts.2020.04.013

These discussions about glial function will be inter-spersed with comments about their apparent relevance to autism. The concluding portion of this presentation will be a detailed review and summation of this diagnosis and prevention proposition.

Abbreviations

AKT protein kinase B
ALS acid labile-subunit
ASD autism spectrum disorder
BBB blood-brain barrier
BDNF brain-derived neurotrophic factor
CNS central nervous system
CSF cerebrospinal fluid
GSK3B glycogen synthase kinase-3-beta
fMRI functional magnetic resonance imaging
IGF insulin-like growth factor
IGFBP IGF binding protein
IGFR IGF receptor
IL interleukin
IRS insulin receptor substrate
mTOR mammalian target of rapamycin
NGF nerve growth factor
OL oligodendrocyte
Olig1 oligodendrocyte transcription factor 1
OPC oligodendrocyte precursor cell
P13K phosphoinositide-3-kinase
PNS peripheral nervous system
PROM premature rupture of membranes
PTEN phosphatase and tensin homolog
SNP single nucleotide polymorphism
VSGA very small for gestational age

1. Background

Beginning in 1943, reports began to appear authored by Leo Kanner and others which described an abnormal form of behavior starting in children around ages 1–4 years.[1] For several years, the phenomenon was considered to be a consequence of dysfunctional parenting. For example, the theory of the "refrigerator mom" blamed the abnormal behavior, which persisted into adulthood, as a lack of affectionate parental attention toward their youngsters displayed by non-tender mothers in particular.[2] Later, this concept was discredited.

Toward the end of the 20th century, attention turned toward genetics to explain this condition. In particular, many sets of monozygotic twins were both observed to display autistic behavior in up to 90% of the cases.[3] Thus, it was assumed that the etiology of *all* cases of autism had a genetic origin. However, extensive studies identified major mutations in only 5–10% of autism occurrences. On the other hand, some children with major genetic defects had some behavioral traits similar to autism but not a complete set of such traits as are found in classical autism.[4]

According to the 5th edition of the Diagnostic and Statistical Manual of Mental Disorders (DSM-5) published in May, 2013, autism was defined by cases which displayed at least three persistent deficits in the areas of

(1) Social and emotional reciprocity,
(2) Non-verbal communication behaviors,
(3) Developing and maintaining appropriate relationships,
(4) Repetitive speech, movements, or employing of objects,
(5) Excessive maintenance of routines, formalized patterns of verbal/nonverbal behavior, or excessive refusal to change,
(6) Highly restricted, unusual, fixated interests (e.g., strong preoccupation with atypical objects), and
(7) Hyper- or hypo-reactivity to sensory input or extraordinary interest in appreciated aspects of the environment.

Until that time, variations of autistic behavior were grouped by specific names such as Asperger's syndrome, autistic disorder, and childhood disintegrative disorder, with various levels of seriousness. From that point in time onward, the differing presentations of this condition were taken to be degrees of symptom severity of the same malady. Hence, ASD (autism spectrum disorder) was understood to mean a group of conditions of the same malady but of varying functionality and intensities of sign presentations[5] (see Sections 8 and 10).

In contradistinction, other ailments were defined by specific genetic errors which were like autism spectrum disorder (ASD) in only *some* of the characteristics they displayed (see Table 1). Although thought to be differing representations of ASD, they were reexamined because these distinct Mendelian human diseases made autism seem to be a similar but genetically not identical heterogeneous condition. In many such instances, rare or common varieties of multiple genes make their genomic profile markedly different from other so-called autism-like cases. For example, Rett syndrome has been considered a variation of ASD but displays markedly different characteristics.[6]

Table 1 Some key differences between Rett syndrome and autism.

Characteristic	**Rett syndrome**	**True autism**
Head	Microcephaly	Macrocephaly
Common gender	Female (nearly all)	Male (4:1)
Brain involvement	Frontal cortex	Cerebrum; cerebellum
Neurons affected	Cholinergic	Serotonergic

Based on Riikonen R. *Insulin-like growth factors—neurobiological regulators of brain growth in autism?* In AW, Zimmerman, ed., Autism Current Theories and Evidence, Totowa, NJ: Humana Press; 2008, Chapter 10, 233-244.

"Autism-like" syndromes (with major mutations noted in parentheses but not found in the vast majority of ASD cases) include Rett syndrome (MECP2—methyl CpG binding protein 2), Fragile X (FMR1—fragile X mental retardation), Tuberous Sclerosis (TSC1—tuberous sclerosis 1), and Phelan-McDermid syndrome (SHANK3—SH3 and multiple ankyrin repeat domain 3). By comparison, MECP2 has been identified in about 1% of all cases of true autism.[4]

IGF-1 (insulin-like growth factor-1) is a single-chain polypeptide containing 70 amino acids. The level of free (active) IGF-1, typically about 1% of the total in humans, is governed by:

(a) Total IGF synthesis—largely controlled by growth hormone,
(b) IGF binding proteins (the most common in the CNS being IGFBP-2, 3, 4, and 5), and IGFBP proteases,
(c) Drugs that inhibit protein-binding of IGF,
(d) IGF cell membrane receptors (IGFR),
(e) Amount of the acid-labile subunit ALS, and
(f) Supply of nutrients.[7–9]

Most IGF-1 (>98%) in the blood is bound to IGFBP-3 and ALS. After release of the binding protein by matrix metalloproteinase-9 (MMP9), the insulin-like growth factor-1 crosses the blood-brain barrier (BBB) through an interaction with endothelial transporter lipoprotein-related receptor-1 (LRP1).[10,11]

IGF-1 is produced primarily in the liver (after birth) and in the placenta (before birth) for endocrine distribution. This polypeptide can also be synthesized locally for paracrine/autocrine provision. IGF-1 acts through tyrosine kinase-bearing receptors (IGFR) and plays a key role in brain development in youngsters. IGF-1 increases dendritic branching of layer two pyramidal neurons and the formation of neo-cortical connections.

This agent has significant effects on fetal and neonatal brain growth (especially with myelination). Particularly sensitive to IGF-1 deficiency are oligodendrocytes, cerebellar Purkinje cells, and motor neurons. IGF-1 stimulates DNA synthesis, neurite growth, neurotransmitter secretion, bone maintenance, control of neurotrophic factors, apoptosis inhibition, and protection of oligodendrocyte precursors.[7]

Overall, the roles of IGF-1 in normal fetal and neonatal nerve development include preventing neuronal injury, reducing neuronal degeneration, and increasing myelination. IGF-1 is involved in neurogenesis and synaptic genesis, and exerts effects in brain repair after mild/moderate injury. IGF-1 is neuroprotective when made available within 2 h of a hypoxic event.[7,11]

2. Glia

As the emphasis on major mutations associated with true autism occurrences decreased in studies about ASD, increasing attention was paid to structural and chemical deviations found in these cases, especially in the brain development of the newborn. To comprehend this shift in focus, it is important to review the factors and interrelations of glia to the disease process. The human neuro-glial cells to be considered here are oligodendrocytes, astrocytes, microglia, and Schwann cells primarily.

Glial cells are part of the immune system, maintaining chemicals needed for synaptic function, producing myelin which insulates the neural axons, synthesizing cerebrospinal fluid, controlling inflammation, and eliminating waste products. Glial cells do not participate directly in electrical impulse transmission in the nerves or in synaptic interactions between axons and dendrites but do affect the speed of neural signal passage. Glia lack axons or dendrites and play no role in adhering nerve cells together. (The word "glia" is derived from an ancient term meaning glue, since early scientists in this area believed that adhesion was the primary function of these cells).[12]

Glia do adjust the ionic environment of nerves in order to transmit electrical signals in the neurons more effectively. Glial cells assist the healing of nerve injury. As will be noted below, glia found in the central nervous system (CNS) bear specific differences from glia in the peripheral nervous system (PNS) in this regard.

Although the figure is debated, the ratio of glia cells to neurons in the brain appears to be roughly 1:1.[13] Special importance in the forthcoming discussions of their functions in autism is the role of oligodendrocytes in the CNS in particular in maintaining and strengthening a fixed nerve connectivity position once created and joined in the young brain; thereby, they reduce the potential of dysconnectivity occurring in infants. As will be described, this is highly dependent on the supply of insulin-like growth factor (IGF-1).

Glial cells in the PNS often promote the regeneration of lost neural function, although such recovery is not found in the CNS. In the central nervous system, nerve damage control is usually limited to apoptotic removal of affected components by astrocytes and microglia, in circumstances somewhat similar to reparable cases in the PNS.[12]

Infrequently, glia can serve as the origin of highly malignant glioblastomas, with no evidence of lower grade precursors. About 95% of cases cannot be related to any specific cause, such as heredity. Apoptosis does not occur in these aggressive tumors, which are difficult to treat or control. It is very uncommon for glioblastomas to spread beyond the brain.[12]

The glia to be examined here are listed in Table 2 in Section 7:

Some of these glia cells will now be discussed as they may relate to the genesis of autism.

3. Astrocytes and microglia (CNS)

Astrocytes and microglia are significantly involved in synapse maturation, maintenance of neural function, and synapse formation in the central nervous system. An astrocyte is typically in contact with many neurons simultaneously and could directly affect thousands of synapses.

The genesis of autism may be related to the release of cytokines such as TNFα, IL-1β, IL-2, IL-6, and IL-10 during infections such as Covid-19. Patterson proposed that such factors could lead to reduced prenatal production of insulin-like growth factor-1 (IGF-1) by the placenta during and following maternal infection, especially first trimester influenza.[14] In that case, the blood-brain barrier (BBB) becomes more permeable, possibly leading to neuro-inflammation. In this situation, microglia and astrocytes become more reactive. This reactive gliosis may affect the pruning of synapses and dendrites, especially by microglia.[12]

Microglia are among the brain's major responses to infection. They also direct post-infection reshaping of the brain cells. Microglia clear debris created by such an assault. Microglia act to restore some affected intra- and inter-cellular connections. Autistic disabilities may result from too few or too many removed connections post-infection in some cases. The role of microglia also appears to be involved with proper brain reconnections. In summary, microglial and pro-inflammatory cytokine activation control brain cell rewiring, reshaping, matching, and restructuring following an insult. Such an attempt may result in brain dysfunction as found in autism, as happens in faulty pruning processes.[15–17]

Microglial progenitors occupy the developing spaces prior to the completion of the blood-brain barrier. In the very young, vitamin D deficiency may induce errors of repair and development by poorly mediated pro-inflammatory release from microglia. Neural dysfunction might be the consequence of aberrant microglia after the insults of imperfect synaptogenesis or dendritic augmentation which they were unable to prevent.[18]

Microglia are the primary macrophages of the CNS. Fetal neo-neurons may experience faulty epigenetic alterations by being exposed to nutritional deficiencies, chemical insults, or stress hormones. Microglia are unusually active in injured patients, especially those destined to develop persistent abnormalities. Primitive neural phagocytes uniquely originate in the embryonic yolk sac. These early microglia colonize the whole CNS. Later, they are derived from lone cells. These glia are found at all stages of brain development.[19]

About 50% of all neurons in the developing brain undergo programmed cell death. Corpses are phagocytized by microglia throughout life. Excluding extraordinary health situations, personal longevity occurs inversely to the quantity of IGF-1 present. For reasons yet to be determined, microglia can promote neuron and neural progenitor cell survival in some situations. Alternatively, microglia phagocytize exuberant synaptic connections by pruning.[20]

Microglia are abnormally active in autism patients, especially those destined to develop overtly this malady between 12 and 36 months of age. Inefficient dendritic pruning may explain enlarged head circumferences in some young autistic children. As noted earlier, maternal infections or inflammatory immune activation in pregnant women may actuate microglial factors and promote the release of pro-inflammatory cytokines.[21,22]

Microglia are able to produce IGF-1. This may reduce the number of babies destined to become autistic (see discussion below). Also, such glial cells engulf dead and dying neural cells everywhere in the developing brain. These glia regulate neuron system vascularization, synaptic connectivity, axon outgrowth, and synaptic maturation, as well as appear in the CNS before the development of blood vessels.[23]

Astrocytes are closely associated with synaptic function. They have segment control of impulse transmission and are essential for the establishment of synaptic connectivity. By their interaction with neuronal synapses, this modulates interneuron function.[24]

Astrocytes secrete thrombospondins, which increase the total number of neuronal synapses. These cells are important to the current timing of synapse formation. The total number of neurotransmitter receptors determines the postsynaptic strength.[25]

Astrocytes have phagocytic capacity for removing nonfunctional synapses. This is largely dependent on two receptors, MEGF10 and MERTK. Thus, in summary, the astrocyte-synapse interaction has several aspects: formation, function, and elimination. A number of autism-like neurological disorders are linked to astrocyte dysfunction: amyotrophic lateral sclerosis, Rett syndrome, Fragile X syndrome, and other psychiatric conditions.[26]

With increased neuronal activity, the demand for blood flow increases. Vasoactive metabolites of arachidonic acid are released from astrocyte endfeet onto the blood vessels. Prostaglandin E2 (PGE2) and epoxyeicosatrienoic acid (EET) dilate vessels. In contrast, 20-hydroxyeicosatetraenoic acid (20-HETE) and ATP constrict the vessels.[27]

Astrocytes supply the bulk of caloric need of the brain, control compensation of interstitial fluid, provide substrates and precursors for biosynthesis, and recycle neurotransmitters. These gliotransmitters include adenosine triphosphate, glutamate, homocysteic acid, taurine, atrial natriuretic factor, tumor necrosis factor-alpha, and gamma-aminobutyric acid. An astrocyte is polarized, with processes at one end making contact with the blood supply, and the other end creating contact with adjacent synapses.[28]

Biochemically, astrocytes perform most of the Kreb cycle utilization of blood-delivered glucose to provide energy to the neurons. Also, astrocytes can store chemical bond energy in the form of glycogen. Similarly, oligodendrocytes rely on aerobic glycolysis. Calcium ions have fundamental roles in this energy-providing process as well. Hence, astrocytes and the blood-brain barrier (BBB) act together as a governor and a barrier with size-selective permeability.[29]

An example of the filter role of the BBB is the case of IGF-1 (see Section 7 for details). The level of IGF in an individual's cerebrospinal fluid (CSF) is directly related to its concentration in the blood; increase in systemic IGF-1 increases its passage into the cerebrospinal fluid. In cerebellar granule cell cultures, added IGF-1 increases cell survival, and withdrawal results in neuronal death. Young children (<4 y.o.) with autism have reduced levels of IGF in their CSF, as noted earlier. This is thought to be the cause of reduced numbers of Purkinje cells in the cerebellum.[30]

This selective role of the BBB can be affected by hypoxia/ischemia and stroke. Astrocyte dysfunction (astrogliosis) in the face of pathologic challenges in the form of injury and disease from CNS insults can extend to microglia and oligodendrocyte progenitors. In such cases, the glia cells can then generate negative effects via cytokines, chemokines, prostaglandins, and other agents. Perhaps the most detrimental effect of astrogliosis is inhibition of axon regeneration. In the presence of astrogliosis, genetic polymorphism (to be discussed later) could change the future of the astrocytes as well. Blindness and paralysis may result in extreme cases. On the other hand, control of inflammation, BBB repair, and neuroprotection depend on the roles played by astrocytes so as to preserve viable tissue and function. [31,32]

4. Oligodendrocytes (CNS)

Ventricular zones of the brain are the sites of origin of oligodendrocyte precursor cells (OPC). They migrate from there to the developing CNS, where they become active oligodendrocyte (OL) units. The rate of OL production can accelerate to replace lost myelin in the presence of injury, advanced age, or disease. The ability to transition from OPC to OL begins shortly before birth and continues for several months thereafter. The mammalian OPCs produce platelet-derived regulators to survive and proliferate.[33]

Of particular interest to the latter discussion here on autism is the manner in which IGF-1 affects proliferation, differentiation, and survival of OLs. Additional agents also able to promote this activity include neuregulins, Wnts, neurotrophin-2, chemokines, and thyroid hormone, among others. OPCs thrive *in vitro* with media containing platelet-derived growth factor (Pdgf). Thyroid hormone promotes differentiation and Pdgf sustains this activity.[33]

In adults, the number of OPCs stays reasonably constant until a demyelination occurs. OL development continues lifelong. This is necessary for normal brain function as well as repair following periodic demyelination events, as needed. It is a result of the balance between inhibitory extracellular signals, such as Notch signaling and differentiation promoters like Olig2. Following OL differentiation, genes controlling myelin-associated proteins such as PLP (proteolipid protein), MAG (myelin-associated glycoprotein), CNP (2′:3′-cyclic nucleotide-3′-phosphodiesterase), and MBP (myelin basic protein) are strongly induced. Myelin-regulator factor (Myrf) is generated as needed to aid in OL maturation and myelination. Epigenesis is also involved in this control process, including histone modification, chromatin remodeling, noncoding RNAs, and DNA methylation.[34]

Myelin production is the main role of oligodendrocytes. Because of its high lipid content (80% vs 20% protein), myelin is poorly hydrated but very stable for prolonged periods. The myelin sheets encircle the axons of nerves several times, to enhance packing density. In this way, Van der Waal's forces strongly promote the potency of this unit. The surface area of each flattened myelin unit is estimated to total up to 50×10^3 μm^2.[2,34]

The sequence of axonal wrapping with myelin is:

(1) Proliferation and migration of OPCs,
(2) Axon-glia recognition,
(3) Differentiation of OPCs to OLs,
(4) Axonal wrapping, and
(5) Myelin compaction.

In addition, oligodendrocytes and axons are joined metabolically. Aggregate cultures of fetal rat brain treated with IGF show marked increases of CNP (2′,3′-cyclic nucleotide-3′-phosphodiesterase) and oligodendrocytes. The emergence of enhanced MBP (myelin basic protein) can also serve as a marker for this transition.[35]

Oligodendrocytes provide support and insulation to CNS neurons. This is achieved by developing a myelin sheath around the neuronal axon. This covering material insulates the neuronal axons, accelerates the passage of nerve impulses from source to objective, and sustains/supports the functional pathways of these units once interneuron connections are formed. In this manner, dysconnectivity in the brain is minimized. The increased velocity over myelinated axons is due to reduced ion leakage, improved synaptic function, saltatory impulse movement, and decreased cell membrane capacitance.[35] (The neurons in the PNS are myelinated by Schwann cells and will be discussed separately.)

Functionality of the axonal portion of the neurons depends in large part if it is myelinated, and, if so, to what extent. This can be identified by the level of axonal myelination. Two-thirds of all the axons in the body are unmyelinated. Such axons in the CNS have radii of 0.05–0.6 μm and conduction speeds of $1.8\sqrt{a}$, where a is the axonal radius. In contrast, myelinated axons typically have overall radii of 0.5–10 μm and conduction speeds of 12(a + b), where b is the myelin sheath thickness. The impulse velocities in unmyelinated axons are typically 0.5–10 m/s; in myelinated axons they can be up to 150 m/s. A typical adult brain has about 100 billion neurons.[36]

Myelination of new neurons in the CNS of the human fetus begins in 24–25 weeks post-conception. The rate peaks at 1-year postpartum and tapers off gradually until about 25–30 years of life. Oligodendrocyte precursors arise from pluripotent cells in neurogenesis. OPCs are highly migratory, proliferative, hardy, and vigorously myelogenic. However, if the level of IGF-1 is reduced, the differentiation of OPCs to generate functional oligodendrocytes is diminished, and rapid cell death ensues. Oligodendrocyte loss results in demyelination. This can lead to impaired neurologic functions such as those changes seen in multiple sclerosis and autism.[35]

In the CNS, functional oligodendrocytes, a type of neuroglia, can be linked to as many as 50 axons for the purpose of creating and maintaining myelogenesis. In the fetal brain, the first wave of OPC formation occurs in the medial ganglionic eminence; the second wave develops in the medial forebrain; and the third wave, near the time of birth, extends from the dorsal and subventricular zones. The dysconnectivity characteristic of autism typically affects the posterior-to-anterior neural pathways, among others, in the brain.[34]

Oligodendrocytes are the last cells to be produced in the CNS. The overall role for oligodendrocytes is a supporting and trophic one, like all the other glia. In this function, they produce GDNF (glial cell line-derived neurotrophic factor), BDNF (brain-derived neurotrophic factor), and supplemental IGF-1.[35]

The impulses carried by the myelinated neurons have spatial insulating gaps, the nodes of Ranvier, which are voltage-gated sodium channel clusters. By promoting saltatory node-to-node impulse passage, the velocity is increased by as much as 100x. Areas of the brain populated by dense collections of myelinated neurons are known as white matter, whereas the nonmyelinated areas are called gray matter. People with high IQs have been found to have greater white matter than average.[35]

Demyelinating diseases such as multiple sclerosis and various leukodystrophies as well as spinal cord injuries and hypoxic episodes can be a stimulus to reduced myelination. The efficiency of remyelination decreases with age or toxin-induction because of less effective OPC recruitment and differentiation (progenitor failure). Transplanted OPCs can ameliorate dysmyelinated cases. Loss of normal homeostatic functions and increases of toxic actions may help explain the neuro-degradation and pathophysiology in Parkinson's disease, Huntington's disease, Alzheimer's disease, and amyotrophic lateral sclerosis.[35]

5. Schwann cells (PNS)

In the PNS, myelination is carried out by Schwann cells rather than oligodendrocytes. These glial cell bodies here are located inside the nerve itself, unlike OLs. Schwann cells evolve as follows:

Neural crest → precursor → immature → promyelin → mature

In sharp contrast to oligodendrocytes in the CNS, Schwann cell function and survival are strongly dependent on IGF-2 instead of IGF-1.[36,37] Neuregulin (NRG1), an epidermal growth factor, is a key regulator of axon myelination in the PNS. (This will have important relevance in the later discussion about autism.)

Peripheral nerve regeneration is dependent on the presence of viable Schwann cells, glial cells that associate synapses of the PNS with nerve fibers. Myelination of these nerve fibers is essential to achieve accelerated propagation of action potentials by saltatory conduction, much like the nerves of the CNS. Diseases specifically of the myelin itself are limited to the PNS (e.g., multiple sclerosis). The internodal portion of the axon is about 99% of its length overall. In this way, myelin provides a high-resistance, low capacitance neural sheath to enhance impulse speed. Like the CNS oligodendrocyte, myelination of the PNS nerves by Schwann cells is subjected in part to epigenetic regulation.[38]

6. Myelination and IGF-1

One of the first studies to relate IGF-1 to autism utilized a hospital-based collection of juvenile spinal tap surpluses. Although the number of cases was relatively small, a clear difference was found in samples from children aged 1 to 4 years. The aliquots from youngsters diagnosed as autistic by conventional psychological methods displayed IGF-1 concentrations of the CSF that were lower than for neurologically normal children of similar ages (normal mean = 0.4; autistic = 0.2 μg/L).[6]

A number of other reported observations pertained to a plausible connection between autism and IGF levels. For example, a study was carried out on an equal number of African American and Caucasian girls, matched for age, body mass index, socioeconomic status, and pubertal stage. Each was tested for serum IGF-1. The African American girls were found in 2000 to have greater total IGF-1 ($p < 0.001$) and free IGF-1 ($p < 0.01$) than their paired white teammates.[39] In a related study of 8-year-old children in 2002–10, the prevalence of ASD among Caucasians was 6.7 and for African Americans was 5.9 per 1000 (95% CI).[40] This would support the deduction that the higher the IGF-1 level, the lower is the potential for autism.

During pregnancy, placental GH (growth hormone) is believed to be the impetus for the rise in fetal serum IGF-1. Especially during the second trimester, the level of IGF-1 in the amniotic fluid increases steadily. Women carrying female fetuses also display a significantly higher placental GH level in their own blood than those bearing male gestations (e.g., 14% difference at 28 weeks gestation). This is consistent with the lower incidence of autism in female offspring than male. Placental (fetal) GH can be distinguished analytically from maternal pituitary GH.[41,42]

Inadequate placental production of IGF-1 can reduce the transfer of nutrients across the placenta to the fetus. Very-small-for-gestational-age (VSGA) newborns eventually have a higher incidence of autism than children who were normal weight at birth.[43]

Increased risk for autism in the offspring is linked in many cases to antepartum maternal infection. Maternal immune activity (MIA) increases in response to infectious challenges, such as influenza and herpes. As a result, the level of interleukin-6 (IL6) rises, whereby the activity of placental JAK/STAT3 (Janus kinase/signal transducer and activator of transcription proteins) escalates. In turn, this reduces the amount of IGF-1 released by the placenta to the developing fetus (see Fig. 1). As a consequence, autism results more often if chorioamnionitis occurs. Hence, MIA (e.g., with premature rupture of membranes—PROM) could serve as a trigger for lowering the neonate's IGF-1 available to aid in neoneuronal myelination, independent of the maternal level of IGF-1. In line with this, IL6 is markedly elevated in the cord bloods of preterm newborn whose mothers had presented in prodromal labor with PROM vs those with intact membranes. IL6 is increased in the cerebellum of many autistic children.[14]

If not fully myelinated, nerve cell transmission is likely to be much slower than normal. Infant responses to environmental stimuli will be diminished,

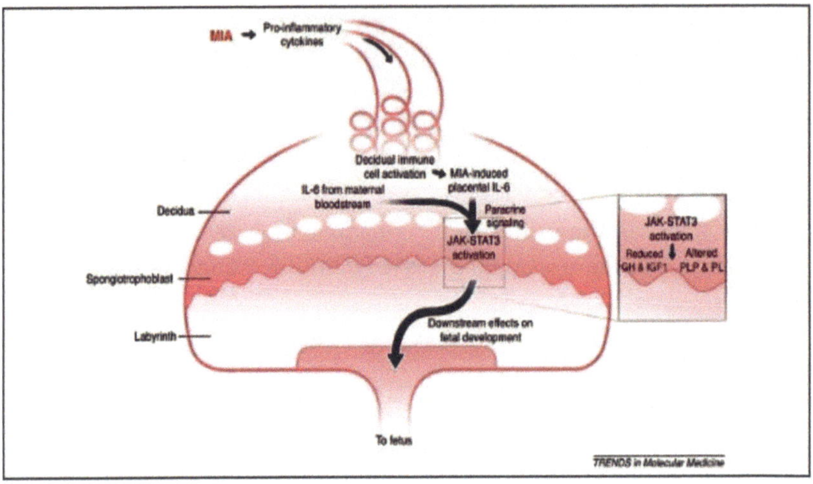

Fig. 1 Maternal immunologic activation, secondary to an inflammatory process, promotes the release of IL6. The cytokine depresses IGF-1 production and release to the fetus. EY Hsiao, pH Patterson. Activation of the maternal immune system induces endocrine changes in the placenta via IL-6. *Published in Brain Behav. Immun. 25 (2011). This diagram is reproduced here with the kind permission of Elsevier Publishers.*

as a result. Reduced IGF-1 in the very young neonates decreases differentiation of oligodendrocytes, causing diminished neuronal myelination, the hallmark of autism in the brain. In autism, brain tissue displays abnormally thin axon myelin coats, as observed by MRI and direct biopsy,[28,44] thereby promoting reduced neural function and impulse transmission. This supports the observation that one of the primary pathologic processes in autism is deficient IGF-mediated myelination in the developing central neural pathways in the white matter of the brain. This becomes overtly evident in the baby's behavior at 1–4 years of age.[7–11,45]

7. Autism etiology

The observations noted so far suggest that supplementation of IGF-1 in the newborn could reduce or eliminate neurologic defects, especially the eventual development of ASD. This condition in particular is to be differentiated from less common syndromic neuro-pathologies which bear some of the characteristics of ASD but are distinguishable from it (see Table 1). For example, repetitive movements are found in both true autism and Rett syndrome (RS); however, as noted earlier, RS is almost unique to girls but ASD is much more common in boys (4:1). RS has a distinguishing major genetic mutation (MECP2), whereas autism does not. Autism can coexist with other genetic disorders such as Downs syndrome in about 10% of the cases.[46]

There is increasing evidence that the etiology of autism is related to disorders of the **IGF1/IGFR/IRS1/PI3K/AKT/mTOR** intracellular signaling pathway. In particular, this conduit effects the translation of IGF-1 to promote myelogenesis. Activation of the tyrosine kinase receptor sites of IGFR, the junction between the extracellular and intracellular milieux, affects a number of signaling pathways, inducing vasomotor and metabolic effects. In contradistinction, phosphorylated AKT is decreased in the autistic brain.[47–49] This could happen if the cellular environment contained a reduced amount of IGF and/or if the IRS1 in the PI3K/AKT chain were polymorphic (rs1801123—see Section 10). As will be noted in Section 11, supplemental dietary IGF-1 could ameliorate either problem.

Another anabolic role of IGF1 in the nervous system relates to an additional application of the PI3K/AKT chain. When stimulated, this chain can also inhibit GSK3B (glycogen synthase kinase-3-beta), which would otherwise prevent the synthesis of glycogen and proteins in the brain.[50] Thus, deficiency of insulin-like growth factor-1 (IGF-1) can reduce anabolic function promoted by this sequence.

Rather than being due to a major genetic defect, the dysconnectivity characteristic of autism appears to be the result of diminished effect of IGF-1 on neonatal neurogenesis and myelogenesis, secondary to impediments in the PI3K/AKT pathway or to insufficient IGF-1 initially.

Autism is currently referred to as ASD—autistic spectrum disorder. Instead of occurring as a conglomerate of several different entities with some shared characteristics, the *spectrum* of ASD may be a single malady with a range of intensities determined by the degree of neonatal IGF-1 deficiency (see Section 10). This would yield a scope of degrees of activity determined by the level of phosphorylation (or inaction) of the PI3K/AKT cellular pathway.[48]

Further supporting the IGF hypothesis is the observation that the rate of cancer in autistic patients is about one-third that of neurologically normal individuals. This may be due to the lower level of IGF in children and adults with ASD. In contradistinction, elevated amounts of IGF are typically accompanied by shortened longevity of life., One of the characteristics of IGF-1 is its mitogenic property. Similarly, the tumor suppressor PTEN also acts to reduce the activity of PI3K, thereby diminishing the active translation of IGF. This could also lessen neuronal myelination in children destined to display autistic behavior later. On the other hand, chemical inhibition of PTEN could enhance myelination.[51–53]

In comparison, the PI3K/AKT/mTOR pathway is a key player in the Schwann cell initiation of myelination. PTEN is a lipid phosphatase with the effect of opposing this PI3K pathway in the PNS environment and preventing the resumption of myelination. Loss of OLs in the CNS, leading to demyelination, can be repaired and restored by remyelination. Such myelin is necessary for axon survival, which requires intact OPCs and oligodendrocytes. Available IGF-1 and Olig1 are utilized for this purpose. Effective remyelination also needs removal of prior myelin debris by glial macrophages. With aging, there is a decrease in OPC recruitment and differentiation.[48]

As noted, nucleotide polymorphic limitation of the cellular IRS-1 factor can retard the translation of the IGF signal. Overall, IGF stimulation of oligodendrocytes is essential for axonal myelination and synaptic development in the prenatal and neonatal stages, especially in the brain. Thus, it is important to enhance the total activity in the CNS and PNS of biologically operational IGF early in a newborn's development. The primary objective is to avoid unalterable dysconnectivity (mis-wiring) in the crucial neonatal stages of brain neural circuit genesis, especially with the long-range posterior-to-anterior connections. In contradistinction, reduced IGF-1 levels in early life can increase the overall lifespan. However, deficiency of this growth factor in the adult negatively affects neurovascular and cognitive function.[51]

As it relates to the avoidance of autism, the level of IGF-1 in the newborn appears to be the most pertinent. During the first 4 years of life, the serum IGF-1 concentration remains reasonably constant unless it is augmented pharmacologically or nutritionally.

Suppressed serum IGF-1 levels, due to a variety of reasons, such as prematurity, can cause dysmyelination of neo-neurons. It is at this stage, especially in the first year of extrauterine life, that irreparable brain mis-wiring could occur at a critical time since the active formation of new nerve tracts is taking place.[46]

In general, IGF-1 promotes protective surveillance of existing brain cells to prevent neuronal derangement. As noted earlier, nuclear polymorphisms can down-regulate the promotion by this factor. Knockout mice which lack the ability to synthesize sufficient IGF-1 have reduced axonal diameters and decreased nerve conduction velocities. In contradistinction, transgenic mice with enhanced IGF1 produce up to 130% the normal amount of myelin. Untreated autistic children between birth and 4 years of age have lower levels of this growth factor in their cerebrospinal fluid than unaffected youngsters.[6] These findings in affected persons are consistent with diminished myelination in early neurogenesis.[50]

The neuronal growth cones (axon precursors) are guided to sites of need (e.g., muscles) by non-diffusible biochemical factors in the vicinity of the target tissues. To refine the path, both chemo-attractive (such as netrins) and chemo-repulsive (such as semaphorins) agents are appropriately placed within the receiving tissues. Once the axon reaches a suitable dendritic or organ target, synapse formation ensues. Such trophic interactions begin before birth and continue postnatally.[46,52,53]

8. Connectivity

With time, neurons in the developing neonate display increasing numbers of synapses. The fittest of them survive and the less functional in the remainder are eliminated by apoptosis. The most effective connections are retained peripherally and in the brain. This is promoted by agents such as nerve growth factor (NGF) and brain-derived neurotrophic factor (BDNF). According to the Hebb's Postulate, active, functional interplay between a presynaptic axon and a postsynaptic target fortifies the synaptic link overall. Such strengthening promotes persistence of the joining of neurons with terminal tissues. Conversely, weakly associated unions eventually resolve and disappear.[53–60] (See figs. 24.1 and 24.2B in Ref. 53.)

Postnatal brain development is dependent on enhancement of functional synapse formation. This is promoted by strong, timely impulses sent to the target via the presynaptic axon. For proper, efficient, coordinated, and rapid neural connectivity (circuitry) to develop between regions of the neonatal brain, early, comprehensive myelination of the participating axons is essential to retain the most functional course. In contradistinction, autistic brains have been observed to display heterogeneity, with various areas of increased and of decreased connectivity. This is especially evident in imaging studies of the corpus callosum, where autistic individuals exhibit changes in white matter myelination related to decreased interhemispheric connectivity. Such studies include fMRI (functional magnetic resonance imaging) and EEG (electroencephalography).[60–62]

Attracted by CX3CR1 chemokine receptor, microglia colonize and become ready for nerve development prior to the appearance of other

Table 2 Summary of glial cell types.

Glia	Site	Main functions
Astrocytes	CNS	Link neurons to blood supply; secrete neurogenic factors, neurotransmitters, and neuromodulators
Microglia	CNS	Prune synapses/dendrites; release CX3CR1/neurotrophic BDNF
Oligodendrocyte	CNS	Produce myelin sheath
Ependymal cells	CNS	Secrete cerebrospinal fluid
Radial cells	CNS	Matrix to assist neo-neuron migration
Schwann cells	PNS	Myelination; clear cellular debris
Satellite cells	PNS	Regulate chemical environment

BDNF, brain-derived neurotrophic factor; CX3CR1, fractalkine receptor in microglia migration. Table based on Barres, BA, Freeman, MR, and Stevens B, eds. *Glia*, Cold Spring Harbor, New York: Cold Spring Harbor Laboratory Press; 2015, sections 1,2,4.

nervous system cells. From there, synaptic junctioning and related axon outgrowth ensue (see Table 2).

Protective myelination in humans reaches its peak growth rate by 1 year of age. Psychosocial behavioral symptoms of autism typically appear between 1 and 4 years after birth. In the brains of autistic children, myelination has been shown by imaging studies and biopsy to be deficient compared to unaffected children.[28,44] Such a diminution would weaken the operation of the axon-target synapse, thereby accounting for the malfunction of areas of the brain involved in comportment. This would include the brain's prefrontal and temporal areas.[60,63]

"Pruning" of excessive neural synapses and dendrites by microglia (autophagy) to enhance the functionality of the retained circuitry has been found in the CNS of children up to the age of 1 year. During this time, the estimated number of synapses increases by a factor of 10. In unaffected individuals, dendrite exuberance normally occurs as the need for high velocity impulse delivery is met by myelinated presynaptic axons. On the other hand, disproportionate processing is the consequence of hypomyelination. Compounding this could be neuronal cell signaling impediments in the PI3K/AKT/mTOR pathway.[32,62]

Initially, brain dysconnectivity in autism was discerned in 2004 to be between frontal and posterior compromised areas. In the absence of effective connectivity (e.g., parietal nerve tract union with frontal zones), functional combinations of nerves would be diminished. Such circuitry disorders could result from abnormalities such as:

(a) arrested OPC development,
(b) OL dysfunction, and
(c) axonal disorders disrupting normal impulse signaling.

As a result of such dysconnectivity and consequential neurological defects, normal brain function would be disrupted, especially in aspects of behavioral issues. This results from myelination deficits causing cerebral white matter hypoplasia, decreased neural connectivity, and impaired or disrupted impulse conduction synchronization between cerebral sites as a consequence of insufficient IGF-1.[63]

Brain connectivity denotes networks that are functionally connected to each other to result in meaningful psychological and physical actions. White matter inflammation can induce a reactive response (of the microglia and/or astrocytes), leading to disordered myelination. Being a prelude to autism, this typically occurs in the first postpartum year without overt signs. The dysconnectivity approach to explain the etiology of autism replaces the previous emphasis on genetic defects as a more germane perspective.[64]

Examination of autistic patients by fMRI has revealed functional underconnectivity in at least five areas (prefrontal, parieto-occipital, motor, somatosensory, and temporal). In contrast is the parallel observation of tempero-thalamic hyperconnectivity in other cephalic areas in the autism group, compared to unaffected individuals of a similar age. The overconnectivity regions are apparently due to incomplete dendritic pruning by glia, especially within medial motor and orbito-frontal zones, whereas underconnectivity is due to hypomyelination in other brain areas. It was concluded that such abnormalities could explain socio-communicative and cognitive impairments in autism.[64,65]

In general, the exposure of children to new concepts, as found in the classroom, acts to stimulate new memory functions. However, therapeutic attempts have so far been marginally successful, at best, to promote correction of neural dysconnectivity in autistic children. Such reduced "normal" brain connectivity in childhood autism is due to aberrant long-range connections of neural networks secondary to hypomyelination.

Autism can be seen as a dysconnection (mis-wiring) syndrome. In other words, children with autism possess a pathology emanating from reduced brain networking at psychosocial levels. As of today, no treatment is available to reverse this error once established. Central to this problem is inadequate myelination able to hold developing nerve circuits fully ligated and aligned together by effective, functional synapses. Similarly, reduced fractional anisotropy near the corpus callosum may be due to reduced myelination of adjacent fiber tracts or to defective early neural migration, synaptogenesis, or dendritic/axonal pruning.[64–67]

Autism becomes symptomatic around ages 1–2 years in many cases. This happens to be approximately contemporaneous with the first 6–14 months of postpartum life when cephalic synaptogenesis, dendritic arborization, and myelination are very active. Dysconnectivity-related behavior would become noticeably apparent near the end of this inaugural period. Generally, hypoconnectivity in autism relates to disrupted neural communication between distant regions (e.g., frontal and parietal lobes bearing complex cognitive and social functions); in contradistinction, hyperconnectivity involves neural contact between local brain regions.

In the first stage of axon/target union, on or about the time of birth, each of many axonal branches of a single neuron may develop synaptic connections with several targets. A large percentage of these multiple rudimentary neuronal innervations are subsequently eliminated. Most susceptible to such early dissolution are those neurons with hypomyelination and low velocity impulse transmission. In such cases, the chance of a functional persistence or reconnection to another site is reduced from the final mature ones.[53]

Because autistic dysconnectivity often encompasses synaptic dysfunction (i.e., synaptic connection problems rather than regional issues), it has been recommended that related studies should consider the possible single nucleotide polymorphisms (SNPs).[63–67]

9. Polymorphism and biomarkers

As of now, autism is largely an irreversible condition once it can be recognized explicitly through psychological testing. In some cases, autistic symptoms may improve to some degree with early psychotherapeutic intervention. The behavior traits characteristic of autism are believed to be the results of brain dysconnectivity developed during the infantile and early childhood years.

ELISA quantitation of deficient serum IGF-1 in umbilical cord blood collected at delivery, for example, may foster early initiation of specific nutrient supplementation to minimize subsequent development of primary symptoms of autism. It might also define where on the spectrum the malady would lie, if left untreated promptly. Antepartum diagnostic cordocentesis could potentially be employed in highly suspicious cases, especially in gravid women with one or more prior affected offspring. Also, fetal SNPs could be identified in the maternal circulation before birth.[67–75]

Most SNPs are silent since 98% occur in the intron or intergeneic regions of the DNA. However, some have been identified in variably symptomatic conditions such as autism. To be classified as a polymorphism, the least common pathognomonic allele must have an occurrence frequency of 1% or more. Each may or may not display overt changes of medical importance. Some of the SNPs possibly relevant to identifying potential autism development include:

(1) Insulin receptor-like growth factor (IRS-1)—G972R
(2) Dual specificity DESP15—rs3746599
(3) IGF1/TasI and IGF1/SmaBI
(4) IGF-1 (Chinese)—rs1520220
(5) IRS-1 (Korean)—rs1801123
(6) FOXP2—rs2056202; rs2292813
(7) GSBRB3—rs878960.

In particular, polymorphic IRS-1 (#5 in the above list) reduces the rate of translation of IGF-1. (See the discussions about the intracellular PI3K/AKT chain noted in Section 8.) Allele frequency analysis determines if a statistically significant association between the incidence of this SNP and autism exists. The presence of this SNP is often correlated with enhanced autism potential in Korean children.[69] However, a SNP allele that is frequent in one ethnic group, as it was in this reported introductory study, may be rarer overall in mixed groups. Thus, investigations of this allele in several locales in the world are necessary before it can be utilized on a regular screening basis.[67–75]

Rett syndrome and Fragile X, for example, which resemble autism in only some physiological and psychosocial characteristics, occur much less frequently than ASD (see Section 2). The present report has enumerated biochemical features of true autism cases, which distinguish them from less common disorders bearing some, but not all, analogous characteristics. One or more early-warning biomarkers detected at birth (or previously) could be anticipatory for the later appearance of authentic autistic disease:

(1) Deficient serum IGF-1
(2) Elevated anti-myelin basic protein
(3) Elevated serotonin
(4) SNP rs1801123
(5) SNP rs878960.

Timely detection could conceivably allow prompt correction of the biochemical flaw(s) before permanent neuro-psychological abnormalities appear.[7,45,69,71] For example, one mathematical approach using three of these blood factors together can help predict at birth which newborn have increased probabilities of displaying autistic characteristics at a later age:

1—IGF1, 2—anti-myelin basic protein (anti-MBP), and 3—serotonin.[45] With these quantities can be calculated an *Autism Index* (AI), where p = weighted probability of variation of the biomarker from normal in autism, and n = absolute percentage (expressed as a decimal) deviation of the biomarker from normal:

$$AI = [p_1n_1 + p_2n_2 + p_3n_3]/0.1$$

AI = 0.00 would be an example of an unaffected newborn, whereas AI more or less than zero would raise the possibility of future appearance of autistic features and the current need for preventative intervention.[46]

10. Autism prevention

Ingestion of milk raises a human's serum level of IGF-1. Human milk increases a baby's IGF-1 level more than bovine milk or formula.[76] In a study performed on adults, it was found that a daily intake of 600 mL of milk for 12 weeks raised the mean serum IGF level by 10% in total.[77,78] On the other hand, vegan women who consume no animal products (e.g., bovine milk) have serum IGF levels that are 13% on average lower than in omnivores.[79] Although ingested IGF is a polypeptide, it apparently survives passage through the stomach because of casein shielding.[80,81]

Special imaging study approaches have shown autistic children with the most unmanageable social interaction skills to have significantly lower global MWF (myelin water fraction) scores when compared with normal control individuals.[82] MRI brain scans of children who have been breast-fed as infants displayed increased white matter in frontal and associated brain regions.[83] The longer a group of babies were breast-fed, the less frequent was the occurrence of autism later.[76] This was again confirmed in a meta-analysis from several sources.[84]

Also, ingestion of vitamin D enhances the benefit of IGF-1 by increasing IGFR receptors. Mothers of autistic children have a 1-in-5 chance of delivering another affected child, whereas those with no previous autistic offspring have a 1-in-60 possibility.[7,85] It is twice as common in siblings as in half-siblings. Vitamin D supplementation of gravid mothers with prior autistic children significantly reduced the chance of bearing another affected child. Such nutritional additives might diminish the possibility of autism in babies in general.

Hence, plausible preventive treatments include:

(1) Human breast milk feeding of the newborn for up to 1 year in order to raise the serum level of IGF-1 of the baby. Children breast-fed exclusively for the first several months of life present a significantly lower incidence of autism than those fed bovine milk or formula. The longer a baby is breast-fed exclusively, the lower is the chance that it will display autistic characteristics later.[76,84] Another means for elevating blood IGF-1 in the neonate is massage therapy, although prolonged treatment in conjunction with autism has not been studied as yet.[86,87]

(2) Oral supplementation of vitamin D increases the circulating level of IGF-1 significantly. Gravid women who were deficient in vitamin D at mid-gestation were twice as likely to deliver children who eventually developed autism than those who were normal in this nutrient. When women who had previously delivered autistic children were given vitamin D supplement in a following gestation, the rate of affected subsequent offspring dropped from 20% to 5%.[88–91]

(3) Modified forms of IGF (such as Des-IGF, IGF_{1-3}, or cGP), one of which can be administered orally, show potential for use in these cases. They more easily cross the BBB (blood-brain barrier) and bind less strongly with IGFBPs (IGF binding proteins), thereby requiring administration of less in the active, unbound form. The ultimate purpose is to raise the level of free IGF-1 and prevent the type of autistic brain changes secondary to hypomyelination found on biopsy.[46,92–97]

$$IGFBP_3\text{-}IGF_1 + cGP \rightarrow IGFBP_3\text{-}cGP + IGF_1$$

(4) Influenza vaccination should be administered seasonally to all non-allergic women contemplating a new pregnancy.[14]

Irreversible CNS dysconnectivity developing with overt symptoms in the first year or two of neonatal life may be the consequence of an uncorrected biochemical deficiency secondary to a nucleotide polymorphism. This is comparable to a partially inhibited enzyme. The initial step of the PI3K chain is the interaction of the IGFR (receptor) with free IGF (see Section 8). The higher the concentration of unbound IGF-1 in the surrounding serum/extracellular milieu, the more rapid would be the signal transfer to IRS1. However, if this IRS1 has a functional polymorphic change from normal IRS1 (e.g., rs1801123), the slower would be the passage of the message to PI3K, the next member of the chain.

Consequently, there may be at least three processes for slowing the PI3K/AKT chain activation of myelination: reduced IGF-1 in the cellular environment or IRS1 polymorphism in the transmembrane chain. Either problem could be compensated kinetically by an increase in the provision of IGF-1, such as with breastfeeding.

(a) Dephosphorylation[98]

(b) IRS1 polymorphism[69]

(c) Selective inhibition of elements of the P13K/AKT pathway[99]

Any of these problems could be compensated kinetically by an increase in the provision of IGF-1, such as with breast-feeding.

Examples of the association of a SNP with autism are rs1801123 and rs878960, as discussed earlier. The observed kinetics of IGF1 translation via the PI3K/AKT chain when comparing affected from neurologically normal individuals would support the purported association of these SNPs in particular with autism. Such an applicable SNP departure from normal (rs878960) could serve as a parameter to assess anticipated symptom severity (i.e., spectrum position) as well. The three alleles of SNP rs878960 in GABRB3 (gamma-aminobutyric acid receptor subunit beta-3) can distinguish severe cases of autism from mild/moderate ones.[71]

Early prevention of autism development could be initiated by measuring IGF-1 in umbilical cord blood at delivery. This would determine the advisability of IGF augmentation for at least one full postpartum year to overcome by biochemical kinetics the neurogenic deficiency of autism. Timely action is necessary because myelination of brain neurons is nearly complete by the child's second birthday.[9,100]

Although the concept of dysconnectivity in autism was proposed as early as 2004, no inclusive mechanism for its origin and prevention has been advanced until now. As described previously in the present discussion, it would appear that reduced myelination due to inadequate provision or deficient translation of IGF-1 by oligodendrocytes explains the basis of dysconnectivity beginning in affected infants. Furthermore, it creates the framework for minimizing the negative effect on connectivity development in the affected newborn.[15,63,96–100]

In Fig. 2 are seen several summarized aspects of postulated faulty metabolism in autistic people, especially young children. Starting from the top, insufficient IGF-1 produced in the liver would diminish the velocity of the PI3K/AKT chain, followed by reduced stimulation of oligodendrocytes

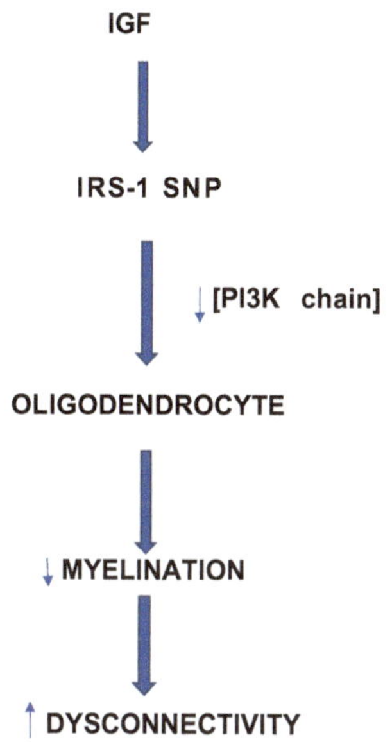

Fig. 2 IGF-1 stimulates oligodendrocytes to supply myelin to the neo-neuron. IGF-1 enhances the velocity of the **IGF1/IGFR/IRS1/PI3K/AKT/mTOR** intracellular chain. In the presence of the polymorphic form of IRS-1 gene or reduced IGF-1, the rate of myelin production is reduced, myelination of new nerves is slowed, and connectivity of new circuits is inadequate to meet the needs of the developing infantile brain. As a result, permanent mis-wiring ensues, especially with long distance connections (e.g., parietal-to-frontal).

to myelinate neo-neurons. The same lowered production would result if there were an IRS-1 SNP in the chain. Impaired guidance of nerve circuit generation in the young brain would lead to overall dysconnectivity.

Related to this is the vitamin D_3 receptor (VDR) gene polymorphism which may explain the enhanced utility of this vitamin in reducing the incidence of autism. Alleles of the VDR gene (T allele at position Taq-1 and α allele at position Apa-I), when present, modulate the ASD incidence for a given level of vitamin D.[101]

The analyses to diagnose and appreciate the various underlying events related to the etiology of autism (Fig. 2) would synopsize the putative cause and effect of this transformed biological pathway. Correction of this insufficiency could be achieved by dietary increase of the amount of IGF-1 to replace deficient amounts produced in the liver or to overcome the decreased kinetics of the PI3K/AKT pathway due to a polymorphic form of one or more components of this reaction sequence. An additional diagnostic test that could be made at delivery is myelin basic protein or its antibody (Tables 3 and 4).[41,45,100]

Table 3 Mean serum IGF-1 level of young children (μg/mL) born at term.

Children (μg/mL) born at term

Age (months)	Girls	Boys
2.5–12	103	91
21–39	132	108
45–51	164	114

Note that within each age grouping, boys have a lower mean level of serum IGF-1 than girls. This may account for the greater occurrence of autism in boys than girls (4–1). See Section 7 and Ref. 42 concerning corresponding fetal placental growth hormone (GH) differences by gender.

Based on Yuksel, B, Nuiozbek, M, Mungan, NO, et al. Serum IGF-1 & IGFBP-3 levels in healthy children between 0 and 6 years of age. *J Clin Res Pediatr Endo*. 2011;3(2):84–88.

Table 4 Associated test results[a].

Neonate/infant source:	Unaffected	Autistic
CSF IGF (age 1–4 years)	Normal	Low
Serum IGF (neonate, VSGA)	Low normal	Low
Serum IGF (via maternal infection)	Normal	Low (↑IL6)
Polymorphic IRS1	Negative	Positive
Phosphorylatd AKT	Positive	Negative
Brain dysconnectivity (fMRI)	Negative	Positive
Neuron impulse velocity	Normal	Reduced
Dendritic mass (MRI)	Normal	Excessive
Cerebral white matter (MRI)	Normal	Hypoplastic
Serum IGF[b]	Normal	Low
Serum serotonin[b]	Normal	Elevated
Anti-myelin basic protein[b]	Normal	Elevated

[a]Based in part on Refs. 7, 12, 14, 46, 93; practicable for evaluation in addition to psychological diagnostic testing in infants (Ref. 5).

[b]See Ref. 46 for a perinatal score which includes these three blood parameters.

This review of glial functions and their bidirectional association with neural activity aids in the understanding of autism's etiology. This is especially true in myelination. The central nervous system network depends to a large extent on the abilities of the glia to support essential roles fundamental to neuronal generation, operation, turnover, and cleanup.[102,103]

Valproic acid (VPA) is an effective anti-epileptic drug. If given to seizure-prone gravidas, it may lead to an increased incidence of autism in their offspring. Studies in laboratory animals exposed to VPA display decreased OL-lineage cells and hypomyelination. If no alternative drug is found effective during the pregnancy, IGF prophylaxis postpartum in the neonate would appear to be indicated.[104]

As reviewed early in this chapter, speculation into the cause(s) of autism which began with the "refrigerator mom" proposal[2] was followed by numerous etiologic proposals. In recent years, ASD has been correlated with major genetic mutations, but such prominent flaws were found in no more than 10% of the autism cases.[4]

Lately, cesarean section has joined the group of putative associations with ASD. The odds ratio distinction of c/sections followed by autism as a postpartum complication (1.34), compared to less often affected vaginally delivered infants, is most pronounced in the gestational range of 39–41 weeks.[105] *Before* primary operative delivery is elected and executed, a term fetus's endurance is tested most while encountering prodromal and active labor stress, especially due to dystocia. The varying risks to the baby are often evidenced by non-reassuring fetal heart rate tracings of fluctuating degrees (mild to severe). This may be due to undiagnosed preexisting IGF deficiency, making hypomyelinated CNS neo-neurons exquisitely sensitive to pre-labor, antepartum, and intrapartum hypoxic impairment as found in stressed deliveries. Affected neurons may be removed by microglia in such cases.

As recommended here for all deliveries, every newborn should be tested for serum IGF level at birth via umbilical cord or heel-stick blood. Neonates scoring positive for deficient IGF-1 should be started immediately on quantified breast-feeding or insulin-like growth factor supplement. (The exact doses remain to be determined.)

11. Challenge

Using the data that 1-in-60 of all neonates develop autism and that 3,853,000 babies were born in the United States in 2017 would average out to another 165 newborn children in America who would develop a form of this disease *every day* of that year. It is estimated that it typically would be in the range of an additional $3,000,000 to raise such an affected child.[7,100] Therefore, research to solve the autism enigma is urgently needed for both humanitarian and medical reasons.

Numerous etiologic theories for autism have been proposed without proven basis (e.g., pollutant exposure,[106] major genetic mutations,[107] and digestive tract changes).[108] The currently entertained model incorporating glia cells, myelination, dysconnectivity, and IGF-1 could explain atypical connectivity. Such a defect is principally found in the frontal networks and corpus callosum, and seems to be most viable inclusive scheme proposed thus far[46,109] (see Section 8 and Fig. 2).

Just as diminished serum IGF-1 parallels poor postpartum brain growth and is a major indicator of depressed CNS development in preterm infants,[103] so too, it relates directly to potential brain dysconnectivity in autistic full-term children. IGF-1 augmentation can diminish brain injury after hypoxic events, much like the benefit of breastfeeding neonates as it relates to autism.

At birth, the baby's cord blood IGF-1 concentration is largely independent of the mother's level.[110] Depressed neonatal serum IGF-1 may result in poor postpartum brain growth, especially in SGA (small for gestational age) babies. It is a predictor of reduced or delayed CNS development, especially in preterm infants.[111] Deficiency of this growth factor can cause autistic brain dysconnectivity if left untreated. What remains to be determined is the umbilical cord serum IGF-1 limit below which aggressive postpartum growth factor replacement is indicated, as well as the minimum breast milk IGF-1 concentration that can be remedial in this regard.

Recent studies have demonstrated encouraging results using IGF derivatives to treat maladies such as Rett and Fragile X syndromes.[112,113] It has been proposed that IGF-1 and IGF derivatives are able to reduce synaptic defects in established cases. Also, these pharmaceuticals may alleviate neurotoxicity present in the AKT/mTOR pathway in affected microglia.[114]

As yet, no study has reported an antepartum or postpartum method which prevents or diminishes the development of classical autism in the infant other than prolonged breastfeeding *from birth* exclusively.[76,83,84,115–117] Although the pathologic processes in autism begin around birth, behavioral symptoms usually do not appear until ages 1–4 years. This could portend the futility of IGF-1 administration to psychologically symptomatic autistic patients, once the malady is well established.

Recent events have centered around the Covid-19 pandemic. To be investigated is if this virus has any effect on the developing fetuses, such as autism.[118]

Acknowledgments

The author wishes to thank: Roberta Zuckerman and David Mankuta, MD, for their helpful discussions and suggestions about the material presented in this chapter; Aviva Adler, librarian, for her cooperative assistance in locating relevant literature; and Prof. D. Teplow and the editorial staff of Elsevier Publications for guidance in preparing this report.

References

1. Kanner L. Autistic disturbances of affective contact. *Nerv Child.* 1943;2:217–250.
2. Co S, Kaufman M. Refrigerator mothers. In: Steinman G, ed. *The Cause of Autism.* New York: Baffin Books Pub; 2014:71–81.
3. Hallmayer J, Cleveland S, Torres A, et al. General heritability and shared environmental factors among twin pairs with autism. *Arch Gen Psychiatry.* 2011;68(11):1095–1102.
4. Din-Lovinescu D, Shackles C, Zomorrodian S. Genetic errors. In: Steinman G, ed. *The Cause of Autism.* New York: Baffin Books Pub; 2014:273–285.
5. Tabag K. DMS-5. In: Steinman G, ed. *The Cause of Autism.* New York: Baffin Books Pub; 2014:39–45.
6. Riikonen R. Insulin-like growth factors—neurobiological regulators of brain growth in autism? In: Zimmerman AW, ed. *Autism Current Theories and Evidence.* Totowa, NJ: Humana Press; 2008:233–244. Chapter 10.
7. Steinman G. Insulin-like growth factor and the etiology of autism. *Med Hypotheses.* 2013;80:475–480.
8. Bianchi VE, Locatelli V, Rissi L. Neurotrophic and neuro-regenerative efforts of GH/IGF1. *Int J Mol Sci.* 2017;18:2441–2466.
9. Delafontaine P, Song Y-H, Li Y. Expression, regulation, and function of IGF-1, IGF-1R, and IGF-1 binding proteins in blood vessels. *Arterioscler Thromb Vasc Biol.* 2004;24:435–444.
10. Buyukkayhan D, Tanzel F, Erselcan T, et al. Umbilical serum insulin-like growth factor (IGF-1) in newborn: effects of gestational age, postnatal age, and nutrition. *Int J Vitam Nutr Res.* 2003;73(5):343–346.
11. Bergman D, Halje M, Nording M, et al. Insulin-like growth factor in development and disease: a mini-review. *Geron.* 2013;59:240–249.
12. Barres BA, Freeman MR, Stevens B, eds. *Glia.* Cold Spring Harbor, NY: Cold Spring Harbor Laboratory Press; 2015. sections 1,2,4.
13. Herculano-Houzel S, Lent R. Isotropic fractionator: a simple, rapid method for the quantification of total cell and neuron numbers in the brain. *J Neurosci.* 2006;25(10):2518–2521.
14. Patterson PH. Maternal infections and immune involvement in autism. *Trends Mol Med.* 2011;17(7):389–394.
15. Riikonen R. Insulin-like growth factors in the pathogenesis of neurological diseases in children. *Int J Mol Sci.* 2017;18:2056–2074.
16. Vargas DL, Nascinbone C, Krishman C. Neurological activation and neuroinflammation in the brain of patients with autism. *Ann Neurol.* 2005;57(1):67–81.
17. Volpe JJ, Inder MB, Terre E, et al. *Neurology of the Newborn.* 6th ed. Elsevier; 2018156–157. 375–77, 407.

18. Petrelli F, Pucci L, Bezzi P. Astrocytes and microglia and their potential link with autism spectrum disorder. *Front Cell Neurosci.* 2016;10(21):1–8.
19. Kim JW, Howg JY, Bae SM. Microglia and autism spectrum disease: overview of current evidence and novel immunomodulatory treatment options. *Clin Psycho Pharm Neurosci.* 2018;16/3:246–252.
20. Chula FZ, Almina AB, Malinovskaya NA, et al. The glial perspective as disorders. *Neurosci Biobehav Rev.* 2014;38:160–172.
21. Bronzvoli MR, Facchinetti R, Ingrassia D, et al. Neuroglia in the autistic brain: evidence from a preclinical model. *Mol Autism.* 2018;9:66.
22. Slovak R, Chakraborty NG. The etiological role of microglia in autism spectrum disorder: a possible route for early intervention. *Am J Immunol.* 2017;13(2):99–106.
23. Kaur C, Rathnasamy G, Ling EA. Biology of microglia in the developing brain. *J Neuropathol Exp Neurol.* 2017;76(9):736–753.
24. Newman EA. New roles for astrocytes regulation of synaptic transmission. *Trends Neurosci.* 2003;26(10):536–542.
25. Santello M, Cali C, Bezzi P. Glia transmission and the tripartite synapse. *Adv Exp Med Biol.* 2012;970:307–331.
26. Chung W-S, Allen NJ, Eroglu C. Astrocytes control synapse formation, function, and elimination. In: Barnes SA, Freeman MR, Stevens B, eds. *Glia.* New York: Cold Spring Harbor Laboratory Press; 2015:17–34.
27. MacVicar BA, Newman EA. Astrocyte regulation of blood flow in the brain. In: Barnes SA, Freeman MR, Stevens B, eds. *Glia.* New York: Cold Spring Harbor Laboratory Press; 2015:35–48.
28. Hoang V, Flores M, Sandoval A. Brain imaging in the cause of autism. In: Steinman G, ed. *The Cause of Autism.* New York: Baffin Books Pub; 2014:82–88.
29. Weber B, Barros LF. The astrocyte: powerhouse and recycling center. In: Barnes SA, Freeman MR, Stevens B, eds. *Glia.* New York: Cold Spring Harbor Laboratory Press; 2015:49–64.
30. Daneman R, Prat A. The blood-brain barrier. In: Barnes SA, Freeman MR, Stevens B, eds. *Glia.* New York: Cold Spring Harbor Laboratory Press; 2015:83–105.
31. Sofroniew MV. Astroliosis. In: Barnes SA, Freeman MR, Stevens B, eds. *Glia.* New York: Cold Spring Harbor Laboratory Press; 2015:107–122.
32. Frost JL, Schafer DP. Microglia: architects of the developing nervous system. *Trends Cell Biol.* 2016;26(8):587–597.
33. Bergies DE, Richardson WD. Oligodendrocyte development and plasticity. In: Barnes SA, Freeman MR, Stevens B, eds. *Glia.* New York: Cold Spring Harbor Laboratory Press; 2015:139–165.
34. Emery B, Lu QR. Transcriptional and epigenetic regulation of oligodendrocyte development and myelination in the CNS. In: Barnes SA, Freeman MR, Stevens B, eds. *Glia.* New York: Cold Spring Harbor Laboratory Press; 2015:167–187.
35. Simons M, Klaus-Armin N. Oligodendrocytes: myelination and axonal support. In: Barnes SA, Freeman MR, Stevens B, eds. *Glia.* New York: Cold Spring Harbor Laboratory Press; 2015:189–203.
36. Herman P. *Physics of the Human Body—Nerve Conduction.* Heidelberg: Springer-Verlag; 2007:721.
37. Jessen KR, Mirsky R, Lloyd AC. Schwann cells: development and role in nerve repair. In: Barnes SA, Freeman MR, Stevens B, eds. *Glia.* New York: Cold Spring Harbor Laboratory Press; 2015:205–219.
38. Rasband MN, Peles E. The nodes of Ranvier: molecular assembly and maintenance. In: Barnes SA, Freeman MR, Stevens B, eds. *Glia.* New York: Cold Spring Harbor Laboratory Press; 2015:221–235.
39. Yanovski JA, Sovik KN, Nguyen TT, et al. Insulin-like growth factors and bone mineral density in African American and white girls. *J Pediatr.* 2000;137(6):826–832.

40. Durkin MS, Maenner MJ, Baio J, et al. Autism spectrum disorder among US children (2002–2010): socioeconomic, racial, and ethnic disparities. *Am J Public Health*. 2017;107(11):1818–1826.
41. Verhaeghe J, Coopmans W, Van Herck E, et al. IGF-I, IGF-II, IGF-binding protein 1, and a peptide in second trimester amniotic fluid are dependent on gestation but do not predict weight at birth. *Pediatr Res*. 1999;46(1):101–108.
42. Chellakooty M, Skibsted L, Skovby SO, et al. Longitudinal study of serum placental GH in 455 normal pregnancies: correlation to gestational age, fetal gender, and weight. *Clin Endocrinol Metab*. 2002;87(6):2734–2739.
43. Pinto-Martin JA, Levy SE, Feldman JF, et al. Prevalence of autism spectrum disorder in adolescents born weighing <2000 grams. *Pediatrics*. 2011;128(4):2010–2016.
44. Zikopoulos B, Barbas H. Changes in prefrontal axons may disrupt the network in autism. *J Neurosci*. 2010;30(44):14595–14608.
45. Steinman G. Predicting autism at birth. *Med Hypotheses*. 2013;81:21–25.
46. Steinman G, Mankuta D. Molecular biology of autism's etiology—an alternative mechanism. *Med Hypotheses*. 2019;130:109272.
47. Manning BD, Cantley LC. AKT/PKB signaling: navigating downstream. *Cell*. 2007;129(7):1261–1274.
48. Bibollet-Bahena O, Almazan G. IGF-1 stimulated protein synthesis in oligodendrocyte progenitors requires PI3K/mTOR/AKT and MEK/ERK pathways. *J Neurochem*. 2009;109(5):1440–1451.
49. Chen J, Alberts I, Li X. Dysregulation of IGF-1/PI3K/AKT/mTOR signaling pathway in autism spectrum disorders. *Int J Dev Neurosci*. 2014;35:35–41.
50. Bondy CA, Lee W-H, Cheng CM. Insulin-like growth factor (IGF1) and brain development. In: LeRoith D, Zumkeller W, Baxster RC, eds. *Insulin-Like Growth Factors*. New York: Kluwer Pub; 2003. Chap. 8.
51. Danbro BW, Singh R, Zimmerman MB, et al. Autism linked to increased oncogene mutations but decreased cancer rate. *PLoS One*. 2016;11(3):e0149041. https://doi.org/10.1371/journal.pone.0149041.
52. Steinman G. Plausible etiology of brain dysconnectivity in autism—review and prospectus. *Med Hypotheses*. 2015;85:405–407.
53. Purves D, Augustine GS, Fitzpatrick D, et al. *Neuroscience*. Sunderland, MA: Sinauer; 2012:77–106. 517–19, 526–38.
54. Nair A, Treiber JM, Shukla DK, et al. eds. Impaired thalamocortical connectivity in autism spectrum disorder: a study of functional and anatomical connectivity. *Brain*. 2013;136:1942–1955.
55. Oldehinkel M, Mennes M, Marquand A, et al. Altered connectivity between cerebellum, visual, and sensory-motor networks in autism spectrum disorder: results from the EU-AIMS longitudinal European autism project. *Biol Psychiatry Cogn Neurosci Neuroimaging*. 2019;4(3):260–270.
56. Zeng K, Kang J, Oyang G, et al. Disrupted brain network in children with autism spectrum disorder. *Sci Rep*. 2017;7(1):16253.
57. Holiga S, Hipp JF, Chatham D, et al. Patients with autism spectrum disorder display reproducible functional connectivity alterations. *Sci Transl Med*. 2019;11(481):eaat9223.
58. Mohammad-Rezazadeh I, Frohlich J, Loo SK, et al. Brain connectivity in autism spectrum disorder. *Curr Opin Neurol*. 2016;29(2):137–147.
59. Kana RK, Just MA. Autism as a disorder of functional brain connectivity. In: Amara D, Dawson G, Geschwind DH, eds. *Autism Spectrum Disorders*. Oxford: Oxford Univ. Press, Inc; 2011:981–990.
60. Pereira AM, Campos BM, Coan AC, et al. Differences in cortical structure and functional MRI connectivity in high functioning autism. *Front Neurol*. 2018; 9:539–569.

61. Pfisterer U, Khodosevich A. Neuronal survival in the brain: neuron type-specific mechanisms. *Cell Death Dis.* 2017;8:e2643–e2680.
62. Tang G, Gudsnuck K, Kuo S-H, et al. Loss of mTOR-dependent macroautophagy causes autistic-like synaptic pruning deficits. *Neuron.* 2014;83(5):1131–1143.
63. Just MA, Cherkasky VL, Keller TA, et al. Cortical activation and synchronization during sentence comprehension in high-functioning autism: evidence of underconnectivity. *Brain.* 2004;127:1811–1821.
64. Banai R, Pritchard B, Rosner M. Synaptic pathology. In: Steinman G, ed. *The Cause of Autism.* New York: Baffin Book Publishing; 2014:119–127.
65. Farag J, Rudin A. Brain macropathology. In: Steinman G, ed. *The Cause of Autism.* New York: Baffin Book Publishing; 2014:89–95.
66. Herzog D, Williams K. Brain pathology & behavior. In: Steinman G, ed. *The Cause of Autism.* New York: Baffin Book Publishing; 2014:95–101.
67. Szenczuk M, Bajurna M, Zych S, et al. Association of the autism-like growth factor I gene polymorphisms (IGF1/TasI and UGF1/SnaB1) with the growth and subsequent milk yield. *Czeh J Anim Sci.* 2013;58(9):404–411.
68. McGettrick AJ, Feener EP, Kash CR. Human insulin receptor substrate-1 (IRS-1) polymorphism G972R causes IRS-1 to associate with the insulin receptor and inhibit receptor autophosphorylation. *J Biol Chem.* 2005;280(8):6441–6446.
69. Park HJ, Kim SK, Kang WS, et al. Association between IRS1 gene polymorphism and autism spectrum disorder: a pilot case-control study in Korean males. *Int J Mol Sci.* 2016;17:1227–1235.
70. Kassam S, Meyer P, Corfield A, et al. Single nucleotide polymorphisms (SNPs): history, biotechnological outlook and practical applications. *Curr Pharmacogenomics.* 2005;3:1–4.
71. Jiao Y, Chen R, Ke X, et al. Single nucleotide polymorphisms predict symptom severity in autism spectrum disorder. *J Autism Dev Disord.* 2012;42(6):971–983.
72. Arends N, Johnston L, Hokken-Koelega A, et al. Polymorphs in the IGF-I gene: clinical relevance for short children born small for gestational age (SGA). *J Clin Endocrinol Metab.* 2002;87(6):2720–2724.
73. Shazbazi M, Abdulmohammadi R, Ebadi H, et al. Novel functional polymorphism in IGF-1 gene associated with multiple sclerosis: a new insight to MS. *Mult Scler Relat Disord.* 2017;13:33–37.
74. Wang Q, Liu L, Li H, et al. Genetic and dietary determinants of insulin-like growth factor (IGF-1) and IGF binding protein (BP)-3. *PLoS One.* 2014;9(10):e108934.
75. Tian Y, Wang L, Jia M, et al. Association of oligodendrocytes differentiation regular gene DUSP15 with autism. *World J Biol Psychiatry.* 2017;18(2):143–150.
76. Steinman G, Mankuta D. Breastfeeding as a possible deterrent to autism—a clinical perspective. *Med Hypotheses.* 2013;81:999–1001.
77. Ventura ER, Konigorski S, Rohrmann S, et al. Association of dietary intake of milk and dairy products with blood concentrations of insulin-like growth factor 1 (IGF-1) in Bavarian adults. *Eur J Nutr.* 2019. https://doi.org/10.1007/s00394-019-01944-7.
78. Heanley RP, McCarron DA, Dawson-Hughes B. Dietary changes favorably affect bone remodeling in older adults. *J Am Diet Assoc.* 1999;99:1228–1233.
79. Allen NE, Appleby PN, Davey GK, et al. The associations of diet with serum insulin-like growth factor I and its main binding proteins in 292 women meat-eaters, vegetarians, and vegans. *Cancer Epidemiol Biomarkers Prev.* 2000;11:1441–1448.
80. Shen WH, Xu RJ. Stability of insulin-like growth factor I in the gastrointestinal lumen in neonatal pigs. *J Pediatr Gastroenterol Nutr.* 2000;30:299–304.
81. Philipps AF, Dvorak B, Ling PJ, et al. Absorption of milk-borne insulin-like growth factor-I into portal blood of suckling rats. *J Pediatr Gastroenterol Nutr.* 2000;31:128–135.
82. Deoni SC, Zinstok JR, Daly E, et al. White-matter relaxation time and myelin water fraction differences in young adults with autism. *Psychol Med.* 2015;45(4):795–805.

83. Deoni SCL, Dean DCIII, Piryatinsky I, et al. Breastfeeding and early white matter development: a cross-sectional study. *Neuroimage*. 2013;82:77–86.
84. Tseng P-T, Chen Y-W, Stubbs B, et al. Maternal breastfeeding and autism spectrum disorder in children: a systematic review and meta-analysis. *Nutr Neurosci*. 2019;22(5):354–362.
85. Ozonoff S, Young GS, Carter A, et al. Recurrence risk for autism spectrum disorder: a baby siblings research consortium study. *Pediatrics*. 2011;128:e488–e495.
86. Field T, Diego M, Hernandez-Reis M, et al. Insulin and IGF1 (IGF-1) increased in preterm neonates. *J Dev Behav Pediatr*. 2008;29(6):463–466.
87. Procianoy RS, Mendex EW, Silveira RD. Massage therapy improves neurodevelopment outcome at two years correct for very low birth weight infants. *Early Hum Dev*. 2010;86:7–11.
88. Stubbs G, Henley K, Green J. Autism: will vitamin D supplementation during pregnancy and early childhood reduce recurrence rate of autism in newborn siblings? *Med Hypotheses*. 2016;88:74–78.
89. Ameri P, Giusti A, Bososchetti M, et al. Vitamin D increases circulating IGF1 in adults: potential implication for the treatment of GH deficiency. *Eur J Endocrinol*. 2017;169:767–772.
90. Gomez JM. The role of insulin-like growth factor I components in the regulation of vitamin D. *Curr Pharm Biotechnol*. 2006;7:125–132.
91. Cannell JJ. Vitamin D and autism, what's new? *Rev Endocr Metab Disord*. 2017; 18(2):183–189.
92. Steinman G, Mankuta D. Gene polymorphism in the genesis of autism. *BAOJ Neurol*. 2018;4(2):158–161.
93. Ballard F, Wallace JC, Francis GS. Des(1-3)IGF-1: a truncated form of insulin-like growth factor. *Int J Biochem Biol*. 1996;28:1085–1086.
94. Yuksel B, Nuiozbek M, Mungan NO, et al. Serum IGF-1 & IGFBP-3 levels in healthy children between 0 and 6 years of age. *J Clin Res Pediatr Endocrinol*. 2011; 3(2):84–88.
95. Siegel GJ. editor-in-chief*Basic Neurochemistry*. Philadelphia: Lippincott Williams & Wilkins; 199986.
96. Belmonte MK, Allen G, Beckel-Mitchener A, et al. Autism and abnormal development of brain connectivity. *J Neurosci*. 2004;24(42):9228–9231.
97. Steinman G, Mankuta D. The role of oligopeptides in preventing autism. *Med Hypotheses*. 2020;138:109604.
98. Copps KD, White M. Regulation of insulin sensitivity by serine/threonine phosphorylation of insulin receptor substrate proteins IRS1 and IRS2. *Diabetologia*. 2012;55(10):2565–2582. 2–12.
99. Wang Y, Wang W, Dongguo L, et al. IGF-1 alleviates NMDA-induces escitotoxicity in cultured hippocampal neurons agains autophagy via the NR2B/P13K-AKT-mTOR pathway. *J Cell Physiol*. 2014;229:1618–1629.
100. Steinman G. Prenatal identification of autism propensity. *Med Hypotheses*. 2019; 122:210–211.
101. Cieslinska A, Lostyra E, Chwala B, et al. Vitamin D receptor gene polymorphisms associated with childhood autism. *Brain Sci*. 2017;7:115–125.
102. Wass S. Distortions and disconnections: disrupted brain connectivity in autism. *Brain Cogn*. 2001;75:18–28.
103. Araque A, Navarrette M. Glial cells in neuronal network function. *Philos Trans R Soc B*. 2010;365:2375–2381.
104. Graciarena M, Seiffe A, Nait-Oumesmar B, et al. Hypomyelination and oligodendroglial alterations in mouse model of autism spectrum disorder. *Front Cell Neurosci*. 2019;12:517. https://doi.org/10.3389/fncel.2018.00517.

105. Yip BHK, Leonard H, Stock S, et al. Caesarean section and risk of autism across gestational age: a multinational cohort study of 5 million births. *In J Epidemiol.* 2017;46(2):429–439.
106. Joseph J, Sanghvi S. Pesticides; air traffic pollution. In: Steinman G, ed. *The Cause of Autism.* New York: Baffin Book Publishing; 2014:165–174.
107. Kassay B, McGill N, Segal R, et al. Vaccination/MMR. In: Steinman G, ed. *The Cause of Autism.* New York: Baffin Book Publishing; 2014:251–261.
108. Mirigliani E, Forte M. Leaky gut syndrome (opiate/gluten/casein). In: Steinman G, ed. *The Cause of Autism.* New York: Baffin Book Publishing; 2014:184–189.
109. Catani M, Dell'Acqua F, Budisavljevic S, et al. Frontal networks in adults with autism spectrum disorder. *Brain.* 2016;139:616–630.
110. Baumert M, Grabowska Z, Wojciechowsk E, et al. Insulin-like growth factor-1 (IGF-1) serum concentration in umbilical blood of term and preterm neonates. *Med Sci Monit.* 2004;10(Suppl. 2):80–82.
111. Hellstrom A, Ley D, Hansen-Pupp I, et al. Role of insulin-like growth factor 1 in fetal development and in the early postnatal life of premature infants. *Am J Perinatol.* 2016;33(11):1067–1071.
112. *Neuren Pharmaceutical.* Melbourne, Australia: Corporate Presentation; October, 2019.
113. Guan J, Gluckman P, Yang P, et al. Cyclic glycine-proline regulates homeostasis by altering the binding of IGFBP-3 to IGF-1. *Sci Rep.* 2014;4:4388.
114. Riikonen R. Treatment of autistic spectrum disorder with insulin-like growth factors. *Eur J Paediatr Neurol.* 2016;20(6):816–823.
115. Soke GN, Maenner M, Windham G, et al. Association between breastfeeding initiation and duration and autism spectrum disorder in preschool children enrolled in the study to explore early development. *Autism Res.* 2019;12:816–829.
116. Alzaree FA, AbuShady MM, Atti MA, et al. Effect of early breast milk nutrition on serum insulin-like growth factor-1 in preterm infants. *J Med Sci.* 2019;7(1):77–81.
117. Kinney HC, Volpe JJ. Myelination events. In: Volpe JJ, ed. *Neurology of the Newborn.* 6th ed. Elsevier; 2018. Chap. 8.
118. Steinman G, Covid-19 and autism, *Med Hypotheses* (in review).

CHAPTER TWO

Role of environmental factors and epigenetics in autism spectrum disorders

Roheeni Saxena[a], Melika Babadi[b], Hasti Namvarhaghighi[b], Florence I. Roullet[c,*]
[a]Department of Environmental Health Sciences, Mailman School of Public Health, Columbia University, New York, NY, United States
[b]School of Interdisciplinary Science, McMaster University, Hamilton, ON, Canada
[c]Department of Psychiatry and Behavioural Neurosciences, McMaster University, Hamilton, ON, Canada
*Corresponding author: e-mail address: roullet@mcmaster.ca

Contents

Abstract

Autism Spectrum Disorder (ASD) is a neurodevelopmental disorder thought to be caused by predisposing high-risk genes that may be altered during the early development by environmental factors. The impact of maternal challenges during pregnancy on the prevalence of ASD has been widely studied in clinical and animal studies. Here, we review some clinical and pre-clinical evidence that links environmental factors (i.e., infection, air pollution, pesticides, valproic acid and folic acid) and the risk of

Progress in Molecular Biology and Translational Science, Volume 173
ISSN 1877-1173
https://doi.org/10.1016/bs.pmbts.2020.05.002

ASD. Additionally, certain prenatal environmental challenges such as the valproate and folate prenatal exposures allow us to study mechanisms possibly linked to the etiology of ASD, for instance the epigenetic processes. These mechanistic pathways are also presented and discussed in this chapter.

It is widely thought that the etiology of ASD is due to predisposing hereditary high-risk genes altered during the early developmental pre- and/or postnatal periods, and that environmental factors may trigger these modifications. In this context, a lot of attention has been drawn to the impact of maternal challenges during pregnancy on the prevalence of ASD and, more recently, animal studies suggest that certain types of prenatal stressors could in fact impact several generations of offspring. These early environmental challenges and possible mechanisms linked to epigenetic processes will be presented and discussed in this chapter.

1. Context: Genetics yes, but not only

According to the Centers for Disease Control and Prevention (CDC), since 2000, Autism Spectrum Disorders (ASD) incidence has increased from 1/150 births to ~1/68.[1] Current (2017) CDC data shows US ASD prevalence at ~1.47%, and demonstrates that all ethnic, racial, and socioeconomic groups are affected by ASD.[1] If children with ASD are identified early and provided with appropriate intervention, core social symptoms characterized by a phenotype of serious social/communication deficits, and highly repetitive behaviors, may become less severe.[2] However, as patients diagnosed with *severe* ASD age, only 10% are able to live independently.[3] For those who cannot live independently, 12% require hospital care, 46% require residential care, and 19% have limited supervised independence.[3] Quality of life is typically not high for adults with ASD, and few are able to work or maintain friendships.[3]

In the United States, where the incidence of ASD has been increasing rapidly in the last two decades, ASD is associated with ~$11.5–60.9 billion in annual costs (adjusted to 2011 dollars).[4] Therefore, the economic burden of ASD is significant. Additionally, measures that quantify the cost of ASD patient care do not capture indirect effects, such as familial chronic stress, and other important family/community-level costs.

Overall, ASD has a significant societal impact in a variety of ways, including outcomes for individual patients, outcomes for families/communities, and economic costs associated with the care of ASD patients. Therefore, additional research into exposures that may increase the likelihood of ASD outcomes is needed. Over the last several years, ASD has had rapidly increasing incidence and prevalence; its current US prevalence is 1%–2%; approximately 7–14 million Americans have ASD diagnoses, with symptoms ranging from mild to severely debilitating.[4] However, despite the alarming increase in the number of people living with ASD, its overall etiology is poorly understood, though genetic risk, environmental exposures, and gene-environment interactions are known to play important roles.[5–13]

Part of the difficulty is the large variety of phenotypes and associated comorbidities within the ASD spectrum,[14] and it has been proposed that different causes may explain different presentations of the disorder.[15] Additional complexity could arise from the categorization of the three disorders regrouped within the ASD classification, namely, autism, Asperger's disorder and pervasive developmental disorder—not otherwise specified (PDD-NOS). Dr. Peter Szatmari in 2007 explains that, based on the Diagnostic and Statistical Manual of Mental Disorder (DSM)-IV, "the differentiation of autistic from non-ASD children can be done very well but the differentiation among the three types of ASD appears to be much less reliably and accurately carried out even when carried out with semi-structured instruments."

The altered function of several hundred of genes could be implicated in the etiology of ASD.[16,17] The genetic modifications most commonly reported include copy number variation, chromosomal rearrangements, and single-base pair mutations,[17,18] however, despite decades of intensive research in genetics, it has been difficult so far to identify a genetic hallmark common to more than 1% of the ASD population.[19] For instance, the maternal duplications in genetic sequence 15q11–q13 detected in ~1% of autism cases is one of the most common genetic abnormality observed.[20,21]

Despite these difficulties, several research studies indicate that genetics plays an important role with monozygotic (MZ) twins showing up between 60% and 90% concordance,[5] siblings share >50% heritability,[6] and ASD is inherited transgenerationally.[7,19] In a recent study performed on a population of over 2 million people mostly from northern Europe (Denmark, Sweden and Finland) (and partly in Israel), the authors suggested that 80% of the etiology of ASD was of genetic nature and environmental factors only account for a small proportion of the etiology of ASD.[22] This study did not include multiple births.

Indeed, twin studies hold a particular place in the assessment of the (genetic vs environment) etiology of autism.[23,24] Particularly, the comparison between pairs of MZ twins concordant and discordant for ASD outcomes could be particularly useful to weigh the proportion of genetics versus environmental factors.[24,25] Early studies found that the ASD-related concordance rates in MZ twins of the European populations are as high as 90%.[5,26–28] However, these earlier conclusions were contradicted by other findings which indicated that shared and non-shared environmental factors were important factors in the concordance of ASD.[29,30] Current limitations and difficulties of these types of studies include the usually small number of MZ twins discordant for ASD (as compared to other comparison groups such as MZ concordant for ASD and dizygotic (DZ) twins), and the possibility that conclusions drawn from twin studies in general would not be generalizable to non-twin population.[31]

With regard to the genes involved in ASD, neuronal network architecture and communication seem to be key genetic domains. Particularly, the genetic alterations most commonly reported affect the synaptic function, communication and structure, transcriptional regulation, chromatin-remodeling and immunity.[19,32–37] Hence, the prenatal and early postnatal developmental periods have been the target of intense investigations. Indeed, the fetal life is the stage of a major neuronal growth and arrangements.[38] From birth to 2 years of age, each neuron will grow an impressive amount in a pattern known as "dendritic arborization." Each dendrite will eventually share several thousands of connections with the dendrites of other adjacent neurons.[39] In early brain development, the connectivity or neural network of these cells is not complete. A necessary and crucial procedure of synaptic development and pruning (reduction of the neuronal branches) shapes the future central nervous system. Interferences with these processes can induce severe congenital abnormalities of the brain (reviewed in Rice and Barone 2000).[38] This process is particularly important in the context of the etiology of autism (reviewed in Copf 2016),[40] as abnormalities both anatomical and functional of the dendritic arborization could explain the dysfunctional sensory perception and integration by the brain in autism.[41–44] Indeed, while no physiological or genetic hallmarks have been identified, it has been widely reported that ASD brains go through an accelerated volume growth in early childhood,[45] characterized by enlarged frontal and temporal lobes.[46] The frontal lobe is involved in planning, projecting future consequences, conscience, and suppression of socially unacceptable responses, while the temporal lobe is involved in memory, language comprehension, and emotional processing.[47] Abnormalities in these regions may be responsible for the ASD behavioral phenotypes of communication and social interaction deficits. However, though these regions of the brain grow more rapidly in childhood for ASD subjects, this growth plateaus in adolescence[48]; while healthy brains continue to grow during adolescence, ASD brains show a period of arrested growth during these years.[41,48–50] It is possible that this period of arrested growth is similarly linked to ASD behavioral phenotypes. Finally, adult ASD brains show decreased total volume compared to healthy brains.[51] This decrease in overall brain volume, known as cerebral atrophy, may be responsible for some of the language and communication deficits seen in ASD.[52] In addition to these volume-related neuroanatomical phenotypes, as ASD brains grow, they show expansion of gray matter, and impaired maturation of cortical white matter.[50,53–55] White matter (which appears white due to the presence of myelinated axons) contributes primarily to learning and neuronal signal relay function, while gray matter (which contains fewer myelinated axons and more cell bodies, neuropil, dendrites, glial cells, synapses and capillaries) contributes to processing and cognition.[56,57] The differential growth of these two types of brain tissue in ASD brains may contribute to ASD symptomatology.

These important anatomo-physiological features of ASD brains reinforce the importance of exploring challenges which occur early during development affecting directly or indirectly neuronal and synaptic structural and functional pathways.

2. Environmental factors

2.1 Maternal stress during pregnancy and ASD

Numerous types of maternal challenges during pregnancy have been linked to increased risk of ASD from a variety of neurotoxins and pollutants such as pesticides and insecticides,[58] viral and bacterial infections,[59] exposure to metals such as mercury and lead,[59] to low level of electromagnetic radiations.[60] Though these exposures have not directly been linked to specific fetal development time points, exposure to these pollutants and maternal stressors during pregnancy has been linked to ASD outcomes.[61,62] Here we describe a non-exhaustive list of articles and findings related to the following stressors during pregnancy: (1) infection, (2) air pollutants, (3) pesticides (4) psychological stress, (5) anti-epileptic drug valproic acid; and (6) vitamin B9 or folate supplementation.

2.1.1 Infections

In 1978, Dr. Stella Chess and colleagues found a strikingly high prevalence of autism among 243 children with congenital rubella, stating that "a very high number of 18, or 7.4%, fell into the category of autism, whereas the expected rate is 0.7 per 10,000."[63] In 2010, a large study conducted in Denmark by Atladottir et al. examined the association between hospitalization for infection in the perinatal period and/or neonatal or childhood (not during pregnancy) and ASD diagnosis. Data from 1,418,152 children was collected and it was observed that children admitted to the hospital for an infectious as well as non-infectious disease had an increased rate in ASD diagnosis (hazard ratio 1.38 [1.31–1.45], and 1.76 [1.68–1.86]) respectively.[64] Because the associations were not specific to infectious diseases nor specific to ASD (admission to the hospital was also associated with an increase in mental retardation for instance), this study did not clearly support a relation of causality between infections in postnatal and childhood periods and ASD diagnosis.

However in 2012, Atladottir and colleagues found a twofold increased risk of infantile autism following maternal infection during pregnancy with influenza virus, whereas no association was found with common infections such as genital or labial herpes and cystitis.[65]

In 2015, Lee and colleagues published a study performed on a Swedish nationwide cohort of children born between 1984 and 2007, counting a total of over 2 million persons including over 24,000 cases of ASD. The authors concluded that approximately a 30% increase in ASD risk was associated with any inpatient diagnosis of viral or bacterial infection. This was true regardless of the timing (first, second or third trimester) of infection during pregnancy.[66]

Zerbo et al. (2015) studied a Northern Californian clinical database and similarly found an association between maternal bacterial infections during pregnancy and increased risk of ASD in the child. Although maternal infections diagnosed during the second trimester of pregnancy were associated with the highest risk of ASD, generally most infections occurred during the third trimester and most of them were urinary tract infections, amniotic fluid infections at delivery, and major puerperal (following childbirth) infections. In this study, the authors did not find an association between an increased risk of ASD and viral infections.

Recently, Bilbo et al. (2018) reviewed the various studies examining the possible mechanisms and links between maternal infection during pregnancy and ASD incidence, and it seems like the evidence supporting this link is growing.[67]

2.1.2 *Air pollutants*

As reviewed in Costa et al. 2014, air pollutants including metals, PM 2.5 (fine particulate matter with a diameter of less than 2.5 μm), highway air pollution, ozone, and nitrogen oxides have also been associated to ASD.[68] Interestingly, perinatal diesel exhaust exposure has also been significantly associated with ASD, particularly in boys.[69]

2.1.3 *Organochlorides (OC) and organophosphates (OP) pesticides*

Pesticides are defined as "substances used for preventing, destroying, repelling or mitigating pests, intended as insects, rodents, weeds, and a host of other unwanted organisms."[70,71] Pesticide usage has increased 50-fold since 1950 and is expected to reach 3.5 million tons worldwide in 2020.[72,73] Until the 1950s, arsenic-based pesticides were used predominantly.[74] Today's chemical families of pesticides include organochlorines (OC) such as dichlorodiphenyl-trichloroethane (DDT) and organophosphates (OP) and several studies have linked these compounds to neurodevelopmental disorders including autism (reviewed in the article by Kalkbrenner et al. 2014).[75] For instance, in 2007, in a large retrospective study with over 260,000 participants from the California Valley region including 465 cases of ASD, it was found that, during gestation, living in proximity to the usage of OC increases the risk of ASD in children while this risk decreases as distance from field sites increases.[76] Similarly, in 2014, a population-based case–control study (the "Childhood Autism Risks from Genetics and Environment" or "CHARGE" study) including 486 cases of ASD found residential proximity to agricultural pesticides during pregnancy to be associated with ASD.[77] However in 2018, Sagiv and collaborators assessed OP pesticides exposure during pregnancy through the measure of maternal pesticides metabolites in urine as well as residential proximity during pregnancy to OP treated sites, and analyzed associations with ASD-related traits in the children.[78] In this study conducted with the Center for Health Assessment of Mothers and Children of Salinas (CHAMACOS) cohort in the agricultural area of Salinas Valley in California, the authors did not find evidence of associations between residential proximity to OP usage during pregnancy and ASD-related traits. Notably, in this cohort which enrolled a total of 601 women, 84% of the mothers were born in Mexico and conclusions made on association may not be generalizable to other populations with different ethnic distributions.

Overall, the results from these studies are mixed and additional research on various geographical regions and various populations are needed to explore this important aspect of early neurodevelopmental challenges.

In 2007, Dr. Roberts wrote that "experimental and epidemiologic evidence regarding pesticides and pediatric neurodevelopment is strikingly lacking, despite considerable knowledge about pesticide toxicity (particularly neurotoxicity)".[76] This may still be true today.

2.1.4 Psychological stress

The link between ASD prenatal maternal psychological stress has also gained some attention. This type of stressor such as a divorce, financial difficulties, or the death of a relative during pregnancy, and has been assessed through self-reports and questionnaires in prospective and retrospective studies.[79–81] A large prospective study using cohorts with over 4400 ASD cases from Sweden and England found no association between exposure to stressful life events during the prenatal period and increased risk of ASD in children.[82] It must be noted that despite the large size of the cohort, in this particular study, stressful life events were experienced by less than 1% of the population in the Swedish cohort and "very few" in the English one. In contrast, in a study for which the stressor was a natural disaster (severe storms in New Orleans from 1980 to 1995) experienced by many, ASD prevalence among Louisiana children increased significantly with the severity of prenatal storm exposure. In particular, ASD prevalence was found especially increased for the "high exposure" group, the New Orleans children who were in the more sensitive periods of their gestation (middle or end of) when the center of severe hurricanes passed directly through New Orleans.[81] More recently, an Australian study of 174 children with ASD showed that repeated exposure to prenatal stressful life events in the 3 months prior to and during the pregnancy was a significant predictor of ASD-related symptom severity.[83]

Despite these interesting results, further research is needed in this domain of investigation which includes so far a large variety of psychological stressors.

2.1.5 Anti-epileptic drug valproic acid

Valproic acid (VPA) or 2-propylpentanoic acid was first discovered in 1963, and was used for its anticonvulsant properties in an animal model of seizures.[84] It has been used subsequently as an anti-epileptic drug and as a mood stabilizer. Today, in industrialized countries, around 20% of pregnant women with epilepsy receive valproate.[85] As reviewed in Roullet et al. (2013) and Chomiak et al. (2013), there have been numerous reports that anti-epileptic drugs have a clinical impact on fetal brain neurodevelopment when administered to the mother during pregnancy.[86,87] VPA is the most teratogenic of the commonly prescribed anti-epileptic drugs[88,89] and exposure of the fetus to VPA during pregnancy can lead to Fetal Valproate Syndrome (FVS), a rare congenital disorder characterized by spina bifida, atrial septal defect, cleft palate, and in some cases autism.[90,91] VPA is also the drug most commonly associated with ASD as ~9% of the studied children exposed to VPA developed either autism or Asperger's syndrome[92] and VPA monotherapy during pregnancy results in sevenfold greater incidence of ASD as compared to the general population.[93]

2.1.6 Folate (vitamine B9) supplementation

The literature is mixed regarding folate's direct connection to ASD.[94] Some data suggest that folic acid supplementation can reduce ASD incidence,[95–97] whereas other data suggest that increased dietary folate could be associated with increased ASD incidence.[94,98–100] Studies in Norway have shown that folic acid supplementation during pregnancy (to eliminate deficiency) have been linked to lower incidence of ASD in offspring.[96] In contrast, the sharp rise in US-based ASD cases has been temporally linked to the implementation of a folic acid food-fortification program, though it is not clear if this association indicates a causal mechanism between folic acid fortification and ASD.[101] Moreover additional recent research supports the stance that prenatal folic acid supplementation may *increase* the risk of ASD.[102,103] In addition to the importance of developmental folate exposure, folate may also play a role in the postnatal physiology of children and adults with ASD. Some studies of children with ASD have shown abnormal folate and abnormal One-Carbon-Metabolism-linked metabolic patterns throughout childhood.[104,105] Studies have also shown reduced 5-methyl-tetrahydrofolate transport into the cerebrospinal fluid for some ASD phenotypes, due to folate receptor autoimmunity[106]; these observations suggest that folate equilibrium may play a role in ASD symptomatology and pathology, perhaps even in early development. While folate nutritional status is not a primary cause of ASD, gene-nutrient interactions may alter the expression of ASD-linked genes, due to epigenetic effects.[100]

2.2 Developmental sensitivity windows

Human prenatal development is a continuous process that begins with fertilization and concludes with birth (detailed in Moore 2015 and Santro and Yussen 1992).[39,107] During normal development, the fetal organogenesis (formation of organs) takes place during the first trimester and constitutes a particularly sensitive period for exposure to teratogens. The second and third trimesters of prenatal development primarily involve fetal growth, as major structural organs have already been formed. Furthermore, important aspects of neurodevelopment, particularly of the brain and central nervous system, continue throughout these trimesters. The critical period for brain development and consequently developmental/intellectual disability runs from embryonic week 3 through embryonic week 16, and the central nervous system remains susceptible to teratogenic risk through birth. Therefore, neuro-teratogens can affect the brain and central nervous system even during the second and third trimesters, resulting in intellectual impairments, and neurological or behavioral abnormalities.

There is mixed data regarding a possible specific time-period during which maternal challenges increase susceptibility to ASD. Some exposures such as maternal illness appear to be linked to earlier pregnancy trimesters,[108] while others such as air pollution appear to be associated with ASD incidence when they occur in the third trimester.[109] ASD risk associated with proximity to pesticides usage during gestation were higher for exposure during the second and third trimesters of pregnancy with OP chlorpyrifos applications and just before conception or during the third trimester with pyrethroid insecticide applications.[77] Finally, during psychological maternal stressor (natural disasters), an increased prevalence was observed for children exposed during the middle or end of gestation period.[81]

3. Environmental animal models for ASD

Mammalian development follows a pattern of fertilization, embryonic development, and fetal growth such that researchers have been able to design studies examining subtle neurodevelopmental effects, such as behavioral disorders, by subjecting pregnant animals such as rats and mice to chemicals or other exposures known to induce brain defects. By timing these exposures so that they occur on specific prenatal days, it is possible to measure their developmental effects, and study both immediate and long-term consequences on offspring.

3.1 Anti-epileptic drug VPA rodent model

Based primarily on a behavioral neuroscience approach, VPA prenatal exposure has been used to create animal models for the study of autism.[110,111] Since the first studies from Dr. Patricia Rodier describing neuropathological similarities between brains from a human thalidomide case of autism and a valproate rat model for autism[112,113] and the pioneering work of Drs.

Thomas Schneider and Ryszard Przewlocki on prenatal exposure with VPA in the rat model,[110] throughout the years, there has been a wide interest for this unique model. The VPA rodent model for autism exhibits all three levels of validity with the human condition ("construct," i.e., similarity of cause, "face," i.e., similarity of phenotype, and "predictive," i.e., similarity of effect to similar treatment),[114] and these validities have been tested by many independent research teams through a large variety of approaches, behavioral, gene expression, electrophysiological, pharmacological and can be consider as a compelling tool to explore further the etiology of autism and potential underlying molecular mechanisms (reviewed in Roullet et al. 2013 and Tartaglione et al. 2019).[86,115]

3.2 Maternal infections models

Maternal Immune Activation, which consists of mimicking viral or bacterial infections which will trigger an inflammatory response in the pregnant dam, has been studied for autism and schizophrenia research.[116,117] The lipopolysaccharide (LPS) is a group of proteins which forms the bacterial wall. Injection of LPS during pregnancy in the rodent has been used to model autism. The LPS model has been characterized by increased pro-inflammatory cytokines levels.[118] Viral infection is simulated by injection of polyinosinic:polycytidylic acid (Poly I:C), a synthetic analog of double-stranded RNA[119,120] which induces the production of many pro-inflammatory cytokines and interferons.[117]

3.3 Exposure to pesticides models

The impact of prenatal exposure to pesticides on early brain development has been extended to the domain of autism. In 2014, a literature review on 13 studies involving gestational or early postnatal exposure to OP pesticides (chlorpyrifos and its active metabolites) in rats or mice, concluded that there was to date no convincing evidence that prenatal exposure to OP pesticides induces ASD-like symptoms in the rodent.[121]

However, in 2015 and 2016, De Felice and colleagues published two studies on prenatal exposure to an OP insecticide in the BTBR T+tf/J (BTBR) mouse strain, a mouse model of idiopathic autism. They show that prenatal exposure to OP insecticide significantly modifies motor activity in BTBR pups, delaying their motor development and leading to abnormally high rates of ultrasonic vocalization during a pup isolation test, and also further increases the brain oxidative stress levels, already heightened at baseline (i.e., without the prenatal challenge) in the model for autism.[122,123] Beyond the study of the role of pesticides in the incidence of autism, future studies similarly conducted on predisposed models for autism submitted to environmental challenges are needed to advance our understanding of the gene x environment interaction in autism.

3.4 Maternal psychological (social) stress model

It is possible to model the psychological and social stressor in animals. Recently a study revealed that parental stress induced by social isolation before and during pregnancy led to autistic-like features including decreased agonistic behavior and social transmission of flavor preference, reduced plasma oxytocin levels, increased plasma and brain BDNF levels in the offspring.[124]

4. Epigenetics mechanisms: The examples of valproate and folate

Stress during pregnancy could impact the brain development of the children through several mechanisms. For instance, a functional immune system is crucial for the early brain development and maternal immune activation could be a common pathway to environmental challenges such as infections and social stress in pregnancy. Another pathway which could be common to various stressors such as pesticides prenatal exposure are epigenetic processes. Here we will describe how environmental challenges, and in particular valproate and folate, may be linked directly or indirectly to epigenetic mechanisms.

As mentioned earlier, one of the best studied and recognized environmentally induced animal models for autism is the rodent VPA prenatal exposure model. How VPA impacts the brain is still not well understood in the adult and even less so in the developing brain, however there are two interesting known molecular effects of VPA: (1) VPA acts on the GABA-ergic system, (2) VPA modulates an important factor of epigenetic regulation: the protein histone deacetylase.

4.1 Indirectly through the γ-aminobutyric acid (GABA)-ergic system

As early as 1969, it was understood that VPA leads to an increase of GABA in the brain which might inhibit the abnormal neuronal activity in epileptic centers in the brain (reviewed in Hammond et al. 1981 and in Cotariu and Saidman 1991)[137,138] acting most probably through indirect effect on the cerebral enzymes that participate in the regulation of GABA.[139] In the mouse, low doses of VPA (125–290mg/kg) elevate GABA concentrations in the brain at the level of the synaptic terminals[140,141] and valproate increases the level of GABA in the cerebro-spinal fluid in human.[142]

GABA is an amino acid that acts as a neuromodulator. As described in Herlenius and Lagercrantz (2004),[125] "GABA is regarded as the main inhibitory transmitter in the mammalian neuronal system. During brain development, it functions as a trophic factor to influence events such as proliferation, migration, differentiation, synapse maturation, and cell deaths."[126,127] GABA plays a crucial role in the development of the cortex, and early modifications of its neurochemical balance will affect the brain wiring and later the functionality of important cortical systems, such as seen in the visual system.[125,128–130] Importantly in the context of in utero challenges, there is a "GABA shift of actions" from fetal life to birth, as GABA acts as an excitatory agent in the immature brain but as an inhibitor in the mature brain.[129] And this GABA excitatory-inhibitory shift during delivery is abolished in the VPA rat model for autism and the mouse genetic model of fragile X syndrome.[131] Cerebro-spinal levels of GABA in the immature neonate brain are higher (up to 100 times) compared to the adult levels[132] and challenges such as ethanol consumption during pregnancy can interfere with the GABA-ergic system and induce intellectual deficiencies, such as seen in the Fetal Alcohol Syndrome.[133] Craniofacial abnormalities, such as cleft palate, that have been observed in the human fetus after maternal absorption of alcohol during the first trimester have similarly been produced in a transgenic mouse model lacking a certain type of GABA receptors.[134]

GABA has gained a lot of attention in the context of autism research. In a mouse model for Rett syndrome, a genetic neurological disorder which shares some features with autism, Chao et al. (2010)[135] show that dysfunction in GABA signaling mediates autism-like stereotypies. Similarly, in a mouse model for Dravet's syndrome, an epileptic disorder associated with neuropsychological comorbidities, a mutation in SCN1A gene induces impairments in the GABA-ergic system. Notably the behavioral deficits observed in the mutant mouse include autistic-like hyperactivity, stereotyped behavior and social interaction. These social behavioral deficits are reversed by clonazepam, a benzodiazepine type of tranquilizer, which binds to GABAA receptors and increases the effect of GABA.[136] These results further reinforce the critical role of GABA on early brain development and in behavioral features linked to ASD.

Interestingly, there may be a link between GABA and epigenetic regulations of neuroplasticity. For instance, in the retina, GABA impacts the proliferation of neural progenitor cells by direct regulation through phosphorylation of transcription factor cAMP-responsive element-binding protein (CREB), and pharmacological blockade of CREB activity inhibited the effect of GABA on the expression of retinal neuronal marker.[143]

Therefore, the role of the GABA-ergic pathway in the context of autism etiology might be of multiple natures.

4.2 Directly through the inhibition of histone deacetylase

VPA is also directly involved in epigenetic regulation and acts as an inhibitor of histone deacetylase (HDAC).[144] In fact, the teratogenic effect of VPA has been described as a direct consequence to his impact on HDACs.[145] VPA-induced HDAC inhibition also affects the expression MECP2 protein which plays an important role in ASD, as it has been suggested that epigenetic modifications at the level of the MECP2 promoter may have direct implication in the development of ASD.[146] In short, VPA inhibits the activity of HDAC, which in turn impacts the expression of transcription factors Pax 6, a crucial regulator of glutamatergic neuronal differentiation, and MecP2 which regulates gene expression through chromatin remodeling.[147,148]

4.3 The transgenerational epigenetic inheritance

Another strong piece of evidence which indicates that epigenetic mechanisms are involved in VPA-induced autism, is the recent work performed in the animal showing that VPA induces transgenerational epigenetic inherited deficits. This phenomenon also referred to as "third generation effect" describes how an environmental challenge affecting one individual is carried on not (only) to the first-generation offspring but to the following generations.

This phenomenon has been previously described in 1998 in cancerology, when reduced fertility observed in female mice treated with an estrogen medication was not transmitted to their descendants but increased susceptibility to tumor formation was transmitted to subsequent generations.[149]

In 2016, Chang Soon Choi and colleagues showed in mice that VPA-induced behavioral and neuropathological modifications are carried through several generations.[150] The F1 "exposed" generation (offspring of the F0 VPA 'injected' generation) exhibited behavioral phenotypes including impaired sociability and repetitive behavior/hyperactivity which persisted onto the F2 and F3 "exposed" generations, even though the later generations (F1, F2 and F3) were never challenged directly by the VPA. Additionally, the cortices of F1 and F3 showed some imbalanced expressions of excitatory/inhibitory synaptic markers.[141] Tartaglione et al. in 2019 also examined the transgenerational effect of VPA in mice. They analyzed the behaviors and levels of expression of endogenous retroviruses (ERVs) as a marker of defective brain development in three generations (F1–F3, these generations being bred in a different fashion than the ones used by Choi et al. 2016). F1, F2 and F3 VPA generations showed increased motor activity and reduced ultrasonic vocalizations. In addition, VPA increased expression of ERVs in the brain and blood across the three generations.[151]

4.4 Folate and epigenetics mechanisms

Folate has a variety of functions within the body. The pervasiveness of folate's role in metabolism, nervous system function, and as a precursor of S-adenosyl-methionine (SAM), the folate donor for DNA methylation, suggests that there may be a connection between folate status during in utero development and ASD outcomes, via the mediator of DNA methylation.

Specifically, folate is important for One-Carbon Metabolism (OCM), and is required for the formation of SAM, a universal methyl donor that facilitates various methylation reactions, including DNA methylation.[152] In addition to its role in generating SAM, folate is a limiting factor in OCM, and consequently in DNA synthesis/methylation.[153,154] Folate and SAM are known to influence cognitive outcomes via their role in DNA methylation.[155–158] Studies have suggested that folate deficiency/excess may alter methylation[159] of genes implicated in neural development and cognitive function.[155] Additionally, animal studies have demonstrated that folate deficiency can affect brain structure, development, plasticity, and neuron proliferation/differentiation/survival, likely via folate-linked changes in DNA methylation profiles.[155,160–162] Via these mechanisms, developmental deficiency/excess in folate may adversely affect central nervous system (CNS) and brain function.[163] The CNS is most vulnerable to environmental exposures during in utero development, when its plasticity is greatest.[95,164–166]

Interestingly, valproate and folate seem to work in similar ways, both being capable of inducing neuroprotection and/or defects. VPA inhibits intestinal folic acid absorption, impacts folate levels in liver, brain, and plasma[138] and both are involved in neural tube defects processing: whereas prenatal exposure to VPA induces neural tube defects, folate supplementation during pregnancy is usually recommended to prevent them.[167]

It would be interesting to study further their relationship, particularly with regard to the epigenetic balance implicated in certain molecular pathways linked to ASD.

5. Conclusion

There are several challenges during pregnancy linked to the prevalence of ASD and their study could lead us to better understand the complexities and commonalities of pathways in the etiology of ASD. One common aspect developed in this manuscript is the direct and indirect links to epigenetic alterations, but the maternal and fetal neuroimmune responses to a variety of stressors during pregnancy is another crucial domain of research which remains to be explored further. Finally, the time window of interest is not restricted to the gestational period but events happening during the early years of postnatal development, together with the pre-gestational period, up to generations back are now being considered.

References

1. Javurek AB, Suresh D, Spollen WG, et al. Gut dysbiosis and neurobehavioral alterations in rats exposed to silver nanoparticles. *Sci Rep*. 2017;7(1):2822. https://doi.org/10.1038/s41598-017-02880-0.
2. Howlin P. Autism spectrum disorders. *Psychiatry (Abingdon)*. 2006;5(9):320–324.
3. Newschaffer CJ, Croen LA, Daniels J, et al. The epidemiology of autism spectrum disorders. *Annu Rev Public Health*. 2007;28:235–258. https://doi.org/10.1146/annurev.publhealth.28.021406.144007.
4. Centers Disease Control. *Autism Spectrum Disorder (ASD)*. Centers Diease Control; 2017.
5. Bailey A, Le Couteur A, Gottesman I, et al. Autism as a strongly genetic disorder: evidence from a British twin study. *Psychol Med*. 1995;25(1):63–77. https://doi.org/10.1017/s0033291700028099.
6. Klei L, Sanders SJ, Murtha MT, et al. Common genetic variants, acting additively, are a major source of risk for autism. *Mol Autism*. 2012;3(1):9. https://doi.org/10.1186/2040-2392-3-9.
7. Tsang KM, Croen LA, Torres AR, et al. A genome-wide survey of transgenerational genetic effects in autism. *PLoS One*. 2013;8(10):e76978. https://doi.org/10.1371/journal.pone.0076978.
8. Siu MT, Weksberg R. Epigenetics of autism spectrum disorder. *Adv Exp Med Biol*. 2017;978:63–90. https://doi.org/10.1007/978-3-319-53889-1_4.

9. Uher R, Zwicker A. Etiology in psychiatry: embracing the reality of poly-gene-environmental causation of mental illness. *World Psychiatry*. 2017;16(2):121–129. https://doi.org/10.1002/wps.20436.
10. Tran NQV, Miyake K. Neurodevelopmental disorders and environmental toxicants: epigenetics as an underlying mechanism. *Int J Genomics*. 2017;2017:7526592. https://doi.org/10.1155/2017/7526592.
11. Keil KP, Lein PJ. DNA methylation: a mechanism linking environmental chemical exposures to risk of autism spectrum disorders? *Environ Epigenetics*. 2016;2(1): dvv012. https://doi.org/10.1093/eep/dvv012.
12. Rossignol DA, Genuis SJ, Frye RE. Environmental toxicants and autism spectrum disorders: a systematic review. *Transl Psychiatry*. 2014;4:e360. https://doi.org/10.1038/tp.2014.4.
13. Mandy W, Lai M-C. Annual research review: the role of the environment in the developmental psychopathology of autism spectrum condition. *J Child Psychol Psychiatry*. 2016;57(3):271–292. https://doi.org/10.1111/jcpp.12501.
14. Szatmari P, White J, Merikangas KR. The use of genetic epidemiology to guide classification in child and adult psychopathology. *Int Rev Psychiatry*. 2007;19(5):483–496. https://doi.org/10.1080/09540260701563619.
15. Frazier TW, Thompson L, Youngstrom EA, et al. A twin study of heritable and shared environmental contributions to autism. *J Autism Dev Disord*. 2014;44(8):2013–2025. https://doi.org/10.1007/s10803-014-2081-2.
16. Abrahams BS, Arking DE, Campbell DB, et al. SFARI Gene 2.0: a community-driven knowledgebase for the autism spectrum disorders (ASDs). *Mol Autism*. 2013;4(1):36. https://doi.org/10.1186/2040-2392-4-36.
17. de la Torre-Ubieta L, Won H, Stein JL, Geschwind DH. Advancing the understanding of autism disease mechanisms through genetics. *Nat Med*. 2016;22(4):345–361. https://doi.org/10.1038/nm.4071.
18. Devlin B, Scherer SW. Genetic architecture in autism spectrum disorder. *Curr Opin Genet Dev*. 2012;22(3):229–237. https://doi.org/10.1016/j.gde.2012.03.002.
19. Bourgeron T. From the genetic architecture to synaptic plasticity in autism spectrum disorder. *Nat Rev Neurosci*. 2015;16(9):551–563. https://doi.org/10.1038/nrn3992.
20. Schroer RJ, Phelan MC, Michaelis RC, et al. Autism and maternally derived aberrations of chromosome 15q. *Am J Med Genet*. 1998;76(4):327–336. https://doi.org/10.1002/(sici)1096-8628(19980401)76:4less than 327::aid-ajmg8greater than 3.0.co;2-m.
21. Geschwind DH, State MW. Gene hunting in autism spectrum disorder: on the path to precision medicine. *Lancet Neurol*. 2015;14(11):1109–1120. https://doi.org/10.1016/S1474-4422(15)00044-7.
22. Bai D, Yip BHK, Windham GC, et al. Association of Genetic and Environmental Factors with autism in a 5-country cohort. *JAMA Psychiatry*. 2019;76(10):1035–1043. https://doi.org/10.1001/jamapsychiatry.2019.1411.
23. Willfors C, Carlsson T, Anderlid B-M, et al. Medical history of discordant twins and environmental etiologies of autism. *Transl Psychiatry*. 2017;7(1):e1014. https://doi.org/10.1038/tp.2016.269.
24. Imamura A, Morimoto Y, Ono S, et al. Genetic and environmental factors of schizophrenia and autism spectrum disorder: insights from twin studies. *J Neural Transm*. 2020; https://doi.org/10.1007/s00702-020-02188-w. Online ahead of print.
25. Isaksson J, Taylor MJ, Lundin K, Neufeld J, Bolte S. Familial confounding on the ability to read minds: a co-twin control study. *Autism*. 2019;23(8):1948–1956. https://doi.org/10.1177/1362361319836380.
26. Folstein S, Rutter M. Genetic influences and infantile autism. *Nature*. 1977;265(5596):726–728. https://doi.org/10.1038/265726a0.

27. Steffenburg S, Gillberg C, Hellgren L, et al. A twin study of autism in Denmark, Finland, Iceland, Norway and Sweden. *J Child Psychol Psychiatry.* 1989;30(3):405–416. https://doi.org/10.1111/j.1469-7610.1989.tb00254.x.
28. Lichtenstein P, Carlström E, Råstam M, Gillberg C, Anckarsäter H. The genetics of autism spectrum disorders and related neuropsychiatric disorders in childhood. *Am J Psychiatry.* 2010;167(11):1357–1363. https://doi.org/10.1176/appi.ajp.2010.10020223.
29. Sandin S, Lichtenstein P, Kuja-Halkola R, Larsson H, Hultman CM, Reichenberg A. The familial risk of autism. *JAMA.* 2014;311(17):1770–1777. https://doi.org/10.1001/jama.2014.4144.
30. Hallmayer J, Cleveland S, Torres A, et al. Genetic heritability and shared environmental factors among twin pairs with autism. *Arch Gen Psychiatry.* 2011;68(11):1095–1102. https://doi.org/10.1001/archgenpsychiatry.2011.76.
31. Isaksson J, Tammimies K, Neufeld J, et al. EU-AIMS longitudinal European Autism Project (LEAP): The autism twin cohort. *Mol Autism.* 2018;9:26. https://doi.org/10.1186/s13229-018-0212-x.
32. Krumm N, Turner TN, Baker C, et al. Excess of rare, inherited truncating mutations in autism. *Nat Genet.* 2015;47(6):582–588. https://doi.org/10.1038/ng.3303.
33. De Rubeis S, He X, Goldberg AP, et al. Synaptic, transcriptional and chromatin genes disrupted in autism. *Nature.* 2014;515(7526):209–215. https://doi.org/10.1038/nature13772.
34. Waye MMY, Cheng HY. Genetics and epigenetics of autism: a review. *Psychiatry Clin Neurosci.* 2018;72(4):228–244. https://doi.org/10.1111/pcn.12606.
35. Voineagu I, Wang X, Johnston P, et al. Transcriptomic analysis of autistic brain reveals convergent molecular pathology. *Nature.* 2011;474(7351):380–384. https://doi.org/10.1038/nature10110.
36. Parikshak NN, Luo R, Zhang A, et al. Integrative functional genomic analyses implicate specific molecular pathways and circuits in autism. *Cell.* 2013;155(5):1008–1021. https://doi.org/10.1016/j.cell.2013.10.031.
37. Sun W, Poschmann J, Cruz-Herrera Del Rosario R, et al. Histone acetylome-wide association study of autism spectrum disorder. *Cell.* 2016;167(5):1385–1397.e11. https://doi.org/10.1016/j.cell.2016.10.031.
38. Rice D, Barone SJ. Critical periods of vulnerability for the developing nervous system: evidence from humans and animal models. *Environ Health Perspect.* 2000; 108(suppl):511–533. https://doi.org/10.1289/ehp.00108s3511.
39. Santrock J, Yussen S. In: Lange M, ed. *Child Development Fifth Edition.* W. C Brown Publishers; 1992.
40. Copf T. Impairments in dendrite morphogenesis as etiology for neurodevelopmental disorders and implications for therapeutic treatments. *Neurosci Biobehav Rev.* 2016;68:946–978. https://doi.org/10.1016/j.neubiorev.2016.04.008.
41. Courchesne E. Brain development in autism: early overgrowth followed by premature arrest of growth. *Ment Retard Dev Disabil Res Rev.* 2004;10(2):106–111. https://doi.org/10.1002/mrdd.20020.
42. Courchesne E, Pierce K. Brain overgrowth in autism during a critical time in development: implications for frontal pyramidal neuron and interneuron development and connectivity. *Int J Dev Neurosci.* 2005;23(2–3):153–170. https://doi.org/10.1016/j.ijdevneu.2005.01.003.
43. Courchesne E, Pierce K, Schumann CM, et al. Mapping early brain development in autism. *Neuron.* 2007;56(2):399–413. https://doi.org/10.1016/j.neuron.2007.10.016.
44. Rinaldi T, Kulangara K, Antoniello K, Markram H. Elevated NMDA receptor levels and enhanced postsynaptic long-term potentiation induced by prenatal exposure to valproic acid. *Proc Natl Acad Sci USA.* 2007;104(33):13501–13506. https://doi.org/10.1073/pnas.0704391104.

45. Courchesne E. Abnormal early brain development in autism. *Mol Psychiatry*. 2002;7(suppl. 2):S21–S23. https://doi.org/10.1038/sj.mp.4001169.
46. Courchesne E, Campbell K, Solso S. Brain growth across the life span in autism: age-specific changes in anatomical pathology. *Brain Res*. 2011;1380:138–145. https://doi.org/10.1016/j.brainres.2010.09.101.
47. Smith EE, Kosslyn SM. *Cognitive Psychology: Mind and Brain*. Pearson; 2007.
48. Lange N, Travers BG, Bigler ED, et al. Longitudinal volumetric brain changes in autism spectrum disorder ages 6-35 years. *Autism Res*. 2015;8(1):82–93. https://doi.org/10.1002/aur.1427.
49. Lin H-Y, Ni H-C, Lai M-C, Tseng W-YI, Gau SS-F. Regional brain volume differences between males with and without autism spectrum disorder are highly age-dependent. *Mol Autism*. 2015;6:29. https://doi.org/10.1186/s13229-015-0022-3.
50. Hazlett HC, Poe MD, Gerig G, et al. Early brain overgrowth in autism associated with an increase in cortical surface area before age 2 years. *Arch Gen Psychiatry*. 2011;68(5):467–476. https://doi.org/10.1001/archgenpsychiatry.2011.39.
51. Carper RA, Moses P, Tigue ZD, Courchesne E. Cerebral lobes in autism: early hyperplasia and abnormal age effects. *Neuroimage*. 2002;16(4):1038–1051. https://doi.org/10.1006/nimg.2002.1099.
52. National Institute of Neurological Disorders and Stroke. *Cerebral Atrophy Fact Sheet*. National Institute of Neurological Disorders and Stroke; 2017.
53. Casanova MF. White matter volume increase and minicolumns in autism. *Ann Neurol*. 2004;56(3):453. author reply 454. https://doi.org/10.1002/ana.20196.
54. Hardan AY, Jou RJ, Keshavan MS, Varma R, Minshew NJ. Increased frontal cortical folding in autism: a preliminary MRI study. *Psychiatry Res*. 2004;131(3):263–268. https://doi.org/10.1016/j.pscychresns.2004.06.001.
55. Levitt JG, Blanton RE, Smalley S, et al. Cortical sulcal maps in autism. *Cereb Cortex*. 2003;13(7):728–735. https://doi.org/10.1093/cercor/13.7.728.
56. Fields D. White matter matters. *Sci Am*. 2008;3(298):54–61.
57. Blumenfeld H. *Neuroanatomy Through Clinical Cases*. Sunderland, MA: Sinauer Associates; 2010.
58. Furlong MA, Barr DB, Wolff MS, Engel SM. Prenatal exposure to pyrethroid pesticides and childhood behavior and executive functioning. *Neurotoxicology*. 2017;62:231–238. https://doi.org/10.1016/j.neuro.2017.08.005.
59. Ornoy A, Weinstein-Fudim L, Ergaz Z. Prenatal factors associated with autism spectrum disorder (ASD). *Reprod Toxicol*. 2015;56:155–169. https://doi.org/10.1016/j.reprotox.2015.05.007.
60. Thornton IM. Out of time: a possible link between mirror neurons, autism and electromagnetic radiation. *Med Hypotheses*. 2006;67(2):378–382. https://doi.org/10.1016/j.mehy.2006.01.032.
61. Grabrucker AM. Environmental factors in autism. *Front Psych*. 2012;3:118. https://doi.org/10.3389/fpsyt.2012.00118.
62. Matelski L, Van de Water J. Risk factors in autism: thinking outside the brain. *J Autoimmun*. 2016;67:1–7. https://doi.org/10.1016/j.jaut.2015.11.003.
63. Chess S, Fernandez P, Korn S. Behavioral consequences of congenital rubella. *J Pediatr*. 1978;93(4):699–703.
64. Atladóttir HÓ, Thorsen P, Schendel DE, Østergaard L, Lemcke S, Parner ET. Association of hospitalization for infection in childhood with diagnosis of autism spectrum disorders: a Danish cohort study. *Arch Pediatr Adolesc Med*. 2010;164(5):470–477. https://doi.org/10.1001/archpediatrics.2010.9.
65. Atladóttir HÓ, Henriksen TB, Schendel DE, Parner ET. Autism after infection, febrile episodes, and antibiotic use during pregnancy: an exploratory study. *Pediatrics*. 2012;130(6):e1447–e1454. https://doi.org/10.1542/peds.2012-1107.

66. Lee BK, Magnusson C, Gardner RM, et al. Maternal hospitalization with infection during pregnancy and risk of autism spectrum disorders. *Brain Behav Immun.* 2015;44:100–105. https://doi.org/10.1016/j.bbi.2014.09.001.
67. Bilbo SD, Block CL, Bolton JL, Hanamsagar R, Tran PK. Beyond infection—maternal immune activation by environmental factors, microglial development, and relevance for autism spectrum disorders. *Exp Neurol.* 2018;299(pt. A):241–251. https://doi.org/10.1016/j.expneurol.2017.07.002.
68. Costa LG, Cole TB, Coburn J, Chang YC, Dao K, Roque P. Neurotoxicants are in the air: convergence of human, animal, and in vitro studies on the effects of air pollution on the brain. *Biomed Res Int.* 2014. https://doi.org/10.1155/2014/736385.
69. Roberts AL, Lyall K, Hart JE, et al. Perinatal air pollutant exposures and autism spectrum disorder in the children of Nurses' Health Study II participants. *Environ Health Perspect.* 2013;121(8):978–984. https://doi.org/10.1289/ehp.1206187.
70. Costa LG, Giordano G, Guizzetti M, Vitalone A. Neurotoxicity of pesticides: a brief review. *Front Biosci.* 2008;1(13):1240–1249. https://doi.org/10.2741/2758.
71. Echobicon D. Toxic effects of pesticides. In: Klaassen C, ed. *Casarett and Doull's Toxicology. The Basic Science of Poisons.* New York: McGraw-Hill; 2001:763–810.
72. Atwood D, Paisley-Jones C. Pesticides industry sales and usage: 2008–2012 market estimates. In: *Office of Pesticide Programs Office of Chemical Safety and Pollution Prevention.* 2017.
73. Sharma A, Kumar V, Shahzad B, et al. Worldwide pesticide usage and its impacts on ecosystem. *SN Appl Sci.* 2019;1(11):1446. https://doi.org/10.1007/s42452-019-1485-1.
74. Ritter S. Pinpointing trends in pesticide use in 1939. *Chem Eng News.* 2009;87(7).
75. Kalkbrenner AE, Schmidt RJ, Penlesky AC. Environmental chemical exposures and autism spectrum disorders: a review of the epidemiological evidence. *Curr Probl Pediatr Adolesc Health Care.* 2014;44(10):277–318. https://doi.org/10.1016/j.cppeds.2014.06.001.
76. Roberts EM, English PB, Grether JK, Windham GC, Somberg L, Wolff C. Maternal residence near agricultural pesticide applications and autism spectrum disorders among children in the California Central Valley. *Environ Health Perspect.* 2007;115(10):1482–1489. https://doi.org/10.1289/ehp.10168.
77. Shelton JF, Geraghty EM, Tancredi DJ, et al. Neurodevelopmental disorders and prenatal residential proximity to agricultural pesticides: the charge study. *Environ Health Perspect.* 2014;122(10):1103–1109. https://doi.org/10.1289/ehp.1307044.
78. Sagiv SK, Harris MH, Gunier RB, et al. Prenatal organophosphate pesticide exposure and traits related to autism Spectrum disorders in a population living in proximity to agriculture. *Environ Health Perspect.* 2018;126(4):47012. https://doi.org/10.1289/EHP2580.
79. Ward AJ. A comparison and analysis of the presence of family problems during pregnancy of mothers of "autistic" children and mothers of normal children. *Child Psychiatry Hum Dev.* 1990;20(4):279–288. https://doi.org/10.1007/BF00706020.
80. Beversdorf DQ, Manning SE, Hillier A, et al. Timing of prenatal stressors and autism. *J Autism Dev Disord.* 2005;35(4):471–478. https://doi.org/10.1007/s10803-005-5037-8.
81. Kinney DK, Miller AM, Crowley DJ, Huang E, Gerber E. Autism prevalence following prenatal exposure to hurricanes and tropical storms in Louisiana. *J Autism Dev Disord.* 2008;38(3):481–488. https://doi.org/10.1007/s10803-007-0414-0.
82. Rai D, Golding J, Magnusson C, Steer C, Lewis G, Dalman C. Prenatal and early life exposure to stressful life events and risk of autism spectrum disorders: population-based studies in Sweden and England. *PLoS One.* 2012;7(6):1–8. https://doi.org/10.1371/journal.pone.0038893.

83. Varcin KJ, Alvares GA, Uljarević M, Whitehouse AJO. Prenatal maternal stress events and phenotypic outcomes in Autism Spectrum Disorder. *Autism Res.* 2017;10(11):1866–1877. https://doi.org/10.1002/aur.1830.
84. Meunier H, Carraz G, Neunier Y, Eymard P, Aimard M. Pharmacodynamic properties of N-dipropylacetic acid. *Therapie.* 1963;18:435–438.
85. Kulaga S, Sheehy O, Zargarzadeh AH, Moussally K, Berard A. Antiepileptic drug use during pregnancy: perinatal outcomes. *Seizure.* 2011;20(9):667–672. https://doi.org/10.1016/j.seizure.2011.06.012.
86. Roullet FI, Lai JKY, Foster JA. In utero exposure to valproic acid and autism—a current review of clinical and animal studies. *Neurotoxicol Teratol.* 2013;36:47–56. https://doi.org/10.1016/j.ntt.2013.01.004.
87. Chomiak T, Turner N, Hu B. What we have learned about autism spectrum disorder from valproic acid. *Patholog Res Int.* 2013;2013:712758. https://doi.org/10.1155/2013/712758.
88. Hernandez-Diaz S, Smith CR, Shen A, et al. Comparative safety of antiepileptic drugs during pregnancy. *Neurology.* 2012;78(21):1692–1699. https://doi.org/10.1212/WNL.0b013e3182574f39.
89. Vajda FJE, Hitchcock A, Graham J, O'Brien T, Lander C, Eadie M. The Australian register of antiepileptic drugs in pregnancy: the first 1002 pregnancies. *Aust N Z J Obstet Gynaecol.* 2007;47(6):468–474. https://doi.org/10.1111/j.1479-828X.2007.00781.x.
90. Christianson AL, Chesler N, Kromberg JG. Fetal valproate syndrome: clinical and neuro-developmental features in two sibling pairs. *Dev Med Child Neurol.* 1994;36(4):361–369. https://doi.org/10.1111/j.1469-8749.1994.tb11858.x.
91. Williams PG, Hersh JH. A male with fetal valproate syndrome and autism. *Dev Med Child Neurol.* 1997;39(9):632–634. https://doi.org/10.1111/j.1469-8749.1997.tb07500.x.
92. Rasalam AD, Hailey H, Williams JHG, et al. Characteristics of fetal anticonvulsant syndrome associated autistic disorder. *Dev Med Child Neurol.* 2005;47(8):551–555. https://doi.org/10.1017/S0012162205001076.
93. Bromley RL, Mawer G, Clayton-Smith J, Baker GA. Autism spectrum disorders following in utero exposure to antiepileptic drugs. *Neurology.* 2008;71(23):1923–1924. https://doi.org/10.1212/01.wnl.0000339399.64213.1a.
94. DeVilbiss EA, Gardner RM, Newschaffer CJ, Lee BK. Maternal folate status as a risk factor for autism spectrum disorders: a review of existing evidence. *Br J Nutr.* 2015;114(5):663–672. https://doi.org/10.1017/S0007114515002470.
95. Roth C, Magnus P, Schjolberg S, et al. Folic acid supplements in pregnancy and severe language delay in children. *JAMA.* 2011;306(14):1566–1573. https://doi.org/10.1001/jama.2011.1433.
96. Tonks A. Periconceptual folate and autism. *BMJ.* 2013;346:f900. https://doi.org/10.1136/bmj.f900.
97. Schmidt RJ, Tancredi DJ, Ozonoff S, et al. Maternal periconceptional folic acid intake and risk of autism spectrum disorders and developmental delay in the CHARGE (CHildhood autism risks from genetics and environment) case-control study. *Am J Clin Nutr.* 2012;96(1):80–89. https://doi.org/10.3945/ajcn.110.004416.
98. Lyall K, Schmidt RJ, Hertz-Picciotto I. Maternal lifestyle and environmental risk factors for autism spectrum disorders. *Int J Epidemiol.* 2014;43(2):443–464. https://doi.org/10.1093/ije/dyt282.
99. Moretti P, Peters SU, Del Gaudio D, et al. Brief report: autistic symptoms, developmental regression, mental retardation, epilepsy, and dyskinesias in CNS folate deficiency. *J Autism Dev Disord.* 2008;38(6):1170–1177. https://doi.org/10.1007/s10803-007-0492-z.

100. Leeming RJ, Lucock M. Autism: is there a folate connection? *J Inherit Metab Dis.* 2009;32(3):400–402. https://doi.org/10.1007/s10545-009-1093-0.
101. James SJ, Melnyk S, Jernigan S, et al. Metabolic endophenotype and related genotypes are associated with oxidative stress in children with autism. *Am J Med Genet B Neuropsychiatr Genet.* 2006;141B(8):947–956. https://doi.org/10.1002/ajmg.b.30366.
102. Desoto MC, Hitlan RT. Synthetic folic acid supplementation during pregnancy may increase the risk of developing autism. *J Pediatr Biochem.* 2012;2(4):251–261. https://doi.org/10.3233/JPB-120066.
103. Wiens D, DeSoto MC. Is high folic acid intake a risk factor for autism?-A review. *Brain Sci.* 2017;7(11):149. https://doi.org/10.3390/brainsci7110149.
104. James SJ. Autism and folate-dependent one-carbon metabolism: serendipity and critical branch-point decisions in science. *Glob Adv Health Med.* 2013;2(6):48–51. https://doi.org/10.7453/gahmj.2013.088.
105. Frye RE, James SJ. Metabolic pathology of autism in relation to redox metabolism. *Biomark Med.* 2014;8(3):321–330. https://doi.org/10.2217/bmm.13.158.
106. Ramaekers VT, Blau N, Sequeira JM, Nassogne M-C, Quadros EV. Folate receptor autoimmunity and cerebral folate deficiency in low- functioning autism with neurological deficits. *Neuropediatrics.* 2007;38(6):276–281. https://doi.org/10.1055/s-2008-1065354.
107. Moore K. *The Developing Human.* 10th ed. Elsevier; 2015.
108. Atladottir HO, Thorsen P, Ostergaard L, et al. Maternal infection requiring hospitalization during pregnancy and autism spectrum disorders. *J Autism Dev Disord.* 2010;40(12):1423–1430. https://doi.org/10.1007/s10803-010-1006-y.
109. Kalkbrenner AE, Windham GC, Serre ML, et al. Particulate matter exposure, prenatal and postnatal windows of susceptibility, and autism spectrum disorders. *Epidemiology.* 2015;26(1):30–42. https://doi.org/10.1097/EDE.0000000000000173.
110. Schneider T, Przewlocki R. Behavioral alterations in rats prenatally exposed to valproic acid: Animal model of autism. *Neuropsychopharmacology.* 2005;30(1):80–89. https://doi.org/10.1038/sj.npp.1300518.
111. Roullet FI, Wollaston L, Decatanzaro D, Foster JA. Behavioral and molecular changes in the mouse in response to prenatal exposure to the anti-epileptic drug valproic acid. *Neuroscience.* 2010;170(2):514–522. https://doi.org/10.1016/j.neuroscience.2010.06.069.
112. Rodier PM, Ingram JL, Tisdale B, Nelson S, Romano J. Embryological origin for autism: developmental anomalies of the cranial nerve motor nuclei. *J Comp Neurol.* 1996;370(2):247–261. https://doi.org/10.1002/(SICI)1096-9861(19960624)370:2less than 247::AID-CNE8greater than 3.0.CO;2-2.
113. Rodier PM, Ingram JL, Tisdale B, Croog VJ. Linking etiologies in humans and animal models: studies of autism. *Reprod Toxicol.* 1997;11(2–3):417–422. https://doi.org/10.1016/s0890-6238(97)80001-u.
114. Crawley JN. Designing mouse behavioral tasks relevant to autistic-like behaviors. *Ment Retard Dev Disabil Res Rev.* 2004;10(4):248–258. https://doi.org/10.1002/mrdd.20039.
115. Tartaglione AM, Schiavi S, Calamandrei G, Trezza V. Prenatal valproate in rodents as a tool to understand the neural underpinnings of social dysfunctions in autism spectrum disorder. *Neuropharmacology.* 2019;159:107477https://doi.org/10.1016/j.neuropharm.2018.12.024.
116. Minakova E, Warner BB. Maternal immune activation, central nervous system development and behavioral phenotypes. *Birth Defects Res.* 2018;110(20):1539–1550. https://doi.org/10.1002/bdr2.1416.

117. Meyer U. Prenatal poly(I:C) exposure and other developmental immune activation models in rodent systems. *Biol Psychiatry*. 2014;75(4):307–315. https://doi.org/10.1016/j.biopsych.2013.07.011.
118. Custódio CS, Mello BSF, Filho AJMC, et al. Neonatal immune challenge with lipopolysaccharide triggers long-lasting sex- and age-related behavioral and immune/neurotrophic alterations in mice: relevance to autism Spectrum disorders. *Mol Neurobiol*. 2018;55(5):3775–3788. https://doi.org/10.1007/s12035-017-0616-1.
119. Amodeo DA, Lai C-Y, Hassan O, Mukamel EA, Behrens MM, Powell SB. Maternal immune activation impairs cognitive flexibility and alters transcription in frontal cortex. *Neurobiol Dis*. 2019;125:211–218. https://doi.org/10.1016/j.nbd.2019.01.025.
120. Shi L, Fatemi SH, Sidwell RW, Patterson PH. Maternal influenza infection causes marked behavioral and pharmacological changes in the offspring. *J Neurosci*. 2003; 23(1):297–302.
121. Williams AL, Desesso JM. Gestational/perinatal chlorpyrifos exposure is not associated with autistic-like behaviors in rodents. *Crit Rev Toxicol*. 2014;44(6):523–534. https://doi.org/10.3109/10408444.2014.907772.
122. De Felice A, Scattoni ML, Ricceri L, Calamandrei G. Prenatal exposure to a common organophosphate insecticide delays motor development in a mouse model of idiopathic autism. *PLoS One*. 2015;10(3):1–19. https://doi.org/10.1371/journal.pone.0121663.
123. De Felice A, Greco A, Calamandrei G, Minghetti L. Prenatal exposure to the organophosphate insecticide chlorpyrifos enhances brain oxidative stress and prostaglandin E2 synthesis in a mouse model of idiopathic autism. *J Neuroinflammation*. 2016;13(1):1–10. https://doi.org/10.1186/s12974-016-0617-4.
124. Pisu Maria G, Boero G, Garau A, et al. Are preconceptional stressful experiences crucial elements for the aetiology of autism spectrum disorder? Insights from an animal model. *Neuropharmacology*. 2018;2019(157):107686. https://doi.org/10.1016/j.neuropharm.2019.107686.
125. Herlenius E, Lagercrantz H. Development of neurotransmitter systems during critical periods. *Exp Neurol*. 2004;190(suppl):S8–21. https://doi.org/10.1016/j.expneurol.2004.03.027.
126. Owens DF, Kriegstein AR. Developmental neurotransmitters? *Neuron*. 2002;36(6): 989–991. https://doi.org/10.1016/s0896-6273(02)01136-4.
127. Owens DF, Kriegstein AR. Is there more to GABA than synaptic inhibition? *Nat Rev Neurosci*. 2002;3(9):715–727. https://doi.org/10.1038/nrn919.
128. Hensch TK, Fagiolini M, Mataga N, Stryker MP, Baekkeskov S, Kash SF. Local GABA circuit control of experience-dependent plasticity in developing visual cortex. *Science*. 1998;282(5393):1504–1508. https://doi.org/10.1126/science.282.5393.1504.
129. Ben-Ari Y, Khalilov I, Kahle KT, Cherubini E. The GABA excitatory/inhibitory shift in brain maturation and neurological disorders. *Neuroscientist*. 2012;18(5):467–486. https://doi.org/10.1177/1073858412438697.
130. Jelitai M, Madarasz E. The role of GABA in the early neuronal development. *Int Rev Neurobiol*. 2005;71(February):27–62. https://doi.org/10.1016/S0074-7742(05)71002-3.
131. Tyzio R, Nardou R, Ferrari DC, et al. Oxytocin-mediated GABA inhibition during delivery attenuates autism pathogenesis in rodent offspring. *Science*. 2014;343(6171): 675–679. https://doi.org/10.1126/science.1247190.
132. Hedner T, Iversen K, Lundborg P. γ-Aminobutyric acid concentrations in the cerebrospinal fluid of newborn infants. *Early Hum Dev*. 1982;7(1):53–58. https://doi.org/10.1016/0378-3782(82)90007-X.
133. Koren G, Loebstein R, Nulman I. Fetal alcohol syndrome. Role of the family physician. *Can Fam Physician*. 1998;44:38–40.

134. Condie BG, Bain G, Gottlieb DI, Capecchi MR. Cleft palate in mice with a targeted mutation in the gamma-aminobutyric acid-producing enzyme glutamic acid decarboxylase 67. *Proc Natl Acad Sci USA*. 1997;94(21):11451–11455. https://doi.org/10.1073/pnas.94.21.11451.
135. Chao H-T, Chen H, Samaco RC, et al. Dysfunction in GABA signalling mediates autism-like stereotypies and Rett syndrome phenotypes. *Nature*. 2010;468(7321):263–269. https://doi.org/10.1038/nature09582.
136. Han S, Tai C, Westenbroek RE, et al. Autistic-like behaviour in Scn1a+/− mice and rescue by enhanced GABA-mediated neurotransmission. *Nature*. 2012;489(7416):385–390. https://doi.org/10.1038/nature11356.
137. Hammond EJ, Wilder BJ, Bruni J. Central actions of valproic acid in man and in experimental models of epilepsy. *Life Sci*. 1981;29(25):2561–2574. https://doi.org/10.1016/0024-3205(81)90629-9.
138. Cotariu D, Zaidman JL. Developmental toxicity of valproic acid. *Life Sci*. 1991;48(14):1341–1350. https://doi.org/10.1016/0024-3205(91)90430-j.
139. Sawaya MC, Horton RW, Meldrum BS. Effects of anticonvulsant drugs on the cerebral enzymes metabolizing GABA. *Epilepsia*. 1975;16(4):649–655. https://doi.org/10.1111/j.1528-1157.1975.tb04747.x.
140. Loscher W. Concentration of metabolites of valproic acid in plasma of epileptic patients. *Epilepsia*. 1981;22(2):169–178. https://doi.org/10.1111/j.1528-1157.1981.tb04098.x.
141. Pinder RM, Brogden RN, Speight TM, Avery GS. Sodium valproate: a review of its pharmacological properties and therapeutic efficacy in epilepsy. *Drugs*. 1977;13(2):81–123. https://doi.org/10.2165/00003495-197713020-00001.
142. Löscher W. Valproate: a reappraisal of its pharmacodynamic properties and mechanisms of action. *Prog Neurobiol*. 1999;58(1):31–59. https://doi.org/10.1016/S0301-0082(98)00075-6.
143. Ramírez M, Hernández-Montoya J, Sánchez-Serrano SL, et al. GABA-mediated induction of early neuronal markers expression in postnatal rat progenitor cells in culture. *Neuroscience*. 2012;224:210–222. https://doi.org/10.1016/j.neuroscience.2012.08.044.
144. Göttlicher M, Minucci S, Zhu P, et al. Valproic acid defines a novel class of HDAC inhibitors inducing differentiation of transformed cells. *EMBO J*. 2001;20(24):6969–6978. https://doi.org/10.1093/emboj/20.24.6969.
145. Eikel D, Lampen A, Nau H. Teratogenic effects mediated by inhibition of histone deacetylases: evidence from quantitative structure activity relationships of 20 valproic acid derivatives. *Chem Res Toxicol*. 2006;19(2):272–278. https://doi.org/10.1021/tx0502241.
146. Grafodatskaya D, Chung B, Szatmari P, Weksberg R. Autism spectrum disorders and epigenetics. *J Am Acad Child Adolesc Psychiatry*. 2010;49(8):794–809. https://doi.org/10.1016/j.jaac.2010.05.005.
147. Kim KC, Lee DK, Go HS, et al. Pax6-dependent cortical glutamatergic neuronal differentiation regulates autism-like behavior in prenatally valproic acid-exposed rat offspring. *Mol Neurobiol*. 2014;49(1):512–528. https://doi.org/10.1007/s12035-013-8535-2.
148. Kim KC, Choi CS, Kim J-W, et al. MeCP2 modulates sex differences in the postsynaptic development of the valproate animal model of autism. *Mol Neurobiol*. 2016;53(1):40–56. https://doi.org/10.1007/s12035-014-8987-z.
149. Newbold RR, Hanson RB, Jefferson WN, Bullock BC, Haseman J, McLachlan JA. Increased tumors but uncompromised fertility in the female descendants of mice exposed developmentally to diethylstilbestrol. *Carcinogenesis*. 1998;19(9):1655–1663. https://doi.org/10.1093/carcin/19.9.1655.

150. Choi CS, Gonzales EL, Kim KC, et al. The transgenerational inheritance of autism-like phenotypes in mice exposed to valproic acid during pregnancy. *Sci Rep*. 2016;6:36250. https://doi.org/10.1038/srep36250.
151. Tartaglione AM, Cipriani C, Chiarotti F, et al. Early behavioral alterations and increased expression of endogenous retroviruses are inherited across generations in mice prenatally exposed to valproic acid. *Mol Neurobiol*. 2019;56(5):3736–3750. https://doi.org/10.1007/s12035-018-1328-x.
152. Gamble MV, Liu X, Slavkovich V, et al. Folic acid supplementation lowers blood arsenic. *Am J Clin Nutr*. 2007;86(4):1202–1209. https://doi.org/10.1093/ajcn/86.4.1202.
153. Kirkbride JB, Susser E, Kundakovic M, Kresovich JK, Davey Smith G, Relton CL. Prenatal nutrition, epigenetics and schizophrenia risk: can we test causal effects? *Epigenomics*. 2012;4(3):303–315. https://doi.org/10.2217/epi.12.20.
154. Yiu TT, Li W. Pediatric cancer epigenome and the influence of folate. *Epigenomics*. 2015;7(6):961–973. https://doi.org/10.2217/epi.15.42.
155. Irwin RE, Pentieva K, Cassidy T, et al. The interplay between DNA methylation, folate and neurocognitive development. *Epigenomics*. 2016;8(6):863–879. https://doi.org/10.2217/epi-2016-0003.
156. Kriaucionis S, Heintz N. The nuclear DNA base 5-hydroxymethylcytosine is present in Purkinje neurons and the brain. *Science*. 2009;324(5929):929–930. https://doi.org/10.1126/science.1169786.
157. Tahiliani M, Koh KP, Shen Y, et al. Conversion of 5-methylcytosine to 5-hydroxymethylcytosine in mammalian DNA by MLL partner TET1. *Science*. 2009;324(5929):930–935. https://doi.org/10.1126/science.1170116.
158. Lister R, Mukamel EA, Nery JR, et al. Global epigenomic reconfiguration during mammalian brain development. *Science*. 2013;341(6146):1237905. https://doi.org/10.1126/science.1237905.
159. Ivanova E, Chen J-H, Segonds-Pichon A, Ozanne SE, Kelsey G. DNA methylation at differentially methylated regions of imprinted genes is resistant to developmental programming by maternal nutrition. *Epigenetics*. 2012;7(10):1200–1210. https://doi.org/10.4161/epi.22141.
160. Kruman II, Mouton PR, Emokpae RJ, Cutler RG, Mattson MP. Folate deficiency inhibits proliferation of adult hippocampal progenitors. *Neuroreport*. 2005;16(10):1055–1059. https://doi.org/10.1097/00001756-200507130-00005.
161. Blaise SA, Nedelec E, Schroeder H, et al. Gestational vitamin B deficiency leads to homocysteine-associated brain apoptosis and alters neurobehavioral development in rats. *Am J Pathol*. 2007;170(2):667–679. https://doi.org/10.2353/ajpath.2007.060339.
162. Akchiche N, Bossenmeyer-Pourie C, Kerek R, et al. Homocysteinylation of neuronal proteins contributes to folate deficiency- associated alterations of differentiation, vesicular transport, and plasticity in hippocampal neuronal cells. *FASEB J*. 2012;26(10):3980–3992. https://doi.org/10.1096/fj.12-205757.
163. Bryan J, Osendarp S, Hughes D, Calvaresi E, Baghurst K, van Klinken J-W. Nutrients for cognitive development in school-aged children. *Nutr Rev*. 2004;62(8):295–306. https://doi.org/10.1111/j.1753-4887.2004.tb00055.x.
164. Georgieff MK. Nutrition and the developing brain: nutrient priorities and measurement. *Am J Clin Nutr*. 2007;85(2):614S–620S. https://doi.org/10.1093/ajcn/85.2.614S.
165. Villamor E, Rifas-Shiman SL, Gillman MW, Oken E. Maternal intake of methyl-donor nutrients and child cognition at 3 years of age. *Paediatr Perinat Epidemiol*. 2012;26(4):328–335. https://doi.org/10.1111/j.1365-3016.2012.01264.x.

166. Chatzi L, Papadopoulou E, Koutra K, et al. Effect of high doses of folic acid supplementation in early pregnancy on child neurodevelopment at 18 months of age: The mother-child cohort "Rhea" study in Crete, Greece. *Public Health Nutr.* 2012;15(9):1728–1736. https://doi.org/10.1017/S1368980012000067.
167. Dawson JE, Raymond AM, Winn LM. Folic acid and pantothenic acid protection against valproic acid-induced neural tube defects in CD-1 mice. *Toxicol Appl Pharmacol.* 2006;211(2):124–132. https://doi.org/10.1016/j.taap.2005.07.008.

CHAPTER THREE

Genomics, transcriptomics, proteomics and big data analysis in the discovery of new diagnostic markers and targets for therapy development

Dong Ik Park[a,b,*]
[a]Danish Research Institute of Translational Neuroscience (DANDRITE)-Nordic EMBL Partnership for Molecular Medicine, Department of Biomedicine, Aarhus University, Aarhus, Denmark
[b]The Danish National Research Foundation Center, PROMEMO, Department of Biomedicine, Aarhus University, Aarhus, Denmark
*Corresponding author: e-mail address: park@biomed.au.dk

Contents

Abstract

Highly complex endophenotypes and underlying molecular mechanisms have prevented effective diagnosis and treatment of autism spectrum disorder. Despite extensive studies to identify relevant biosignatures, no biomarker and therapeutic targets are available in the current clinical practice. While our current knowledge is still largely incomplete, -omics technology and machine learning-based big data analysis have provided novel insights on the etiology of autism spectrum disorders, elucidating systemic impairments that can be translated into biomarker and therapy target

Progress in Molecular Biology and Translational Science, Volume 173
ISSN 1877-1173
https://doi.org/10.1016/bs.pmbts.2020.04.017

candidates. However, more integrated and sophisticated approaches are vital to realize molecular stratification and individualized treatment strategy. Ultimately, systemic approaches based on -omics and big data analysis will significantly contribute to more effective biomarker and therapy development for autism spectrum disorder.

1. Introduction

Autism spectrum disorder (ASD) is a neurodevelopmental disorder with a wide range of conditions characterized by repetitive behaviors, intellectual disability, and deficits in social interaction, communication and language skills. ASD has a high rate of comorbidity with other psychiatric conditions including depression and anxiety.[1,2] Despite its low life prevalence, the disease is highly heritable and estimated to have a median prevalence of 62 in 10,000 worldwide.[3]

While human genetic studies found that a large number of genes are associated with ASD, no major gene which accounts for more than 1% of the cases was found[4] indicating that ASD is not a single disorder with single cause.[5,6] In fact, ASD includes several diseases previously categorized as separate disorders including autistic disorder, Asperger syndrome (AS) and pervasive developmental disorder not otherwise specified (PDD-NOS). In this regard, identifying molecular mechanisms and diagnosis/treatment targets for ASD has been challenging due to the highly heterogenous genetic and symptomatic architecture.

Omics technology has been a powerful tool to identify biomarkers and pathways which could explain the etiology of the psychiatric disorders. With the help of -omics technology, risk genes and proteins involved in neuronal calcium signaling (CACNA1E, CACNA2D1 and CAMK2 alpha), dopaminergic neurotransmission (DRD2), glutamatergic neurotransmission (GRIA1, GRIA2, GRIK5, GRIN2B and GRM5), presynaptic vesicle trafficking (PCLO and STXBP1) and proinflammatory cytokine response (IFN gamma, CCL4 and CXCL8) have been implicated for MDD.[7–10] Diverse pathways such as angiogenesis and vascular system development (ANGPT2 and STAB1), insulin secretion (INS, IRS1 and IRS2) and endocannabinoid signaling (CNR1, DAGLA and DAGLB), as well as ion channel and transporters (SCN2A and SLC4A1), and synaptic component (RIMS1 and ANK3) were associated with BD.[11,12] In SCZ, susceptible pathways involved in glutamatergic neurotransmission and synaptic plasticity (AMPAR, DPYSL2, DRD2, GNB1, GRIA1,

GRIN2A, GRM3, GLUL, NEFL, NEFM, SRR and VDAC), neuronal development (CTNNA1, GPM6A, SOX2OT), calcium channel (CACNA1C, CACNB2, CACNA11, CALM1, CALM2, CAMK2B, CAMK2D, CAMK2G, PMCA4, S100A6, S100A12), chromatin regulation (PU.1 and RPB2), and immune system and inflammation (alpha-defensins, interleukins and COX2) were identified.[13–16] In addition, several biological pathways including calcium signaling (CAMKMT) and mitochondria-related functions (PREPL, ATP5H, MT-CO1, MT-ND2, MT-ND6, NDUFA2, NDUFS6, COX7A2 and COX7C) were found to be related to anxiety disorders.[17,18]

Despite limitations by polygenicity, consequent variable disease symptoms and comorbidity with other psychiatric conditions, -omics studies have identified valuable ASD biomarker and therapy candidates with the help of data resources such as the Simon Foundation Autism Research Initiative (SFARI) (https://gene.sfari.org/), the autDB (http://autism.mindspec.org/autdb/Welcome.do) and the Autism Sequencing Consortium (ASC) (https://genome.emory.edu/ASC/). Recent advances in big data processing with computational algorithms further accelerate ASD research by helping data integration and interpretation. Here, we present the state-of-the-art progress made in -omics and big data studies, and discuss the future directions that will provide a breakthrough in the ASD biomarker and therapy discovery.

2. Omics and big data studies of autism spectrum disorder

2.1 Immune and inflammation pathways

Dysregulated immune system has been shown to highly contribute to the autism-related neurodevelopmental and behavioral abnormalities. Particularly, in utero immune system dysfunctions including maternal production of autoantibodies reactive to fetal brain, altered immunoglobulin levels and innate/adaptive immune cell response have been suggested to play a critical role in the etiology and pathology of ASD.[19] To support this, numerous omics studies have provided evidence of immune dysfunction and autoimmunity in ASD.

Gandal et al. integrated genomics and transcriptomics data of brain tissues acquired from ASD and healthy individuals.[20] They found that gene, isoform and non-coding RNA expression were significantly dysregulated in several brain regions of ASD patients. The differential expression profiles enriched diverse immune-related pathways including inflammatory response, response to cytokine, innate immune response and cytokine production.

Gupta et al. analyzed transcriptome from postmortem human cortical tissues corresponding to Brodmann Area 10, Brodmann Area 19 (BA19) and Brodmann Area 44.[21] The authors identified 12 co-expression modules using weight gene correlation network analysis. The module upregulated in autism patients significantly enriched activated microglial cells and multiple immune pathways.

Voineagu et al. used transcriptomics to analyze postmortem frontal and temporal cortex tissues from ASD and control cases.[22] Genes involved in immune response, inflammation, astrocyte and activated microglia were upregulated in ASD cortices. These genes enriched a network module highly related to autism disease status. Since further genome wide association study (GWAS) enrichment analysis using published data did not enrich the immune and glial module, the authors suggested the non-genetic etiology for immune dysregulation in ASD.

Abnormal immune system in ASD may be also supported by comorbidity with cancer. By analyzing electronic health record, young ASD adults without intervention was shown to be at high risk for cancer development by midlife.[23] Genome-wide exome sequencing analysis showed that risk genes associated with immune system including ERBB2IP (involved in NF-kB signaling, proinflammatory cytokine secretion and TGF beta signaling) and PAX5 (involved in B cell development, leukemia, acute lymphoblastic leukemia and Wnt signaling pathway) are shared between ASD and cancer.[24] Fores-Martos et al. performed transcriptomics meta-analysis and found that a number of genes related to immune system (CDK2, CDKN1A, COL4A1, COL4A2, DDIT4, F2R, IL4R, ITGA5, MYC, NFKB1 and VEGFA) and oxidative stress (ATP5F1, ATP5J, ATP50, COX7B, CYC1, DLD, OGDHL, PFKM, NDUFAF1, NDUFB2, NDUFB6, NDUFV1 and UQCRFS1) were commonly regulated between ASD and several types of cancer cases.[25]

Redox proteomics analysis identified altered protein oxidation profiles in blood plasma of autistic children.[26] The study found that two carbonylated plasma proteins, Ig kappa chain C and complement component C8 alpha chain, were significantly higher in autistic patients than healthy controls. Protein carbonylation is an irreversible post-translational modification induced by reactive oxygen species. Elevated levels of carbonylated immunoglobulin and complement in the study indicated activated oxidative stress and autoimmune response in ASD.

Momeni et al. performed blood plasma proteome profiling in children with ASD.[27] They found that children with ASD had significantly higher levels of complement factor I, a protease for complement protein C3b. In line with this, C3f, a cleaved C3b, and its fragment C3f-des-arginine levels were found to be elevated in ASD children compared to healthy controls.[28]

Urinary proteomics analysis revealed the differential excretion of kininogen-1 (KNG1), IgG1 heavy chain variable region and mannan-binding lectin serine protease-2 (MASP2) isoform-2 precursor between ASD and typically developing children.[29] KNG1 is a precursor protein for high-molecular-weight kininogen that is suggested to play a role in the inflammatory disorders by stimulating cytokine and chemokine secretion.[30] MASP2 activates complement system by cleavaging the components such as C2 and C4.[31] Thus, the findings supported a crucial role of complement system in ASD.

Junaid et al. used two dimensional gel electrophoresis followed by proteomics identification and direct DNA sequencing of frontal lobe gray matter obtained from ASD patients and controls.[32] The authors identified a single nucleotide polymorphism (SNP) in glyoxalase I (GLO1) gene which causes Ala111Glu substitution in the protein sequence. GLO1 is mainly involved in methylglyoxal detoxification and oxidative stress. It was also shown to be associated with other psychiatric disorders such as anxiety disorders and depression.[33–35] The study revealed that the GLO1 polymorphism results in reduced protein activity and methylglyoxal-derived advanced glycation end product accumulation which may affect ASD-susceptible pathways such as cell growth and differentiation.[36,37]

Vogel Ciernia et al. used a mouse model of maternal allergic asthma (MAA) which induces ASD-like behaviors in offspring.[38] Using microglial fraction isolated from the offspring, they performed whole genome bisulfite sequencing and RNA sequencing to investigate DNA methylation and gene expression, respectively. Differentially methylated regions were associated with a number of cytokine and chemokine-mediated immune pathways and transcription factor binding motifs for early microglial development and immune activation including runt-related transcription factor 1 (RUNX1), PU.1, interferon regulatory factor 8 (IRF8), nuclear factor kappa-B (NF-kB) and MAF BZIP transcription factor B (MAFB). In addition, voltage-gated ion channel genes involved in neuronal connection regulation and microglia sensitivity to environmental signals were differentially expressed between MAA and control mice. The study suggested that maternal immune activation during pregnancy can contribute to ASD susceptibility by affecting fetal synaptic function and development.

2.2 Synapse and neurodevelopment

The ultimate culprit of ASD is more likely to be dysregulated neurodevelopmental process during gestation and early postnatal stage. The resultant functional and structural impairment at the synapse can cause cognitive and psychological dissonance.

Multiple synaptic features such as synaptic vesicle trafficking, neurotransmitter release, GABAergic transmission, and balance between excitatory and inhibitory synaptic transmission were found to be impaired in ASD.[39] Also, abnormalities in dendritic spine density/length, branching and morphology have been extensively associated with ASD.[40]

De Rubeis et al. performed whole exome sequencing and integrated several datasets for de novo, inherited and case-control variations, and de novo missense variants.[41] The study showed that genes affected in ASD are highly enriched for synaptic networks and chromatin remodeling. Several chromatin regulators have been shown to be essential for neural development processes including neural progenitor proliferation, migration, differentiation, synaptogenesis and synaptic pruning.[42] Thus, the findings supported synaptic connectivity and plasticity alterations in ASD.

Callaghan et al. performed whole genome sequencing using whole blood of ASD probands.[43] The authors identified five de novo subject-specific damaging variants using computational prioritization based on predicted damage, population frequency, literature evidence and phenotype concordance. The variants included two novel de novo variants of SCN2A, a well-documented ASD gene.[44] SCN2A encodes voltage-gated sodium channel Nav1.2 which is expressed in the axon initial segment. SCN2A was shown to be important for action potential initiation and propagation.[45]

Integrated genomics and transcriptomics analysis by Pain et al. identified 14 differentially expressed genes between ASD and control cases.[46] The authors found that the protein disulfide-isomerase A6 precursor (PDIA6) involved in protein folding was significantly downregulated in blood of ASD individuals. Gene sets including synaptic vesicle, presynapse, abnormal axon guidance and early cortical development represented nominal significance.

Sanders et al. analyzed de novo CNVs using the Simons Simplex Collection (SSC) cohort, the resource of the SFARI.[47] They integrated published de novo CNV data from the Autism Genome Project, and exome sequencing data from the SSC and the ASC. The integrated analysis confirmed previously reported de novo CNV loci, and identified 65 ASD genes.

Protein-protein interaction (PPI) analysis further revealed that the ASD genes enrich interconnected networks for synapse and chromatin regulation.

Grunwald et al. performed transcriptomics analysis using human induced pluripotent stem cells (iPSCs) and -derived neurons generated from dermal fibroblasts of ASD and SCZ patients.[48] While iPSCs-derived neurons from three ASD patients shared a small number of dysregulated genes, transcriptome profiles were discriminative between the ASD and SCZ patients. The authors found that several genes linked to neural development such as SHH, PTCH1, GREM1, FEZF1 and FEZ-AS1 were significantly altered in the ASD iPSCs-derived neurons compared to those from healthy controls.

Single cell transcriptomics was used to analyze gene expression profiles from doublecortin expressing immature neurons in the dentate gyrus of mouse hippocampus.[49] While the study identified several subpopulations with distinct developmental stages, genes positively associated with neuronal maturation progression from stem-like to neuron-like cells were enriched for autism-related gene sets. This suggested that autism development may be attributed to gene and network dysregulation involved in neuronal maturation process during neurogenesis.

DeRosa et al. examined temporal transcriptome changes during cortical neuron differentiation of iPSCs derived from peripheral blood mononuclear cells (PBMCs) of individuals with idiopathic ASD.[50] The authors compared transcriptome differences between two time points, early (DIV 35) and later (DIV 135) developmental stages. Differentially regulated genes in early stage exhibited expression changes in opposite direction in later stage, indicating dramatic gene expression changes during development. Several biological processes including synaptic activity, calcium signaling and neuronal cell migration were found to be dysregulated in the early stage of ASD iPSCs. Consistent with transcriptomics data, neuronal process migration, spontaneous neural spiking activity and calcium transients were found to be impaired.

Pinto et al. identified 36 pathogenic CNVs in ASD-affected individuals.[51] The CNVs implicated genes linked to ASD-related neurodevelopmental disorders such as CHD2, HDAC4 and GDI1. Several synaptic processes including cell projection, neural development, axonogenesis and neuronal synapse were enriched by rare exonic gene deletions. Integrative analysis with de novo loss-of-function SNVs from four ASD exome sequencing studies further revealed the convergent pathway enrichment in neuronal development and axon guidance as well as microtubule associated protein kinase (MAPK) signaling and chromatin modification/transcriptional regulation.

Hu et al. performed quantitative trait analysis using ASD patient subgroups based on the Autism Diagnostic Interview-Revised (ADI-R) scores from five symptom categories including spoken language skills, non-verbal communication, play skills, social development and insistence on sameness.[52] Quantitative trait locus associated with distinct ASD subgroups enriched several cellular pathways including neurogenesis, axogenesis and long-term synaptic potentiation. Network analysis using genes associated with intronic SNPs identified hubs such as HTR4 and GCH1, indicating a role of serotonergic system in the ASD etiology.

Bralten et al. performed GWAS to identify SNPs associated with autistic traits.[53] The authors measured autistic trait scores from Dutch general population using self-report questionnaire. They used GWAS data from Psychiatric Genomics Consortium autism group to find overlaps between genetic variants for ASD susceptibility and three autistic traits (childhood behavior, rigidity and attention to detail). A number of neurite outgrowth-related genes including MET showed a significant association with rigidity trait.

Using exome sequencing, O'Roak et al. identified recurrent de novo mutations of chromodomain helicase DNA binding protein 8 (CHD8) from ASD probands.[54] CHD8 is a chromatin remodeling factor which regulates Wnt signaling pathways.[55,56] CHD8 deficiency was shown to impair axon and dendrite growth, and delay cortical neuron migration.[57] In addition, CHD8 mutations have been strongly associated with clinical features including macrocephaly, speech delay and psychopathology.[58]

Sugathan et al. performed RNA-seq and ChIP sequencing using iPSC-derived human neural progenitor cells with CHD8 reduction.[59] Several critical pathways including synapse, axon guidance and neuron differentiation were enriched by downregulated genes. Furthermore, most CHD8 binding sites were found to be active transcription start sites, supporting a crucial role of CHD8 in chromatin and transcription regulation.

Genomic DNA methylation study using blood specimens examined genetically defined ASD patient subgroups with 16p11.2 deletion or with $CHD8^{+/-}$ variant.[60] Both genotypes have been strongly associated with behavioral and neurodevelopmental deficits of ASD patients.[58,61] Methylated gene profiles obtained from the subgroups were able to distinguish the subgroups from undefined individuals, showing that ASD subgroup stratification could provide more sensitive and specific diagnostics.

Neurexin (NRXN) is a neuronal presynaptic cell adhesion molecule which is crucial for synaptic function and development.[62,63] One study identified CNV losses of chromosome 2p16 which results in NRXN1 coding exon deletion in families with ASD individuals.[64] Zahir et al. identified de novo NRXN1-alpha deletion from a ASD patient.[65] Feng et al. identified two putative missense structural variants in NRXN1 beta gene from several ASD patients.[66] Lam et al. performed single cell RNA-seq analysis using human iPSC-derived neural stem cells and differentiated cells generated from autism patients carrying bi-alleic NRXN1-alpha deletion.[67] The authors found that NRXN1-alpha deletion resulted in astroglia generation, depressed calcium signaling and impaired excitatory neuron maturation.

Single nucleus RNA sequencing using ASD patient cortical tissues revealed that genes important for synaptic function and brain development such as NRXN1, STX1A, SYN2, TCF25, SOX5 and RBFOX3 are differentially regulated in upper-layer excitatory neurons, microglia and protoplasmic astrocytes.[68] The authors examined correlation of cell type-specific gene expression changes with the clinical ASD symptom severity. Transcriptomic changes in L2/3 neurons and microglia were found to be the most predictive of symptom severity. By comparing single cell profiles, they further revealed that cortico-cortical projection neurons across layers are commonly affected in multiple patients, suggesting that upper-layer cortical circuit dysfunction may be relevant to ASD pathology.

Provenzano et al. used microarray to analyze hippocampal transcriptome of two ASD mouse models, BTBR T^+ $Itpr3^{tf}$/J (BTBR) and Engrailed-2 knockout mice, to identify conserved ASD molecular signatures.[69] While more than 150 genes were commonly expressed between the two groups, differentially expressed gene profiles common to both mouse models enriched biological processes such as regulation of ion transmembrane transport and synaptic transmission.

Parikshak et al. performed ribosomal RNA-depleted RNA sequencing using postmortem human frontal cortex, temporal cortex and cerebellum of ASD and control subjects.[70] Significant alterations in long non-coding RNAs (lncRNAs) expression and neuron-specific alternative splicing were found in ASD cortices. In addition, multiple lncRNAs were predicted to interact with microRNA-protein complexes that are differentially regulated between control and ASD cortices, and with fragile X mental retardation protein (FMRP) that is strongly associated with intellectual disability and ASD.[71] The study revealed that cortical patterning between the frontal and temporal cortices were attenuated in ASD. In addition, transcription factor binding site enrichment analysis further found that SOX5, a mammalian corticogenesis regulator, might be involved in cortical patterning dysregulation in ASD. The co-expression network analysis showed that age-related microglial and synaptic function were changed, suggesting that ASD genetic risks may affect cortical gene expression.

Lee et al. performed transcriptomics analysis using zebrafish embryos and larvae treated with valproic acid as a model of ASD.[72] They found that ASD-associated genes such as ADSL, MBD5, SHANK3 and TSC1b were differentially regulated between the control and drug-treated larvae. Several synapse-related processes including generation of a signal involved in cell-cell signaling, synaptic transmission and transmission of nerve impulse were enriched for genes altered in a concentration-dependent manner. Therefore, the authors suggested that valproic acid-treated zebrafish can be a promising alternative animal model for ASD research.

Multiple clustering algorithms based on scores from the ADI-R were used to identify ASD proband subgroups.[73] While the clustering analysis identified four phenotype clusters with distinct behavioral deficits, the authors further performed DNA microarray using lymphoblastoid cell lines from three of the subgroups (groups with severely impaired language, mild phenotype and savant skills).[74] Differentially expressed genes in severely language-impaired subgroup were involved in biological processes including circadian rhythm and apoptosis/cell death of neuroglia, astrocytes and neurons. Interestingly, differentially expressed genes common to all three ASD subgroups were found to be in uncharacterized intronic or intergenic regions associated with cellular response to androgen. Seven of the transcripts showed significant expression changes in response to dihydroxytestosterone, suggesting that the transcripts responsive to androgen may be related to a higher rate of ASD prevalence in males.

Transcriptomics and proteomics profiling of the BTBR mouse hippocampus identified several differentially regulated genes and proteins such as blood-brain-derived neurotrophic factor (BDNF) and SH3 and multiple ankyrin repeat domains 3 (SHANK3).[75] The altered genes and proteins represented several pathways including axon guidance, endocytosis, and regulation of actin cytoskeleton, indicating synaptic dysfunction in the BTBR mouse hippocampus.

Broek et al. performed targeted proteomics analysis using postmortem brain tissues collected from ASD patients and typical controls.[76] The authors found that proteins involved in axon myelination and synaptic regulation were oppositely regulated between prefrontal cortex and cerebellum. The study indicated that distinct brain regions can be differentially affected in ASD.

Meta-analysis of 14 different studies including more than 2700 participants found that peripheral BDNF levels are significantly higher in ASD patients compared to controls.[77] Despite inconsistent reports on BDNF levels, the results concluded that peripheral BDNF alteration can be significantly implicated in ASD.

Synaptic interactome studies using co-immunoprecipitation (Co-IP) coupled mass spectrometry-based proteomics identified key PPIs. One interactome study with developing mouse telencephalon synaptosome revealed that p140Cap protein interacts with more than 300 proteins highly associated with multiple psychiatric disorders including ASD, SCZ, and BD.[78] Another study using developing mouse neocortical synaptosome found that MET receptor tyrosine kinase interaction networks include several key synaptic proteins such as SHANK3, synaptic Ras GTPase-activating protein 1 (SYNGAP1) and glutamate ionotropic receptor NMDA type subunit 2B (GRIN2B).[79] Mejia et al. immunoprecipitated tuberous sclerosis 1 (TSC1) interactome in mouse neuro2a cells and primary rat cortical neurons. The authors found a novel interaction between huntingtin associated protein 1 (HAP1) and TSC1. Furthermore, HAP1 was found to regulate mammalian target of rapamycin complex 1 (mTORC1) signaling and axon-dendrite morphogenesis/positioning in brain.[80]

Baucum et al. performed Co-IP followed by quantitative proteomics to identify CaMKII interactome from mouse forebrains.[81] The study employed the CaMKIIα T286A knock-in mutation which results in reduced CaMKIIα and β phosphorylation. The mutation was found to alter CaMKII interaction with ASD-related synaptic scaffolding proteins including SHANK3, disk large-associated protein 2 (DLGAP2) and SYNGAP1. Since dysregulated CaMKII signaling has been linked to Angelman syndrome which exhibits features overlapping with ASD, the authors concluded that altered PPIs in CaMKII networks may contribute to the ASD etiology.

2.3 Mitochondrial energy metabolism

Mitochondria are key organelles for cellular energy generation. Most energy produced in the brain is consumed on synaptic mechanisms mediating synaptic transmission.[82] Mitochondria were also shown to mediate critical neurodevelopmental processes with a high energy demand such as genetic reprogramming during neuronal differentiation.[83] It has been shown that mitochondria are transported to the active synapse and support dendritic spine density and synaptogenesis[84] and synaptic transmission.[85] Thus, mitochondrial impairment is strongly implicated in dysregulation of synaptic vesicle trafficking, neurotransmitter release and concomitant synaptic transmission in ASD.

Gordon et al. analyzed transcriptome from cortices and hippocampi of three mouse lines, Df(h15q13)/+, Df(h22q11)/+ and Df (h1q21)/+, carrying ASD and SCZ-associated mutations.[86] While several cortical and hippocampal co-expression networks were found to be shared among the three mouse models, one cortical module associated with neuronal mitochondria and firing rate showed gene expression patterns overlapping with transcriptomic changes in ASD and SCZ postmortem human brains. The common gene expression patterns in mice and humans suggested that neuronal bioenergetics can be a crucial mechanism for ASD.

He et al. analyzed 11 transcriptomics datasets of different human tissues from ASD and control individuals.[87] Downregulated genes in ASD patients' brain and blood specimens significantly enriched several pathways including mitochondria-related functions and oxidative phosphorylation. The authors further analyzed eight transcriptomics datasets from ASD rodent models to validate the findings from the human subjects. While all differentially expressed genes in the animal models were found to be upregulated, similar biological processes were enriched between the ASD animal and humans.

Transcriptomic meta-analysis mentioned in Section 2.1 also found that a number of genes in mitochondrial processes including oxidative phosphorylation, mitochondrial electron transport and ATP synthesis were downregulated in ASD.[25]

While Broek et al. identified differential synaptic protein expression patterns between prefrontal cortex and cerebellum as mentioned in Section 2.2, the authors also found that creatine kinase B-type (CKB) levels were altered in ASD.[76] Creatine kinases play a central role in energy storage and distribution by catalyzing phosphate transfer between ATP and creatine phosphate. CKB was shown to affect formation and maintenance of mossy fiber connections in the hippocampus.[88] Glial CKB was associated with high energy demands related to ion homeostasis and TCA cycle metabolite/neurotransmitter trafficking with neurons.[89] Thus, altered CKB levels in ASD brains suggest that disturbed CKB-mediated energy metabolism may be involved in synaptic dysfunction.

Shen et al. performed proteomics analysis using PBMCs of autistic and healthy children.[90] The authors reported that 17 proteins involved in carbon metabolism, pyruvate metabolism, TCA cycle and respiratory electron transport were significantly altered in the ASD PBMCs.

West et al. interrogated blood plasma metabolome of ASD and healthy children.[91] The authors revealed that mitochondrial metabolites including isoleucine, glutaric acid, aspartate, glutamate and TCA cycle-associated molecules (citric acid and succinic acid) are significantly altered in children with ASD. Levels of dehydroepiandrosterone sulfate which affects oxidative energy metabolism by altering mitochondrial respiratory chain complex contents and activities[92,93] were found to be different between autistic and healthy children.

Metabolomics analysis of the prefrontal cortex gray matter obtained from ASD individuals showed that significantly altered metabolites enrich glutathione metabolism and TCA cycle.[94] Transcriptomics data analysis in the human prefrontal and temporal cortices further found that genes for ASD-related metabolites were elevated in autistic individuals.

Blood plasma metabolome was analyzed in healthy controls and children with ASD, idiopathic-developmental delay and Down syndrome.[95] The study revealed that metabolites involved in one-carbon metabolism (serine and glycine) and TCA cycle (alanine and ornithine) were significantly elevated only in children affected by ASD. Since serine and glycine are agonists for NMDA and glycine receptors, respectively, the results suggest the close relationship between energy metabolism and synaptic activity.

2.4 Lipid transport and metabolism

Lipids are essential components of cellular membranes and vesicles. Most cholesterol in the brain are required in myelin sheath, neurons and astrocytes. Particularly, more than 70% of rodent brain cholesterol is in myelin,[96] indicating its major role in synaptic transmission regulation. Lipid rafts which contains high concentration of cholesterol and glycosphingolipids were shown to play a major role in maintenance of synapses and dendritic spines.[97] Lipids are also crucial for brain development. Mammalian brain lipids including cholesterol, phospholipids, cerebrosides, sulfatides and gangliosides highly increase during development. Several studies have demonstrated a role of lipid transport and metabolism in CNS development.[98,99] Therefore, perturbed lipid homeostasis can directly hamper critical synaptic functions which can lead to cognitive, behavioral and emotional impairment in ASD.

Gudenas et al. re-analyzed previously acquired RNA-seq data to identify differentially expressed lncRNAs in ASD brain cortices.[100] More than 200 lncRNAs were found to be altered in ASD cortices. Five among the identified lncRNAs were neighboring and antisense to ASD risk genes including

RAPGEF4, DLX6, STXBP5, KLC2 and DMXL2, suggesting that the lncRNAs may be cis-regulatory elements. Furthermore, differentially expressed lncRNAs enriched functional expression modules related to lipid transport along with synaptic transmission, immune response, drug response and nucleic acid metabolism, indicating a role of lipids at the nexus of other essential pathways.

Corbett et al. performed proteomics analysis using sera obtained from autistic and healthy children.[101] The authors separated the autism children into low-functioning autism (LFA) and high-functioning autism (HFA) subgroups based on intelligence quotient. While apolipoprotein B100 precursor (APOB100), complement factors and fibronectin 1 isoform 1 preproprotein (FN1) levels were altered in general autistic children, comparison between the autism subgroups revealed that apolipoprotein A-IV (APOA4) and APOB100 levels were significantly lower in LFA than HFA sera. Since most brain cholesterols are locally produced by glial cells,[102] apolipoprotein-mediated lipid transport can play a crucial role in diverse neuronal pathways.

Yang et al. analyzed blood sera of Han Chinese children with ASD to identify peptide biomarkers.[103] The authors found that eight peptide peaks were higher in ASD than control sera. The peptides were found to be derived from proteins such as fatty acid binding protein 1 (FABP1) and apolipoprotein C-I precursor (APOC1) involved in lipid metabolism regulation.

Steeb et al. used multiplex immunoassay profiling and proteomics to analyze sera of adults with AS.[104] The authors found that several lipid transport and metabolism-related proteins including apolipoprotein A1 (APOA1), apolipoprotein E (APOE), apolipoprotein C2 (APOC2), adiponectin (ADIPO), fetuin-B precursor (FETUB) and D-glucuronyl C5-epimerase (GLCE) were altered specifically in female AS patients. In addition, sex hormone binding globulin (SHBG) expression patterns between men and women were found to be opposite. SHBG is a major transporter of sex steroids such as testosterone and other androgens which were suggested to contribute to the ASD development.[105] Based on their results, the authors demonstrated that gender subgroup stratification may be required in the studies of AS biomarkers and drug targets.

Wang et al. found the serum metabolomic profile differences between autistic and healthy individuals.[106] The authors revealed that lipids including sphingosine 1-phosphate (S1P), docosahexaenoic acid (DHA) and docosapentaenoic acid were significantly altered in ASD cohorts. S1P and

DHA were significantly correlated with the autism behavior checklist (ABC) score, representing biomarker candidates for ASD diagnosis.

2.5 Gut microbiome

The microbiota-gut-brain axis coordinates communication and interaction between enteric and central nervous system, therefore affecting emotional and cognitive functions in the host. Gut microbes provide neuroactive compounds and immune modulators including serotonin, dopamine, γ-aminobutyric acid (GABA), acetylcholine, histamine and short-chain fatty acids (SCFAs). In addition, gut microbiome was shown to regulate blood-brain barrier formation and hypothalamic-pituitary-adrenal stress response.[107] increasing evidence strongly suggests that abnormal gut microbiome could serve as one of major contributors to ASD development.[108,109] Various gastrointestinal symptoms in autistic individuals also support a significant link between gut microbiome and ASD.[110]

de Theije et al. analyzed microbiota composition of caeca samples from mouse pups exposed to in utero valproic acid.[111] Pyrosequencing analysis revealed that prenatal valproic acid exposure resulted in transgenerational gut microbiota changes in the offspring. Microbiota differences of drug-exposed male offspring were significantly correlated with increased caecal butyrate and decreased ileal serotonin levels.

Ming et al. performed metabolomics analysis to identify urinary biosignatures for ASD. They found that microbial metabolites including 2-(4-hydroxyphenyl)propionate, taurocholenate sulfate, 3-(3-hydroxyphenyl)propionate and 5-aminovalerate (5-AV) were significantly altered in children with ASD.[112]

Yap et al. analyzed urinary metabolite profiles of ASD patients using ^{1}H NMR spectroscopy.[113] The authors found that gut microbial and mammalian cometabolites including dimethylamine, hippurate and phenyacetylglutamine were significantly different between ASD and control subjects. The metabolites were suggested to be involved in interconnected pathways among gut microbiome, energy metabolism and methylamine, suggesting the systemic effects of gut microbiome.[114]

Kaluzna-Czaplinska interrogated urinary metabolome profiles to find organic acid level differences between autistic and control children.[115] The author suggested that gut bacterial metabolism might affect several metabolites including hydroxyphenylacetate and hippurate.

Emond et al. examined urinary metabolome to investigate biomarker candidates for ASD.[116] Nineteen metabolites including succinate, glycolate, hippurate, 3-hydroxyphenylacetate, hydroxyacetate, 1H-indole-3-acetate, phosphate, palmitate, stearate and 3-methyladipate were found to be discriminative between autistic and healthy children, thus representing potential biomarkers for ASD diagnosis. Altered levels of microbe-derived metabolites such as hippurate, 3-hydroxyphenylacetate, 3-hydroxyhippurate and 1H-indole-3-acetate implicated significant effects of gut microbiome on the host metabolome.

De Angelis et al. analyzed fecal metabolome of children with PPD-NOS and ASD.[117] The authors found that beneficial metabolites such as free amino acids (FAAs) and SCFAs were significantly decreased in feces of PDD-NOS and ASD compared to healthy individual. They also revealed significant correlations between metabolically active bacteria abundance and metabolites levels. *Faecalibacterium*, *Ruminococcus* and *Bifidobacterium* genera were positively correlated with total SCFA levels. Total FAA levels were correlated with *Bacteroides* genus.

Sharon et al. performed metabolomics using colon contents from mice harboring ASD human gut microbiome.[108] Adult offspring from mice transplanted with gut microbiome of ASD patient donors showed significantly altered autistic behaviors, and microbiome and metabolome profiles. Weak $GABA_A$ receptor agonists including 5-AV and taurine were found to be lower in mice with ASD microbiome. 5-AV treatment significantly decreased cortical neuron excitability in BTBR mice. Taurine treatment was found to delay inhibitory neurotransmission switch in response to GABA exposure in primary rat cortical neurons. Therefore, the study suggested that neuroactive metabolite alterations by ASD gut microbiome may affect the host neuronal activity. Transcriptomics analysis using mouse prefrontal cortex and striatum further revealed differential expression of diacylglycerol lipase beta which is required for axonal growth and guidance.[118] In addition, alternative splicing events relevant for several ribosome binding protein targets were significantly altered in ASD mouse brains.

Kang et al. analyzed fecal microbial metabolites in children with ASD using ^{1}H NMR spectroscopy.[119] The authors found that isopropanol, *p*-cresol and GABA concentrations were significantly altered in ASD compared to healthy individuals. In addition, they showed that Fischer discriminant analysis model using caprate, nicotinate, glutamine, thymine and aspartate can separate ASD from control cases, therefore suggesting that the metabolite group may be used as biomarkers.

2.6 Ubiquitin-proteasome system

The ubiquitin-proteasome system was shown to control synaptic structure and activity by regulating synaptic protein composition that is important for synaptic formation and maintenance such as SNARE complex proteins, AMPAR and PSD95.[120,121] Thus, perturbed proteolysis processes including ubiquitin mis-conjugation to the protein of target and subsequent failure of proteasomal degradation can lead to synaptic dysfunction, implicating a crucial role of the defective machinery in ASD.

Yi et al. found the link between ubiquitin and Wnt signaling pathways in ASD. The authors compared published databases for ubiquitin protein ligase E3A (UBE3A) interactome and Wnt signaling regulators.[122] While multiple UBE3A interacting proteins were found to be negative Wnt signaling regulators, they quantitated ubiquitinated peptide profiles to identify UBE3A substrates. They revealed that ASD-associated de novo $UBE3A^{T485S}$ mutation depleted the original UBE3A interactors and substrates including proteasome 19S regulatory subunits (PSMD1, PSMD2, PSMD11, PSMC2, and PSMC5) and 20S core subunits (PSMA1, PSMA2, and PSMB1).

Glessner et al. interrogated whole genome CNVs which confer an ASD risk.[123] The authors found that neuronal development-associated genes such as NRXN1, CNTN4, ASTN2 and NLGN1 had a higher CNV frequency in ASD compared to control cases. Interestingly, CNVs associated with ubiquitin ligase genes including UBE3A, PARK2, RFWD2 and FBXO40 were found only in ASD patients, suggesting a role of ubiquitin-proteasome system for ASD susceptibility.

2.7 Epigenetic regulation

Epigenetic regulation including DNA CpG methylation and histone modifications determines chromatin architecture, transcription factor accessibility and gene expression levels. The important role of epigenetic regulation has been highlighted in ASD. Multiple ASD-associated epigenetic changes in PRRT1, FAM181A, CHFR, AURKA, MAP8KIP3, NALP1L5, TET, MECP2, DNMT1 and MTHFR were suggested to affect crucial synaptic and neurodevelopmental processes.[124]

Whole exome sequencing using whole blood obtained from 10-year-old autistic male identified a de novo mutation in HIST1H1E which encodes histone H1.4 or H1E proteins involved in heterochromatin formation.[125] The authors performed a systemic review using autism genetics databases including SFARI and autDB. The analysis revealed that 42 genes related to ASD are directly associated with epigenetic machinery.

Ladd-Acosta et al. examined global DNA methylation profiles using postmortem brain tissues including dorsolateral prefrontal cortex, temporal cortex and cerebellum from autism and control cases.[126] Four differentially methylated regions near PRRT1, 11orf21, ZFP57 and SDHAP3 genes were found commonly across three brain regions. Ellis et al. analyzed DNA methylation using postmortem BA19 brain tissues from autistic and control subjects.[127] The authors found that several CpH sites were hypermethylated in autism brains. Interestingly, 10 histone methylations were only enriched in brains, but not in lymphoblastoid cell lines from the corresponding patients, indicating that epigenetic dysregulation can be tissue-specific.

2.8 Big data analysis with machine learning approach

Big data analysis empowered by machine learning methods has achieved promising advances in ASD research. Machine learning based on large-scale omics and clinical diagnosis data has assisted studies aimed at more precise diagnosis.

Gong et al. applied the KF algorithm to predict new ASD susceptibility genes.[128] The algorithm used datasets generated from biomedical text mining. The authors identified a number of ASD candidate genes for several members of orthodenticle homeobox, 5-hydroxytryptamine receptor (HTR), GABA receptor (GABAR) and protocadherin (PCDH).

Jiao et al. applied machine learning techniques to assess whether SNPs can predict ASD symptom severity.[129] They analyzed 29 SNPs of nine ASD-associated genes in ASD children subgroups with different symptom severity. SNP-based diagnostic models were generated using three different methods including decision stumps, alternating decision trees and FlexTrees algorithms. The study found the SNP rs878960 in GABRB3, a gene encoding GABAR subunit beta-3, to be predictive for symptom severity. Thus, the authors suggested that SNPs in ASD genes can be used as accurate classifiers for ASD symptom severity.

Oh et al. used the published microarray data obtained from peripheral leukocytes of ASD patients.[130] The authors established a predictive diagnosis model based on 19 differentially expressed genes to distinguish autistic and typically developing individuals. While the 19-probe set showed a high accuracy to discriminate the two groups, pyruvate kinase muscle isozyme involved in glycolysis was found to be the best discriminant.

Xiong et al. used deep learning algorithms to compute a score for RNA splicing by DNA variants in ASD.[131] The authors analyzed whole genome sequence of brain tissues from the Autism Tissue Program to identify causal SNPs for splicing misregulation. They found that genes with predicted mis-splicing were enriched in several ASD-relevant pathways including synaptic transmission, neuron projection, mitotic cell cycle, embryonic development and CNS development.

To predict functional and pathogenic de novo mutations in ASD, Zhou et al. analyzed whole genome of ASD simplex families using deep convolutional-network-based framework.[132] Non-coding mutations in ASD probands were predicted to affect brain-specific genes. In addition, genes in synapse and chromatin-related clusters showed similar expression patterns with those with high-impact proband mutations. Thus, the study discovered the causal contribution of non-coding mutations in the ASD etiology.

3. Conclusions and future directions in the biomarker and therapy discovery based on -omics and big data analysis

While -omics and big data analysis have profoundly advanced ASD research by providing information on affected biosignatures and pathways (Fig. 1), none has been validated as reliable clinical biomarkers and therapy targets. This may be the result of the current clinical diagnosis strategy based on standard tests including the Diagnostic and Standard Manual of Mental Disorders, the Childhood Autism Rating Scale, the ABC, the Gilliam Autism Rating Scale, the ADI-R and the Checklist for Autism in Toddlers. Since the standard tests are only based on phenotypic and behavioral traits, underlying biological mechanisms need to be considered for more precise diagnosis and treatment. Therefore, Research Domain Criteria (RDoC) has been suggested to be an alternative framework for biological classification system. The RDoC can address more biological and functional domains related to ASD. For instance, Hennessey et al. attempted to classify amygdala dysfunction in ASD using five RDoC domains including Negative Valence Systems, Positive Valence Systems, Cognitive Systems, Social Processes and Arousal and Regulatory Systems.[133] Systemic molecular profiling using -omics and big data analysis will strongly assist RDoC-based diagnosis strategy.

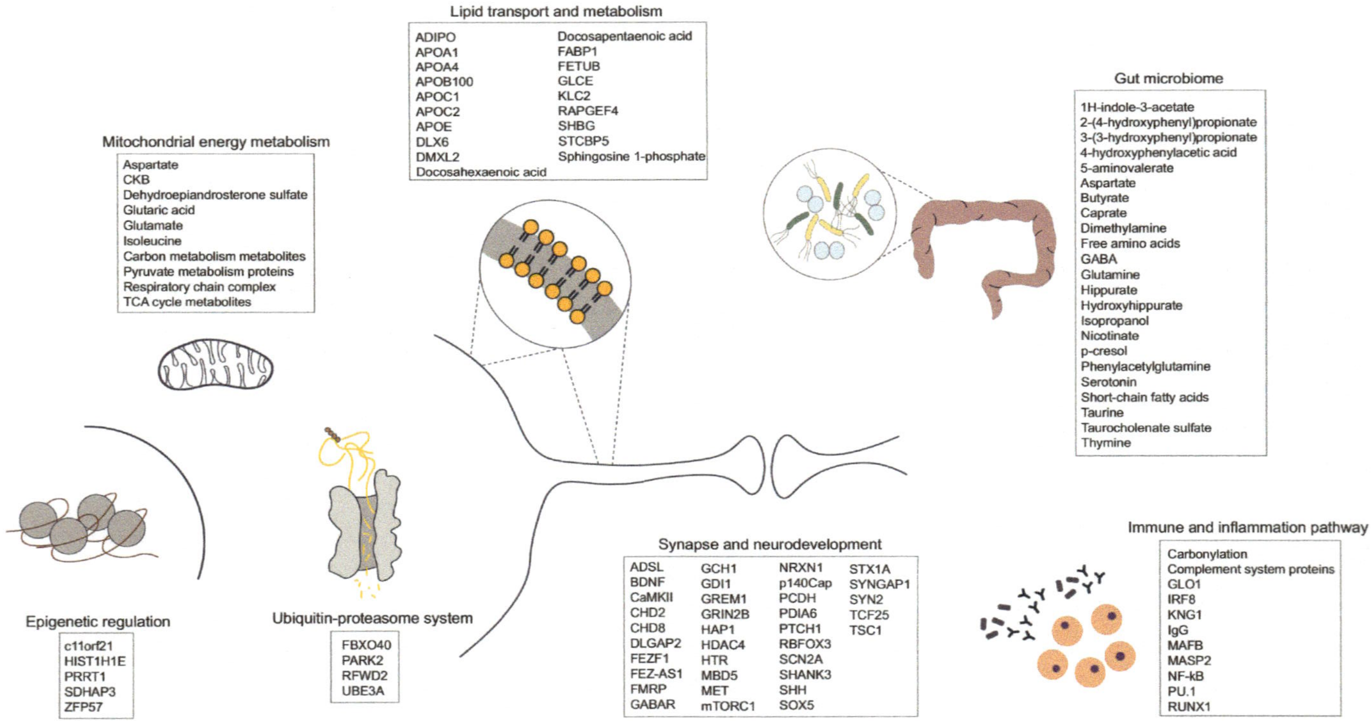

Fig. 1 The overview of biosignatures and pathways identified from -omics and big data analysis for autism spectrum disorder.

More research on ASD subgroup stratification based on molecular profiles is strongly required to overcome the current diagnostic and therapeutic limitations. The current studies have used clinical phenotypes and characterized single genotypes to stratify the subgroups. However, stratification efforts based on -omics and other big data are largely missing. The molecular subgrouping will not only provide information on underlying mechanisms specific to relevant subgroups, but also assist customized pharmacological interventions. While effective treatment for ASD core symptoms has been very challenging, several classes of drugs including atypical antipsychotics, selective serotonin reuptake inhibitors, tricyclic antidepressants, anticonvulsants, NMDA receptor antagonists, acetylcholinesterase inhibitors, psychostimulants, adrenergic alpha-2 receptor agonists and opiate antagonists have been shown to attenuate associated behavioral symptoms such as hyperactivity, anxiety, repetitive behavior, aggression and self-injury.[134] With the help of advanced machine learning methods, integrated multi-omics data may further help to improve pharmacological treatment efficiency.

Ultimately, there is an absolute necessity to establish the integrated molecular guideline based on databases acquired from -omics and other biological measurements. The molecular guideline databases will significantly advance clinical diagnosis and treatment of ASD.

Conflict of interest

The author declares no conflict of interest.

References

1. DeFilippis M. Depression in children and adolescents with autism spectrum disorder. *Children (Basel)*. 2018;5(9):E112. https://doi.org/10.3390/children5090112. [published Online First: Epub Date].
2. van Steensel FJ, Bogels SM, Perrin S. Anxiety disorders in children and adolescents with autistic spectrum disorders: a meta-analysis. *Clin Child Fam Psychol Rev*. 2011; 14(3):302–317. https://doi.org/10.1007/s10567-011-0097-0. [published Online First: Epub Date].
3. Elsabbagh M, Divan G, Koh YJ, et al. Global prevalence of autism and other pervasive developmental disorders. *Autism Res*. 2012;5(3):160–179. https://doi.org/10.1002/aur.239. [published Online First: Epub Date].
4. Huguet G, Ey E, Bourgeron T. The genetic landscapes of autism spectrum disorders. *Annu Rev Genomics Hum Genet*. 2013;14:191–213. https://doi.org/10.1146/annurev-genom-091212-153431. [published Online First: Epub Date].
5. Geschwind DH, Levitt P. Autism spectrum disorders: developmental disconnection syndromes. *Curr Opin Neurobiol*. 2007;17(1):103–111. https://doi.org/10.1016/j.conb.2007.01.009. [published Online First: Epub Date].

6. Happe F, Ronald A, Plomin R. Time to give up on a single explanation for autism. *Nat Neurosci*. 2006;9(10):1218–1220. https://doi.org/10.1038/nn1770. [published Online First: Epub Date].
7. Pantazatos SP, Huang YY, Rosoklija GB, Dwork AJ, Arango V, Mann JJ. Whole-transcriptome brain expression and exon-usage profiling in major depression and suicide: evidence for altered glial, endothelial and ATPase activity. *Mol Psychiatry*. 2017;22(5):760–773. https://doi.org/10.1038/mp.2016.130. [published Online First: Epub Date].
8. Silva-Costa LC, Carlson PT, Guest PC, de Almeida V, Martins-de-Souza D. Proteomic markers for depression. *Adv Exp Med Biol*. 2019;1118:191–206. https://doi.org/10.1007/978-3-030-05542-4_10. [published Online First: Epub Date].
9. Woo HI, Lim SW, Myung W, Kim DK, Lee SY. Differentially expressed genes related to major depressive disorder and antidepressant response: genome-wide gene expression analysis. *Exp Mol Med*. 2018;50(8):92. https://doi.org/10.1038/s12276-018-0123-0. [published Online First: Epub Date].
10. Wray NR, Ripke S, Mattheisen M, et al. Genome-wide association analyses identify 44 risk variants and refine the genetic architecture of major depression. *Nat Genet*. 2018;50(5):668–681. https://doi.org/10.1038/s41588-018-0090-3. [published Online First: Epub Date].
11. Luykx JJ, Giuliani F, Giuliani G, Veldink J. Coding and non-coding RNA abnormalities in bipolar disorder. *Genes (Basel)*. 2019;10(11):E946. https://doi.org/10.3390/genes10110946. [published Online First: Epub Date].
12. Stahl EA, Breen G, Forstner AJ, et al. Genome-wide association study identifies 30 loci associated with bipolar disorder. *Nat Genet*. 2019;51(5):793–803. https://doi.org/10.1038/s41588-019-0397-8. [published Online First: Epub Date].
13. Schizophrenia Working Group of the Psychiatric Genomics Consortium, Ripke S, et al. Biological insights from 108 schizophrenia-associated genetic loci. *Nature*. 2014;511(7510):421–427. https://doi.org/10.1038/nature13595. [published Online First: Epub Date].
14. Gusev A, Mancuso N, Won H, et al. Transcriptome-wide association study of schizophrenia and chromatin activity yields mechanistic disease insights. *Nat Genet*. 2018;50(4):538–548. https://doi.org/10.1038/s41588-018-0092-1. [published Online First: Epub Date].
15. Jaffe AE, Straub RE, Shin JH, et al. Developmental and genetic regulation of the human cortex transcriptome illuminate schizophrenia pathogenesis. *Nat Neurosci*. 2018;21(8):1117–1125. https://doi.org/10.1038/s41593-018-0197-y. [published Online First: Epub Date].
16. Nascimento JM, Martins-de-Souza D. The proteome of schizophrenia. *NPJ Schizophr*. 2015;1(1):14003. https://doi.org/10.1038/npjschz.2014.3. [published Online First: Epub Date].
17. Misiewicz Z, Iurato S, Kulesskaya N, et al. Multi-omics analysis identifies mitochondrial pathways associated with anxiety-related behavior. *PLoS Genet*. 2019;15(9):e1008358. https://doi.org/10.1371/journal.pgen.1008358. [published Online First: Epub Date].
18. Otowa T, Hek K, Lee M, et al. Meta-analysis of genome-wide association studies of anxiety disorders. *Mol Psychiatry*. 2016;21(10):1391–1399. https://doi.org/10.1038/mp.2015.197[published Online First: Epub Date].
19. Hughes HK, Mills Ko E, Rose D, Ashwood P. Immune dysfunction and autoimmunity as pathological mechanisms in autism spectrum disorders. *Front Cell Neurosci*. 2018;12:405. https://doi.org/10.3389/fncel.2018.00405[published Online First: Epub Date].
20. Gandal MJ, Zhang P, Hadjimichael E, et al. Transcriptome-wide isoform-level dysregulation in ASD, schizophrenia, and bipolar disorder. *Science*. 2018;362(6420):eaat8127. https://doi.org/10.1126/science.aat8127. [published Online First: Epub Date].

21. Gupta S, Ellis SE, Ashar FN, et al. Transcriptome analysis reveals dysregulation of innate immune response genes and neuronal activity-dependent genes in autism. *Nat Commun.* 2014;5:5748. https://doi.org/10.1038/ncomms6748. [published Online First: Epub Date].
22. Voineagu I, Wang X, Johnston P, et al. Transcriptomic analysis of autistic brain reveals convergent molecular pathology. *Nature.* 2011;474(7351):380–384. https://doi.org/10.1038/nature10110. [published Online First: Epub Date].
23. Tyler CV, Schramm SC, Karafa M, Tang AS, Jain AK. Chronic disease risks in young adults with autism spectrum disorder: forewarned is forearmed. *Am J Intellect Dev Disabil.* 2011;116(5):371–380. https://doi.org/10.1352/1944-7558-116.5.371. [published Online First: Epub Date].
24. Crawley JN, Heyer WD, LaSalle JM. Autism and cancer share risk genes, pathways, and drug targets. *Trends Genet.* 2016;32(3):139–146. https://doi.org/10.1016/j.tig.2016.01.001. [published Online First: Epub Date].
25. Fores-Martos J, Catala-Lopez F, Sanchez-Valle J, et al. Transcriptomic metaanalyses of autistic brains reveals shared gene expression and biological pathway abnormalities with cancer. *Mol Autism.* 2019;10:17. https://doi.org/10.1186/s13229-019-0262-8. [published Online First: Epub Date].
26. Feng C, Chen Y, Pan J, et al. Redox proteomic identification of carbonylated proteins in autism plasma: insight into oxidative stress and its related biomarkers in autism. *Clin Proteomics.* 2017;14:2. https://doi.org/10.1186/s12014-017-9138-0. [published Online First: Epub Date].
27. Momeni N, Brudin L, Behnia F, et al. High complement factor I activity in the plasma of children with autism spectrum disorders. *Autism Res Treat.* 2012; 2012:868576. https://doi.org/10.1155/2012/868576. [published Online First: Epub Date].
28. Momeni N, Bergquist J, Brudin L, et al. A novel blood-based biomarker for detection of autism spectrum disorders. *Transl Psychiatry.* 2012;2:e91. https://doi.org/10.1038/tp.2012.19. [published Online First: Epub Date].
29. Suganya V, Geetha A, Sujatha S. Urine proteome analysis to evaluate protein biomarkers in children with autism. *Clin Chim Acta.* 2015;450:210–219. https://doi.org/10.1016/j.cca.2015.08.015. [published Online First: Epub Date].
30. Khan MM, Bradford HN, Isordia-Salas I, et al. High-molecular-weight kininogen fragments stimulate the secretion of cytokines and chemokines through uPAR, Mac-1, and gC1qR in monocytes. *Arterioscler Thromb Vasc Biol.* 2006;26(10): 2260–2266. https://doi.org/10.1161/01.ATV.0000240290.70852.c0. [published Online First: Epub Date].
31. Wallis R, Dodds AW, Mitchell DA, Sim RB, Reid KB, Schwaeble WJ. Molecular interactions between MASP-2, C4, and C2 and their activation fragments leading to complement activation via the lectin pathway. *J Biol Chem.* 2007;282(11): 7844–7851. https://doi.org/10.1074/jbc.M606326200. [published Online First: Epub Date].
32. Junaid MA, Kowal D, Barua M, Pullarkat PS, Sklower Brooks S, Pullarkat RK. Proteomic studies identified a single nucleotide polymorphism in glyoxalase I as autism susceptibility factor. *Am J Med Genet A.* 2004;131(1):11–17. https://doi.org/10.1002/ajmg.a.30349. [published Online First: Epub Date].
33. Distler MG, Plant LD, Sokoloff G, et al. Glyoxalase 1 increases anxiety by reducing GABAA receptor agonist methylglyoxal. *J Clin Invest.* 2012; 122(6):2306–2315. https://doi.org/10.1172/jci61319. [published Online First: Epub Date].
34. Hovatta I, Tennant RS, Helton R, et al. Glyoxalase 1 and glutathione reductase 1 regulate anxiety in mice. *Nature.* 2005;438(7068):662–666. https://doi.org/10.1038/nature04250. [published Online First: Epub Date].

35. Williams R, Lim JE, Harr B, et al. A common and unstable copy number variant is associated with differences in Glo1 expression and anxiety-like behavior. *PLoS One*. 2009;4(3):e4649. https://doi.org/10.1371/journal.pone.0004649. [published Online First: Epub Date].
36. Guo Y, Wang P, Sun H, Cai R, Xia W, Wang S. Advanced glycation end product-induced astrocytic differentiation of cultured neurospheres through inhibition of Notch-Hes1 pathway-mediated neurogenesis. *Int J Mol Sci*. 2013;15(1):159–170. https://doi.org/10.3390/ijms15010159. [published Online First: Epub Date].
37. Kim J, Wan CK, Joc S, Shaikh SB, Nicholson LF. The role of receptor for advanced glycation end products (RAGE) in neuronal differentiation. *J Neurosci Res*. 2012;90(6):1136–1147. https://doi.org/10.1002/jnr.23014. [published Online First: Epub Date].
38. Vogel Ciernia A, Careaga M, LaSalle JM, Ashwood P. Microglia from offspring of dams with allergic asthma exhibit epigenomic alterations in genes dysregulated in autism. *Glia*. 2018;66(3):505–521. https://doi.org/10.1002/glia.23261. [published Online First: Epub Date].
39. Giovedi S, Corradi A, Fassio A, Benfenati F. Involvement of synaptic genes in the pathogenesis of autism spectrum disorders: the case of synapsins. *Front Pediatr*. 2014;2:94. https://doi.org/10.3389/fped.2014.00094. [published Online First: Epub Date].
40. Martinez-Cerdeno V. Dendrite and spine modifications in autism and related neurodevelopmental disorders in patients and animal models. *Dev Neurobiol*. 2017;77(4):393–404. https://doi.org/10.1002/dneu.22417. [published Online First: Epub Date].
41. De Rubeis S, He X, Goldberg AP, et al. Synaptic, transcriptional and chromatin genes disrupted in autism. *Nature*. 2014;515(7526):209–215. https://doi.org/10.1038/nature13772. [published Online First: Epub Date].
42. Ronan JL, Wu W, Crabtree GR. From neural development to cognition: unexpected roles for chromatin. *Nat Rev Genet*. 2013;14(5):347–359. https://doi.org/10.1038/nrg3413. [published Online First: Epub Date].
43. Callaghan DB, Rogic S, Tan PPC, et al. Whole genome sequencing and variant discovery in the ASPIRE autism spectrum disorder cohort. *Clin Genet*. 2019;96(3):199–206. https://doi.org/10.1111/cge.13556. [published Online First: Epub Date].
44. Lim ET, Uddin M, De Rubeis S, et al. Rates, distribution and implications of postzygotic mosaic mutations in autism spectrum disorder. *Nat Neurosci*. 2017;20(9):1217–1224. https://doi.org/10.1038/nn.4598. [published Online First: Epub Date].
45. Bender KJ, Trussell LO. The physiology of the axon initial segment. *Annu Rev Neurosci*. 2012;35:249–265. https://doi.org/10.1146/annurev-neuro-062111-150339. [published Online First: Epub Date].
46. Pain O, Pocklington AJ, Holmans PA, et al. Novel insight into the etiology of autism spectrum disorder gained by integrating expression data with genome-wide association statistics. *Biol Psychiatry*. 2019;86(4):265–273. https://doi.org/10.1016/j.biopsych.2019.04.034. [published Online First: Epub Date].
47. Sanders SJ, He X, Willsey AJ, et al. Insights into autism spectrum disorder genomic architecture and biology from 71 risk loci. *Neuron*. 2015;87(6):1215–1233. https://doi.org/10.1016/j.neuron.2015.09.016. [published Online First: Epub Date].
48. Grunwald LM, Stock R, Haag K, et al. Comparative characterization of human induced pluripotent stem cells (hiPSC) derived from patients with schizophrenia and autism. *Transl Psychiatry*. 2019;9(1):179. https://doi.org/10.1038/s41398-019-0517-3. [published Online First: Epub Date].

49. Gao Y, Wang F, Eisinger BE, Kelnhofer LE, Jobe EM, Zhao X. Integrative single-cell transcriptomics reveals molecular networks defining neuronal maturation during postnatal neurogenesis. *Cereb Cortex*. 2017;27(3):2064–2077. https://doi.org/10.1093/cercor/bhw040. [published Online First: Epub Date].
50. DeRosa BA, El Hokayem J, Artimovich E, et al. Convergent pathways in idiopathic autism revealed by time course transcriptomic analysis of patient-derived neurons. *Sci Rep* 2018;8(1):8423. https://doi.org/10.1038/s41598-018-26495-1. [published Online First: Epub Date].
51. Pinto D, Delaby E, Merico D, et al. Convergence of genes and cellular pathways dysregulated in autism spectrum disorders. *Am J Hum Genet*. 2014;94(5):677–694. https://doi.org/10.1016/j.ajhg.2014.03.018. [published Online First: Epub Date].
52. Hu VW, Addington A, Hyman A. Novel autism subtype-dependent genetic variants are revealed by quantitative trait and subphenotype association analyses of published GWAS data. *PLoS One*. 2011;6(4):e19067. https://doi.org/10.1371/journal.pone.0019067. [published Online First: Epub Date].
53. Bralten J, van Hulzen KJ, Martens MB, et al. Autism spectrum disorders and autistic traits share genetics and biology. *Mol Psychiatry*. 2018;23(5):1205–1212. https://doi.org/10.1038/mp.2017.98. [published Online First: Epub Date].
54. O'Roak BJ, Vives L, Fu W, et al. Multiplex targeted sequencing identifies recurrently mutated genes in autism spectrum disorders. *Science*. 2012;338(6114):1619–1622. https://doi.org/10.1126/science.1227764. [published Online First: Epub Date].
55. Durak O, Gao F, Kaeser-Woo YJ, et al. Chd8 mediates cortical neurogenesis via transcriptional regulation of cell cycle and Wnt signaling. *Nat Neurosci*. 2016;19(11):1477–1488. https://doi.org/10.1038/nn.4400. [published Online First: Epub Date].
56. Nishiyama M, Skoultchi AI, Nakayama KI. Histone H1 recruitment by CHD8 is essential for suppression of the Wnt-beta-catenin signaling pathway. *Mol Cell Biol*. 2012;32(2):501–512. https://doi.org/10.1128/mcb.06409-11. [published Online First: Epub Date].
57. Xu Q, Liu YY, Wang X, et al. Autism-associated CHD8 deficiency impairs axon development and migration of cortical neurons. *Mol Autism*. 2018;9:65. https://doi.org/10.1186/s13229-018-0244-2. [published Online First: Epub Date].
58. Merner N, Forgeot d'Arc B, Bell SC, et al. A de novo frameshift mutation in chromodomain helicase DNA-binding domain 8 (CHD8): case report and literature review. *Am J Med Genet A*. 2016;170a(5):1225–1235. https://doi.org/10.1002/ajmg.a.37566. [published Online First: Epub Date].
59. Sugathan A, Biagioli M, Golzio C, et al. CHD8 regulates neurodevelopmental pathways associated with autism spectrum disorder in neural progenitors. *Proc Natl Acad Sci U S A*. 2014;111(42):E4468–E4477. https://doi.org/10.1073/pnas.1405266111. [published Online First: Epub Date].
60. Siu MT, Butcher DT, Turinsky AL, et al. Functional DNA methylation signatures for autism spectrum disorder genomic risk loci: 16p11.2 deletions and CHD8 variants. *Clin Epigenetics*. 2019;11(1):103. https://doi.org/10.1186/s13148-019-0684-3. [published Online First: Epub Date].
61. Fernandez BA, Roberts W, Chung B, et al. Phenotypic spectrum associated with de novo and inherited deletions and duplications at 16p11.2 in individuals ascertained for diagnosis of autism spectrum disorder. *J Med Genet*. 2010;47(3):195–203. https://doi.org/10.1136/jmg.2009.069369. [published Online First: Epub Date].
62. Craig AM, Kang Y. Neurexin-neuroligin signaling in synapse development. *Curr Opin Neurobiol*. 2007;17(1):43–52. https://doi.org/10.1016/j.conb.2007.01.011. [published Online First: Epub Date].

63. Sudhof TC. Neuroligins and neurexins link synaptic function to cognitive disease. *Nature*. 2008;455(7215):903–911. https://doi.org/10.1038/nature07456. [published Online First: Epub Date].
64. Szatmari P, Paterson AD, Zwaigenbaum L, et al. Mapping autism risk loci using genetic linkage and chromosomal rearrangements. *Nat Genet*. 2007;39(3):319–328. https://doi.org/10.1038/ng1985. [published Online First: Epub Date].
65. Zahir FR, Baross A, Delaney AD, et al. A patient with vertebral, cognitive and behavioural abnormalities and a de novo deletion of NRXN1alpha. *J Med Genet*. 2008;45(4):239–243. https://doi.org/10.1136/jmg.2007.054437. [published Online First: Epub Date].
66. Feng J, Schroer R, Yan J, et al. High frequency of neurexin 1beta signal peptide structural variants in patients with autism. *Neurosci Lett*. 2006;409(1):10–13. https://doi.org/10.1016/j.neulet.2006.08.017. [published Online First: Epub Date].
67. Lam M, Moslem M, Bryois J, et al. Single cell analysis of autism patient with bi-allelic NRXN1-alpha deletion reveals skewed fate choice in neural progenitors and impaired neuronal functionality. *Exp Cell Res*. 2019;383(1):111469. https://doi.org/10.1016/j.yexcr.2019.06.014. [published Online First: Epub Date].
68. Velmeshev D, Schirmer L, Jung D, et al. Single-cell genomics identifies cell type-specific molecular changes in autism. *Science*. 2019;364(6441):685–689. https://doi.org/10.1126/science.aav8130. [published Online First: Epub Date].
69. Provenzano G, Corradi Z, Monsorno K, et al. comparative gene expression analysis of two mouse models of autism: transcriptome profiling of the BTBR and En2 (−/−) hippocampus. *Front Neurosci*. 2016;10:396. https://doi.org/10.3389/fnins.2016.00396. [published Online First: Epub Date].
70. Parikshak NN, Swarup V, Belgard TG, et al. Genome-wide changes in lncRNA, splicing, and regional gene expression patterns in autism. *Nature*. 2016;540(7633):423–427. https://doi.org/10.1038/nature20612. [published Online First: Epub Date].
71. McLennan Y, Polussa J, Tassone F, Hagerman R. Fragile x syndrome. *Curr Genomics*. 2011;12(3):216–224. https://doi.org/10.2174/138920211795677886. [published Online First: Epub Date].
72. Lee S, Chun HS, Lee J, et al. Plausibility of the zebrafish embryos/larvae as an alternative animal model for autism: a comparison study of transcriptome changes. *PLoS One*. 2018;13(9): e0203543. https://doi.org/10.1371/journal.pone.0203543. [published Online First: Epub Date].
73. Hu VW, Steinberg ME. Novel clustering of items from the autism diagnostic interview-revised to define phenotypes within autism spectrum disorders. *Autism Res*. 2009;2(2):67–77. https://doi.org/10.1002/aur.72. [published Online First: Epub Date].
74. Hu VW, Sarachana T, Kim KS, et al. Gene expression profiling differentiates autism case-controls and phenotypic variants of autism spectrum disorders: evidence for circadian rhythm dysfunction in severe autism. *Autism Res*. 2009;2(2):78–97. https://doi.org/10.1002/aur.73. [published Online First: Epub Date].
75. Daimon CM, Jasien JM, Wood 3rd WH, et al. Hippocampal transcriptomic and proteomic alterations in the BTBR mouse model of autism spectrum disorder. *Front Physiol*. 2015;6:324. https://doi.org/10.3389/fphys.2015.00324. [published Online First: Epub Date].
76. Broek JA, Guest PC, Rahmoune H, Bahn S. Proteomic analysis of post mortem brain tissue from autism patients: evidence for opposite changes in prefrontal cortex and cerebellum in synaptic connectivity-related proteins. *Mol Autism*. 2014;5:41. https://doi.org/10.1186/2040-2392-5-41. [published Online First: Epub Date].
77. Zheng Z, Zhang L, Zhu T, Huang J, Qu Y, Mu D. Peripheral brain-derived neurotrophic factor in autism spectrum disorder: a systematic review and meta-analysis. *Sci Rep*. 2016;6:31241. https://doi.org/10.1038/srep31241. [published Online First: Epub Date].

78. Alfieri A, Sorokina O, Adrait A, et al. Synaptic interactome mining reveals p140Cap as a new hub for PSD proteins involved in psychiatric and neurological disorders. *Front Mol Neurosci.* 2017;10:212. https://doi.org/10.3389/fnmol.2017.00212. [published Online First: Epub Date].
79. Xie Z, Li J, Baker J, Eagleson KL, Coba MP, Levitt P. Receptor tyrosine kinase MET interactome and neurodevelopmental disorder partners at the developing synapse. *Biol Psychiatry.* 2016;80(12):933–942. https://doi.org/10.1016/j.biopsych.2016.02.022. [published Online First: Epub Date].
80. Mejia LA, Litterman N, Ikeuchi Y, et al. A novel Hap1-Tsc1 interaction regulates neuronal mTORC1 signaling and morphogenesis in the brain. *J Neurosci.* 2013; 33(46):18015–18021. https://doi.org/10.1523/jneurosci.2290-13.2013. [published Online First: Epub Date].
81. Baucum 2nd AJ, Shonesy BC, Rose KL, Colbran RJ. Quantitative proteomics analysis of CaMKII phosphorylation and the CaMKII interactome in the mouse forebrain. *ACS Chem Nerosci.* 2015;6(4):615–631. https://doi.org/10.1021/cn500337u. [published Online First: Epub Date].
82. Harris JJ, Jolivet R, Attwell D. Synaptic energy use and supply. *Neuron.* 2012;75(5):762–777. https://doi.org/10.1016/j.neuron.2012.08.019. [published Online First: Epub Date].
83. Zheng X, Boyer L, Jin M, et al. Metabolic reprogramming during neuronal differentiation from aerobic glycolysis to neuronal oxidative phosphorylation. *Elife.* 2016;5: e13374. https://doi.org/10.7554/eLife.13374. [published Online First: Epub Date].
84. Li Z, Okamoto K, Hayashi Y, Sheng M. The importance of dendritic mitochondria in the morphogenesis and plasticity of spines and synapses. *Cell.* 2004;119(6):873–887. https://doi.org/10.1016/j.cell.2004.11.003. [published Online First: Epub Date].
85. Levy M, Faas GC, Saggau P, Craigen WJ, Sweatt JD. Mitochondrial regulation of synaptic plasticity in the hippocampus. *J Biol Chem.* 2003;278(20):17727–17734. https://doi.org/10.1074/jbc.M212878200. [published Online First: Epub Date].
86. Gordon A, Forsingdal A, Klewe IV, et al. Transcriptomic networks implicate neuronal energetic abnormalities in three mouse models harboring autism and schizophrenia-associated mutations. *Mol Psychiatry.* 2019;https://doi.org/10.1038/s41380-019-0576-0. [published Online First: Epub Date].
87. He Y, Zhou Y, Ma W, Wang J. An integrated transcriptomic analysis of autism spectrum disorder. *Sci Rep.* 2019;9(1):11818. https://doi.org/10.1038/s41598-019-48160-x. [published Online First: Epub Date].
88. Jost CR, Van Der Zee CE, In 't Zandt HJ, et al. Creatine kinase B-driven energy transfer in the brain is important for habituation and spatial learning behaviour, mossy fibre field size and determination of seizure susceptibility. *Eur J Neurosci.* 2002;15(10):1692–1706. https://doi.org/10.1046/j.1460-9568.2002.02001.x. [published Online First: Epub Date].
89. Hemmer W, Wallimann T. Functional aspects of creatine kinase in brain. *Dev Neurosci.* 1993;15(3–5):249–260. https://doi.org/10.1159/000111342. [published Online First: Epub Date].
90. Shen L, Feng C, Zhang K, et al. Proteomics study of peripheral blood mononuclear cells (PBMCs) in autistic children. *Front Cell Neurosci.* 2019;13:105. https://doi.org/10.3389/fncel.2019.00105[published Online First: Epub Date].
91. West PR, Amaral DG, Bais P, et al. Metabolomics as a tool for discovery of biomarkers of autism spectrum disorder in the blood plasma of children. *PLoS One.* 2014;9(11): e112445. https://doi.org/10.1371/journal.pone.0112445. [published Online First: Epub Date].
92. Patel MA, Katyare SS. Treatment with dehydroepiandrosterone (DHEA) stimulates oxidative energy metabolism in the cerebral mitochondria. A comparative study of effects in old and young adult rats. *Neurosci Lett.* 2006;402(1–2):131–136. https://doi.org/10.1016/j.neulet.2006.03.057. [published Online First: Epub Date].

93. Safiulina D, Peet N, Seppet E, Zharkovsky A, Kaasik A. Dehydroepiandrosterone inhibits complex I of the mitochondrial respiratory chain and is neurotoxic in vitro and in vivo at high concentrations. *Toxicol Sci.* 2006;93(2):348–356. https://doi.org/10.1093/toxsci/kfl064. [published Online First: Epub Date].
94. Kurochkin I, Khrameeva E, Tkachev A, et al. Metabolome signature of autism in the human prefrontal cortex. *Commun Biol.* 2019;2:234. https://doi.org/10.1038/s42003-019-0485-4. [published Online First: Epub Date].
95. Orozco JS, Hertz-Picciotto I, Abbeduto L, Slupsky CM. Metabolomics analysis of children with autism, idiopathic-developmental delays, and Down syndrome. *Transl Psychiatry.* 2019;9(1):243. https://doi.org/10.1038/s41398-019-0578-3. [published Online First: Epub Date].
96. Dietschy JM, Turley SD. Thematic review series: brain Lipids. Cholesterol metabolism in the central nervous system during early development and in the mature animal. *J Lipid Res.* 2004;45(8):1375–1397. https://doi.org/10.1194/jlr.R400004-JLR200. [published Online First: Epub Date].
97. Hering H, Lin CC, Sheng M. Lipid rafts in the maintenance of synapses, dendritic spines, and surface AMPA receptor stability. *J Neurosci.* 2003;23(8):3262–3271.
98. Betsholtz C. Lipid transport and human brain development. *Nat Genet.* 2015;47(7):699–701. https://doi.org/10.1038/ng.3348. [published Online First: Epub Date].
99. Liu Q, Trotter J, Zhang J, et al. Neuronal LRP1 knockout in adult mice leads to impaired brain lipid metabolism and progressive, age-dependent synapse loss and neurodegeneration. *J Neurosci.* 2010;30(50):17068–17078. https://doi.org/10.1523/jneurosci.4067-10.2010. [published Online First: Epub Date].
100. Gudenas BL, Srivastava AK, Wang L. Integrative genomic analyses for identification and prioritization of long non-coding RNAs associated with autism. *PLoS One.* 2017;12(5): e0178532. https://doi.org/10.1371/journal.pone.0178532. [published Online First: Epub Date].
101. Corbett BA, Kantor AB, Schulman H, et al. A proteomic study of serum from children with autism showing differential expression of apolipoproteins and complement proteins. *Mol Psychiatry.* 2007;12(3):292–306. https://doi.org/10.1038/sj.mp.4001943. [published Online First: Epub Date].
102. Dietschy JM. Central nervous system: cholesterol turnover, brain development and neurodegeneration. *Biol Chem.* 2009;390(4):287–293. https://doi.org/10.1515/bc.2009.035. [published Online First: Epub Date].
103. Yang J, Chen Y, Xiong X, et al. Peptidome analysis reveals novel serum biomarkers for children with autism spectrum disorder in China. *Proteomics Clin Appl.* 2018;12(5). e1700164. https://doi.org/10.1002/prca.201700164. [published Online First: Epub Date].
104. Steeb H, Ramsey JM, Guest PC, et al. Serum proteomic analysis identifies sex-specific differences in lipid metabolism and inflammation profiles in adults diagnosed with Asperger syndrome. *Mol Autism.* 2014;5(1):4. https://doi.org/10.1186/2040-2392-5-4. [publishedOnline First: Epub Date].
105. Ingudomnukul E, Baron-Cohen S, Wheelwright S, Knickmeyer R. Elevated rates of testosterone-related disorders in women with autism spectrum conditions. *Horm Behav.* 2007;51(5):597–604. https://doi.org/10.1016/j.yhbeh.2007.02.001. [published Online First: Epub Date].
106. Wang H, Liang S, Wang M, et al. Potential serum biomarkers from a metabolomics study of autism. *J Psychiatry Neurosci.* 2016;41(1):27–37. https://doi.org/10.1503/jpn.140009. [published Online First: Epub Date].
107. Rogers GB, Keating DJ, Young RL, Wong ML, Licinio J, Wesselingh S. From gut dysbiosis to altered brain function and mental illness: mechanisms and pathways. *Mol Psychiatry.* 2016;21(6):738–748. https://doi.org/10.1038/mp.2016.50. [published Online First: Epub Date].

108. Sharon G, Cruz NJ, Kang DW, et al. Human gut microbiota from autism spectrum disorder promote behavioral symptoms in mice. *Cell*. 2019;177(6). https://doi.org/10.1016/j.cell.2019.05.0041600–18.e17. [published Online First: Epub Date].
109. Stilling RM, Moloney GM, Ryan FJ, et al. Social interaction-induced activation of RNA splicing in the amygdala of microbiome-deficient mice. *Elife*. 2018;7:e33070. https://doi.org/10.7554/eLife.33070. [published Online First: Epub Date].
110. Li Q, Han Y, Dy ABC, Hagerman RJ. The gut microbiota and autism spectrum Disorders. *Front Cell Neurosci*. 2017;11:120. https://doi.org/10.3389/fncel.2017.00120. [published Online First: Epub Date].
111. de Theije CG, Wopereis H, Ramadan M, et al. Altered gut microbiota and activity in a murine model of autism spectrum disorders. *Brain Behav Immun*. 2014;37:197–206. https://doi.org/10.1016/j.bbi.2013.12.005. [published Online First: Epub Date].
112. Ming X, Stein TP, Barnes V, Rhodes N, Guo L. Metabolic perturbance in autism spectrum disorders: a metabolomics study. *J Proteome Res*. 2012;11(12):5856–5862. https://doi.org/10.1021/pr300910n. [published Online First: Epub Date].
113. Yap IK, Angley M, Veselkov KA, Holmes E, Lindon JC, Nicholson JK. Urinary metabolic phenotyping differentiates children with autism from their unaffected siblings and age-matched controls. *J Proteome Res*. 2010;9(6):2996–3004. https://doi.org/10.1021/pr901188e. [published Online First: Epub Date].
114. Mussap M, Noto A, Fanos V. Metabolomics of autism spectrum disorders: early insights regarding mammalian-microbial cometabolites. *Expert Rev Mol Diagn*. 2016;16(8):869–881. https://doi.org/10.1080/14737159.2016.1202765. [published Online First: Epub Date].
115. Kaluzna-Czaplinska J. Noninvasive urinary organic acids test to assess biochemical and nutritional individuality in autistic children. *Clin Biochem*. 2011;44(8–9):686–691. https://doi.org/10.1016/j.clinbiochem.2011.01.015. [published Online First: Epub Date].
116. Emond P, Mavel S, Aidoud N, et al. GC-MS-based urine metabolic profiling of autism spectrum disorders. *Anal Bioanal Chem*. 2013;405(15):5291–5300. https://doi.org/10.1007/s00216-013-6934-x. [published Online First: Epub Date].
117. De Angelis M, Piccolo M, Vannini L, et al. Fecal microbiota and metabolome of children with autism and pervasive developmental disorder not otherwise specified. *PLoS One*. 2013;8(10): e76993. https://doi.org/10.1371/journal.pone.0076993. [published Online First: Epub Date].
118. Brittis PA, Silver J, Walsh FS, Doherty P. Fibroblast growth factor receptor function is required for the orderly projection of ganglion cell axons in the developing mammalian retina. *Mol Cell Neurosci*. 1996;8(2–3):120–128. https://doi.org/10.1006/mcne.1996.0051. [published Online First: Epub Date].
119. Kang DW, Ilhan ZE, Isern NG, et al. Differences in fecal microbial metabolites and microbiota of children with autism spectrum disorders. *Anaerobe*. 2018;49:121–131. https://doi.org/10.1016/j.anaerobe.2017.12.007. [published Online First: Epub Date].
120. Colledge M, Snyder EM, Crozier RA, et al. Ubiquitination regulates PSD-95 degradation and AMPA receptor surface expression. *Neuron*. 2003;40(3):595–607. https://doi.org/10.1016/s0896-6273(03)00687-1. [published Online First: Epub Date].
121. Saldate JJ, Shiau J, Cazares VA, Stuenkel EL. The ubiquitin-proteasome system functionally links neuronal Tomosyn-1 to dendritic morphology. *J Biol Chem*. 2018;293(7):2232–2246. https://doi.org/10.1074/jbc.M117.815514. [published Online First: Epub Date].
122. Yi JJ, Paranjape SR, Walker MP, et al. The autism-linked UBE3A T485A mutant E3 ubiquitin ligase activates the Wnt/beta-catenin pathway by inhibiting the proteasome. *J Biol Chem*. 2017;292(30):12503–12515. https://doi.org/10.1074/jbc.M117.788448. [published Online First: Epub Date].

123. Glessner JT, Wang K, Cai G, et al. Autism genome-wide copy number variation reveals ubiquitin and neuronal genes. *Nature*. 2009;459(7246):569–573. https://doi.org/10.1038/nature07953. [published Online First: Epub Date].
124. Eshraghi AA, Liu G, Kay SS, et al. Epigenetics and autism spectrum disorder: is there a correlation? *Front Cell Neurosci*. 2018;12:78. https://doi.org/10.3389/fncel.2018.00078. [published Online First: Epub Date].
125. Duffney LJ, Valdez P, Tremblay MW, et al. Epigenetics and autism spectrum disorder: a report of an autism case with mutation in H1 linker histone HIST1H1E and literature review. *Am J Med Genet B Neuropsychiatr Genet*. 2018;177(4):426–433. https://doi.org/10.1002/ajmg.b.32631. [published Online First: Epub Date].
126. Ladd-Acosta C, Hansen KD, Briem E, Fallin MD, Kaufmann WE, Feinberg AP. Common DNA methylation alterations in multiple brain regions in autism. *Mol Psychiatry*. 2014;19(8):862–871. https://doi.org/10.1038/mp.2013.114. [published Online First: Epub Date].
127. Ellis SE, Gupta S, Moes A, West AB, Arking DE. Exaggerated CpH methylation in the autism-affected brain. *Mol Autism*. 2017;8:6. https://doi.org/10.1186/s13229-017-0119-y. [published Online First: Epub Date].
128. Gong L, Yan Y, Xie J, Liu H, Sun X. Prediction of autism susceptibility genes based on association rules. *J Neurosci Res*. 2012;90(6):1119–1125. https://doi.org/10.1002/jnr.23015. [published Online First: Epub Date].
129. Jiao Y, Chen R, Ke X, et al. Single nucleotide polymorphisms predict symptom severity of autism spectrum disorder. *J Autism Dev Disord*. 2012;42(6):971–983. https://doi.org/10.1007/s10803-011-1327-5. [published Online First: Epub Date].
130. Oh DH, Kim IB, Kim SH, Ahn DH. Predicting autism spectrum disorder using blood-based gene expression signatures and machine learning. *Clin Psychopharmacol Neurosci*. 2017;15(1):47–52. https://doi.org/10.9758/cpn.2017.15.1.47. [published Online First: Epub Date].
131. Xiong HY, Alipanahi B, Lee LJ, et al. RNA splicing. The human splicing code reveals new insights into the genetic determinants of disease. *Science*. 2015;347(6218):1254806. https://doi.org/10.1126/science.1254806. [published Online First: Epub Date].
132. Zhou J, Park CY, Theesfeld CL, et al. Whole-genome deep-learning analysis identifies contribution of noncoding mutations to autism risk. *Nat Genet*. 2019;51(6):973–980. https://doi.org/10.1038/s41588-019-0420-0. [published Online First: Epub Date].
133. Hennessey T, Andari E, Rainnie DG. RDoC-based categorization of amygdala functions and its implications in autism. *Neurosci Biobehav Rev*. 2018;90:115–129. https://doi.org/10.1016/j.neubiorev.2018.04.007. [published Online First: Epub Date].
134. Eissa N, Al-Houqani M, Sadeq A, Ojha SK, Sasse A, Sadek B. Current enlightenment about etiology and pharmacological treatment of autism spectrum disorder. *Front Neurosci*. 2018;12:304. https://doi.org/10.3389/fnins.2018.00304. [published Online First: Epub Date].

CHAPTER FOUR

Autism spectrum disorder risk prediction: A systematic review of behavioral and neural investigations

Xiujuan Geng[a,*], Xin Kang[a,b,*], Patrick C.M. Wong[a,b,c]
[a]Brain and Mind Institute, The Chinese University of Hong Kong, Shatin, Hong Kong
[b]Department of Linguistics and Modern Languages, The Chinese University of Hong Kong, Shatin, Hong Kong
[c]Department of Otorhinolaryngology, Head and Neck Surgery, The Chinese University of Hong Kong, Shatin, Hong Kong
[*]Corresponding authors: e-mail address: gengx@cuhk.edu.hk; xin.kang@cuhk.edu.hk

Contents

Abstract

A reliable diagnosis of autism spectrum disorder (ASD) is difficult to make until after toddlerhood. Detection in an earlier age enables early intervention, which is typically more effective. Recent studies of the development of brain and behavior in infants

Progress in Molecular Biology and Translational Science, Volume 173
ISSN 1877-1173
https://doi.org/10.1016/bs.pmbts.2020.04.015

and toddlers have provided important insights in the diagnosis of autism. This extensive review focuses on published studies of predicting the diagnosis of autism during infancy and toddlerhood younger than 3 years using behavioral and neuroimaging approaches. After screening a total of 782 papers, 17 neuroimaging and 43 behavioral studies were reviewed. The features for prediction consist of behavioral measures using screening tools, observational and experimental methods, brain volumetric measures, and neural functional activation and connectivity patterns. The classification approaches include logistic regression, linear discriminant function, decision trees, support vector machine, and deep learning based methods. Prediction performance has large variance across different studies. For behavioral studies, the sensitivity varies from 20% to 100%, and specificity ranges from 48% to 100%. The accuracy rates range from 61% to 94% in neuroimaging studies. Possible factors contributing to this inconsistency may be partially due to the heterogeneity of ASD, different targeted populations (i.e., high-risk group for ASD and general population), age when the features were collected, and validation procedures. The translation to clinical practice requires extensive further research including external validation with large sample size and optimized feature selection. The use of multi-modal features, e.g., combination of neuroimaging and behavior, is worth further investigation to improve the prediction accuracy.

1. Introduction

Autism spectrum disorder (ASD) is a complex neurodevelopmental disorder and is behaviorally characterized by the presence of social communication deficits and restricted and repetitive behaviors in childhood, according to the Diagnostic and Statistical Manual of Mental Disorders, Fifth Edition (DSM-5).[1] ASD affects 1%–2% of all children in developed countries and regions including North America and Europe.[2] Early detection of ASD allows for early intervention, which has been shown to be effective in decreasing impairments, influencing developmental trajectories,[3–6] and resulting in more positive long-term outcomes for children.[7–9] Behavioral symptoms at the first year of life are not specific to ASD and cannot differentiate infants who will develop ASD from who have developmental delays.[10] Therefore a confirmed and reliable diagnosis of ASD is not typically made until 4–5 years.[11–13] Recently a few reviews have summarized the potential early behavioral and neural signs of ASD,[14–16] which may assist in detecting the risk of ASD at the pre-symptomatic stage. Confirmed genetic associations with autism have also been reported.[17,18] Yet few studies have used genetics for detecting ASD. Therefore, in this review we surveyed behavioral and neural investigations on the early prediction of ASD.

1.1 Early behavioral signs of ASD

In the past two decades, interest in studying early signs of autism and the timing of their onset has grown considerably. Although the mean age of diagnosis is around 4–5 years,[12,13] studies have suggested that early behavioral signs of ASD may be detected before clinical diagnosis as early as before the first year of life.[19–23] Studies in the 1990s mainly reported evidence collected using parents' retrospective reports,[24,25] home videos[26,27] and case studies.[28] Although retrospective studies provide important insights into early manifestations of autism, they may be subjected to reporting and sampling biases.[29,30] In recent years, there has been an increasing number of prospective longitudinal studies that use screening tools, non-invasive observational and experimental techniques (e.g., eye-tracking technology). Most of these studies focus on infants at 'high-risk' (e.g., siblings of children with ASD). Previous studies have shown that 10%–20% of children whose older siblings are probands with autism will meet diagnostic criteria for the disorder themselves.[31,32] There are also studies that examine infants and toddlers with delays in motor, language, and cognitive development and the general population. Motor skills,[33] visual reception,[34] language,[35] and eye gaze patterns of social scenes and faces[36,37] show different developmental patterns in the first year of life in infants who later meet criteria for ASD. All these findings highlight the developmental nature of ASD and also hold promise for earlier detection of ASD before formal diagnosis. In the past decade, standardized assessment tools based on human coded behaviors and parents' interviews have been developed and used by more and more clinicians, including the Autism Diagnostic Observation Schedule, second edition (ADOS-2),[38] the Autism Diagnostic Interview Revised (ADI-R).[39] These diagnostic tools have been standardized and shown to have diagnostic accuracy in children as young as 24 months.[40] Despite accumulating evidence on the prodromal signs of ASD among younger children, few behavior studies except those on screening tools have examined how well these behaviors of infants may predict future ASD diagnosis.

1.2 Early neuroimaging derived signs of ASD

1.2.1 Morphological signs using structural and diffusion MRI

Recent studies of early brain development with MRI have demonstrated neural abnormalities across the first 2 years of life for the individuals who will develop autism after 2 years or older. One of the most consistent findings is the enlarged brain size, i.e., overgrowth of cerebral volume at the end of the first and second year of life, yet the enlargement is not present at birth.[41–45] Frontal, parietal and temporal lobes also appear to have the overgrowth pattern.[46,47] Subcortical regions, amygdala (mainly in girls with ASD),[45,48,49] hippocampus[43,45,48] and components of the basal ganglia, including the caudate nucleus, globus pallidus, and putamen show increased volumes in toddlers with ASD.[50,51] A longitudinal study using infant data has suggested that brain growth, specifically, the cortical surface area, begins to diverge from a typical trajectory starting from 6 months of age.[52]

Converging studies using diffusion MRI have indicated increased fractional anisotropy (FA) and over-connectivity in infants and toddlers who have or are later diagnosed with ASD.[53–57] Aberrant white matter integrity has been detected as early as 6 months of age in multiple fiber tracts.[54,58] The white matter integrity in genu and cerebellar peduncles has also been shown to be associated with abnormal sensory responsiveness at 2 years of age, which is a clinical domain affected in ASD.[55] The findings of increased FA of ASD in infants and toddlers across multiple independent studies are in contrast to reports among older children and adults with autism where FA is relatively lower in individuals with autism.[59] Longitudinal studies on developmental trajectories suggest that FA may be elevated during the first years of age, but become lower in individuals with ASD.[55,58] A recent study in tuberous sclerosis complex (TSC) patients, however, has revealed that TSC patients with ASD exhibited reduced FA in white matter regions along several fiber bundles over the first 2 years of life compared to TSC subjects without ASD.[60]

1.2.2 Neural activities with functional MRI

Functional MRI (fMRI) can be used for estimating brain activation under tasks across the entire brain based on blood oxygen levels. Task-evoked fMRI studies have been conducted to explore neural areas and circuits responsible for core defects in autism, such as language production. Redcay et al. have reported atypical functional activations in language related brain regions, and the patterns are associated with receptive and expressive language function levels.[61–63] One of their early studies used auditory-evoked fMRI in young children with autism during natural sleep. Significant differences in regional activation patterns were detected in response to speech sounds between autism and control toddlers.[61] Later, the authors applied the same task paradigm and examined neural responses in children aged 12–48 months with ASD and controls.[64] Less activation in the left superior temporal gyrus in children with ASD and right lateralization in response to speech sounds were found, which is contrast to left lateralization observed in controls. Lombardo et al.[63] examined activation patterns in regions associated with language function in toddlers with ASD and found that ASD toddlers with poorer language a year later showed less activation in bilateral temporal and frontal regions of interest relative to groups including children with ASD who had relatively typical language scores.

Resting-state fMRI is used to study intrinsic neural fluctuations and functional connectivity. Altered functional connectivity has also been found in the development of ASD. Dinstein et al.[62] have shown weaker interhemispheric synchronization in ASD toddlers. Reduced resting-state functional connectivity between amygdala and several other brain regions has been reported in children with ASD with average age of 3.5 years.[65]

1.2.3 Neurophysiological signs measured by EEG

Electroencephalography (EEG) records the electrical activity of a large population of synchronous neurons in the cerebral cortex. Although the source localization is challenging, ASD pathophysiology may involve early disruptions in the brain's oscillatory rhythms that can be captured by EEG. It has the potential to explore the neural functional differences of early autism phenotype.[66] A growing body of work has demonstrated the sensitivity of EEG to developmental changes between high- and low-risk infants,[67,68] between high-risk who develop ASD (HR-ASD) and high-risk who did not (HR-neg),[69] and associations between EEG measures and ASD symptoms.[68] Orekhova et al.[70] reported elevated phase-lagged alpha-range connectivity in 14-month-old HR-ASD compared to both low-risk infants and HR-neg infants. The alpha wave suppression is thought to reflect visual and attentional processes.[71,72] Components of event related potentials (ERPs) evoked in response to dynamic eye gaze during 6–10 months old were reported to be associated with autism diagnosed at 36 months.[73] Jones et al.[74] measured ERPs to faces and objects of infants at 6 and 12 months, and found that HR-ASD (high-risk who met criteria of ASD at 24 months) had shorter epochs of visual attention, faster but less prolonged activation to faces, and delayed sensitization responses to faces at 6 months. These findings are consistent with disrupted engagement of sustained attention to social stimuli in children with ASD.

2. Methods

2.1 Scope of the review

The emerging findings of potential risk markers for autism measured by behavior and neuroimaging in early years of life have encouraged the work on predicting the diagnosis of ASD in early ages. We conducted a literature search on all studies that used behavioral or neuroimaging measures as predictors for the classification on the diagnosis of ASD in infants or toddlers younger than 3 years. The corpus of studies we surveyed was identified through systematic searches of PubMed and Web of Science using variations related to autism (autism OR autistic OR ASD), infants OR toddlers (infants OR infancy OR toddler* OR early children), brain imaging (MRI OR magnetic resonance OR neuroimaging OR brain imaging OR EEG OR ERP OR magnetoencephalography (MEG) OR functional near infrared spectroscopy (fNIRS)) or behaviors (behav* OR eye), and variations related to prediction (predict* OR detect* OR classif* OR accuracy OR sensitivity OR specificity). The search was concluded on the 1 Jan, 2020. We have also searched the references in related review papers recently published to add missing literature. After the first round searching and removing the duplicated ones, we have ended up with 583 papers using neuroimaging techniques and 199 behavioral papers. We further screened the papers based on the following criteria: average ages of participants were below 3 years old; there is diagnosis of ASD; used neuroimaging including MRI, EEG and other imaging methods, (e.g., fNIRS, MEG), or behavioral methods including screening tools and behavioral observation or experiments; have the prediction or classification performance reported; longitudinal or cross-sectional studies. We did not include retrospective studies that have been covered extensively previously.[75]

2.2 Modalities and features

The studies we surveyed were categorized based on the modalities of techniques used for the prediction. Modalities used in related literature include: structural, diffusion and functional MRI, EEG, fNIRS, eye tracking, screening tools, and behavioral assessments. All these measures have their unique advantages and disadvantages in assessing biological or psychological states. For instance, functional MRI can image the fluctuations under stimuli or rest of the entire brain. Yet it has a poor temporal resolution. In contrast, EEG has very high temporal resolution of brain activity but low spatial resolution. Screening and behavioral assessments may be a more direct evaluation of the symptoms expressed in ASD, yet they are less associated with neurobiological basis compared to neuroimaging measures.

For the non-imaging behavioral literature, we start with reviewing screening tools that employed parent's reports and/or structured play. We then move on to observational and experimental behavioral studies in the home and lab settings. The first category of neuroimaging studies reviewed is using structural or diffusion MRI for prediction. The derived neural features can be brain volumes of different tissue types, cortical thickness and surface area, and white matter characteristics. Next, we focused on predictions using functional MRI, both functional connectivity at rest and functional activation patterns under stimuli can be computed as potential features for the diagnosis of ASD. Another modality reviewed is to use EEG including both ERP recorded under stimuli and EEG signals recorded at rest. A total of five categories were included in this review based on the modalities of features: screening tools, observational and experimental behavior measures, structural MRI, functional MRI, and EEG. There was no prediction study of early diagnosis of ASD using other neuroimaging techniques such as MEG or fNIRS.

2.3 Analytical methods

Prediction of ASD diagnosis using behavioral and neural biological features has begun to gain prominence in the ASD literature over a decade ago. Some conventional classification methods used in prediction of ASD diagnosis include: logistic regression classifier, linear discriminant function, and binary cutoff at a threshold optimized by receiver operating characteristic (ROC) curve evaluation. Most behavioral studies reviewed here used these types of classification methods. Machine learning methods are also widely used in prediction of autism, consisting of k-nearest neighbor, Bayesian classification, decision trees such as random forest and ensemble trees, support vector machine (SVM), and deep learning network. More advanced approaches such as deep learning are characterized by many layers of artificial neurons, which require large sample size for automatic construction of features.[76]

3. Results

3.1 Screening tools

Although early screening tools are available for parents and clinicians, there is still debate whether it is beneficial to use these tools for early diagnosis and intervention. The American Academy of Pediatrics (AAP) have recommended universal screening for ASDs at 18 and 24 months, and additional selective screening for toddlers with identified concerns.[40] However, the US Preventive Services Task Force (UDPSTF) claims that there is insufficient evidence of benefits for using universal ASD screening tools for young children when their parents or clinicians have raised no concerns of ASD.[77] They challenged the validity of these screening tools due to limitations of the evidence in size, study design, and heterogeneity and called for more research.

Among 32 studies on prospective screening tools of ASD, 20 of them have reported systematic evaluation of screening tools for the general population regardless of risk status.[78] Eleven studies examined the high-risk samples including infants reported with developmental concerns or having an older sibling with ASD diagnosis.[10] One study examined both high-risk and general population samples.[79] Two studies investigated the stability of diagnostic instruments when children were first examined between 12 and 27 months and re-examined between 27 and 61 months.[80,81] We first reviewed studies that have investigated the general population and included a large sample size before reviewing studies involving the high-risk population. For both populations, we summarized the studies involving a large sample size ($N > 10{,}000$), a higher accuracy rate, or younger age of children (below 12 months) in the main content. All screening tools in this review can be found in Table 1.

3.1.1 Screening tools for the general population

The Checklist for Autism in Toddlers (CHAT),[78] is a screening tool intended for clinical, research, and education use on 18-month-old toddlers from community samples. With a sample size of 16,235 children, the cutoff criteria of CHAT demonstrate a sensitivity of 20% and specificity of 99%, suggesting it would probably miss 80% of children who are later diagnosed with ASD but may be used as a "rule in" screening tool to identify children who may not need extra concerns, but not a "rule out" assessment. The Modified Checklist for Autism in Toddlers—Revised with Follow-up (M-CHAT-R/F),[82] adapted from the CHAT, was reported to have increased sensitivity of 73% and specificity of 89% when 3 was used as the cutoff score with a sample size of 16, 071 children.

Early Screening of Autistic Traits Questionnaire (ESAT), a parent/caregiver-completed checklist, also had a large sample size (N=31,724) of 14–15-month children.[83] However, the true positive value of this test

Table 1 List of studies using behavioral methods.

Modalities	Study	Target group	*N*	Sex (male)	Age diagnosed	Method for diagnosis	Features for assessment	Age assessed	Analytical method	Accuracy
Screening	Pierce et al.[85]	GP	*N*=10,479 infants; 137 pediatricians	Not clear	32–36 months	Clinical judgment based on ADOS-T between 12 and 18 months and ADI-R at 32–36 months.	Communication and Symbolic Behavior Scales Developmental Profile Infant-Toddler Checklist (CSBS-DP-IT-Checklist)	10–16 months; and followed every 6 months until 36 months.	True positive results/true positive results + false positive results	PPV=0.75
Screening	Wu et al.[145]	GP	*N*=119 (ASD=57; DD=62)	*N*=47 (ASD); *N*=44 (DD)	24 and 35 months	DSM-5 criteria	Screening Tool for Autism in 2-Year-Olds (STAT)	16 and 24 months (Time 1); 35 and 46 months (Time 2)	ROC	AUC=0.87 (Time 1); AUC=0.82 (Time 2)
Screening	Fulcer et al.[96]	GP	*N*=158 (ASD=79; TD=79)	*N*=64	23–71 months	ADOS-G and DSM-IV-TR criteria.	Repetitive Behavior Scale-Revised (RBS-R)	23–71 months	ROC	AUC=0.903
Screening	Gray et al.[146]	GP	*N*=207	*N*=172	3 years	DSM-IV criteria and information from ADI-R, ADOS	Development Behavior Checklist-Early Screen (DBC-ES)	20–51 months	ROC	SEN=0.83; SPE=0.48

Continued

Table 1 List of studies using behavioral methods.—cont'd

Modalities	Study	Target group	*N*	Sex (male)	Age diagnosed	Method for diagnosis	Features for assessment	Age assessed	Analytical method	Accuracy
Screening	Wong et al.[97]	GP	$N=211$ (autism = 53, PDD-NOS = 33, and non-autism = 125)	$N=156$	18–24 months	Clinical diagnosis based on DSM-IV and AOI-R.	Checklist for Autism in Toddlers (CHAT) - 23: Chinese checklist combining the M-CHAT (self-administered questionnaire) and Section B of the CHAT (direct observation by trained investigators).	18–24 months	ROC	Part A: SEN = 0.931; SPE = 0.768; Part B: SEN = 0.736, SPE = 0.912
Screening	Smith et al.[147]	GP	$N=217$ (Study 1); $N=232$ (Study 2)	$N=160$ (Study 1); $N=124$ (Study 2)	48 months	NA	Parent's Observational Screen of Social Interactions (POSI); Modified Checklist for Autism in Toddlers (M-CHAT)	18–48 months (Study 1); 16–36 months (Study 2)	ROC	Study1: SEN of POSI = 0.89; M-CHAT = 0.71; SPE of POSI = 0.54; M-CHAT = 0.62; Study 2: SEN of POSI = 0.83; M-CHAT = 0.50; SPE of POSI = 0.75; M-CHAT = 0.84
Screening	Dietz et al.[83]	GP	$N=31{,}724$	$N=19{,}986$	34–64 months	Diagnoses were made by three child psychiatrists using DSM-IV criteria	Early Screening of Autistic Traits Questionnaire (ESAT)	14–15 months	Chi-square analysis	PPV = 0.25

Screening	Kruizinga et al.[89]	GP	$N=3127$	$N=1564$	on average 3.5 months after screening	ADOS + ADI-R	Brief Infant-Toddler Social Emotional Assessment (BITSEA)	12–40 months	ROC	AUC = 0.90 (Problem scale); AUC = 0.93 (Competence scale)
Screening	Scambler et al.[148]	GP	$N=44$	$N=28$	4–5 years	Clinical diagnoses based on ADI-R, ADOS, and DSM-IV	Checklist for Autism in Toddlers (CHAT)	Time 1: 2–3 years; Time 2: 4 and 6 years.	Percentage	83% of the sample predicted by original CHAT criteria; 93% predicted by Denver modification
Screening	Giserman et al.[149]	GP	$N=512$ (223 ASD; 289 non-ASD)	$N=343$	Not clear	ASD was determined by DSM-IV criteria for autism or PDD-NOS, ADOS, and expert clinical judgment	Brief Infant-Toddler Social Emotional Assessment (BITSEA)	15–48 months	ROC	AUC = 0.92; SEN = 0.8; SPE = 0.79
Screening	Nah et al.[150]	GP	$N=55$	$N=44$	2 and 6 years	Best estimate clinical (BEC) DSM-IV-TR diagnosis	Autism detection in early childhood (ADEC); Childhood Autism Rating Scale (CARS)	19–42 months	ROC	2-Year Follow-up: ADEC AUC = 0.92; CARS AUC = 0.81; 6-Year Follow-up: ADEC AUC = 0.85; CARS AUC = 0.72
Screening	Dereu et al.[86]	GP	$N=6808$	$N=3553$	Not clear	Diagnosed in autism clinics or diagnostic centers	Checklist of Early Signs of Developmental Disorders (CESDD)	3–39 months	ROC	AUC = 0.90 (95% CI = 0.84–0.97); SEN = 0.68; SPE = 0.96; PPV = 0.10; NPV = 0.99

Continued

Table 1 List of studies using behavioral methods.—cont'd

Modalities	Study	Target group	N	Sex (male)	Age diagnosed	Method for diagnosis	Features for assessment	Age assessed	Analytical method	Accuracy
Screening	Choueiri and Wagner[151]	GP	$N=61$	$N=41$	Not clear	Diagnosed by clinicians using DSM-IV criteria as the benchmark	Rapid Interactive Screening Test for Autism in Toddlers (RITA-T)	18–36 months	ROC	SEN=1.00; SPE=0.84; PPV=0.88; NPV=1
Screening	Matson et al.[88]	GP	$N=957$	$N=670$	Not clear	Clinical judgment using DSM-IV-TR algorithm for Autistic Disorder.	Baby and Infant Screen for aUtIsm Traits (BISCUIT)	17–37 months	Logistical regression	Variances account=~90% (Study 1); Accuracy=0.89 (Study 2)
Screening	Wetherby et al.[87]	GP	$N=5385$	$N=2752$	3 years or older	Best estimate diagnosis based on MSEL, VABS, developmental history, ADOS, and Social Communication Questionnaire (SCQ)	Infant-Toddler Checklist (ITC)	6–24 months	Logistic regression	Both PPV & NPV >0.70
Screening	Robins et al.[82]	GP	$N=16,071$	$N=8056$	Not clear	Clinical diagnosis following DSM-IV-TR	Modified Checklist for Autism in Toddlers–Revised with follow-up (M-CHAT-R/F)	18–30 months	ROC	SEN=0.73; SPE=0.89

Screening	Baird et al.[78]	GP	N=16,235	N=9918	3 1/2 years; 5 1/2 years; 7–8 years	Diagnosed by clinicians or the research team	Checklist for Autism in Toddlers (CHAT): parental report and health practitioner observation through direct testing	18 months	Combined medium- and high-risk threshold	SEN=0.20; SPE=0.99; PPV=0.29
Screening	Barbaro et al.[84]	GP	N=20,770	Not clear	24 months	ADOS and ADI-R together with clinical judgment.	Social Attention and Communication Study (SACS) Items	8, 12, 18 and 24 months	ROC	SEN= 0.69–0.838; SPE =0.998–0.999; PPV=0.81
Screening	Turner-Brown et al.[152]	GP	N=699	N=442	3 years	DSM-IV criteria and information of Developmental Concerns Questionnaire (DCQ), MSEL, ADOS, VSAB-2.	First Year Inventory (FYI)	12 months	ROC	PPV=0.31
Screening	Stone et al.[153]	GP	N=71	N=44	24 months	ADOS scores and clinical diagnostic criteria in DSM-IV-TR	Screening Tool for Autism in Toddlers (STAT)	24–35 months	ROC	SEN=0.95; SPE=0.73; PPV=0.56, NPV=0.97

Continued

Table 1 List of studies using behavioral methods.—cont'd

Modalities	Study	Target group	*N*	Sex (male)	Age diagnosed	Method for diagnosis	Features for assessment	Age assessed	Analytical method	Accuracy
Screening	Feldman et al.[154]	HR	HR-ASD = 108	$N = 74$	3 years	ADI-R by phone interview	Parent Observation of Early Markers Scale (POEMS)	3, 6, 9, 12, 18, and 24 months	ROC	Overall SEN = 0.74; SEN at 3, 6, 9, 12, 18, and 24 months of age was 0.50, 0.25, 0.57, 0.71, 0.89 and 1.00, respectively. Overall SPE = 0.73. SPE at 3, 6, 9, 12, 18, and 24 months of age was 0.87, 0.82, 0.84, 0.68, 0.65 and 0.70, respectively. Overall PPV = 0.21. PPV at 3, 6, 9, 12, 18, and 24 months of age was 0.25, 0.10, 0.29, 0.16, 0.20, 0.24, respectively
Screening	Lee et al.[92]	HR	$N = 121$ (HR = 86; Control = 35)	$N = 670$	36 months	Clinical diagnoses of ASD according to DSM-IV-TR and results from the ADOS and ADI-R.	First Year Inventory (Version 2.0)	12 months	ROC	AUC = 0.63; SEN = 0.34; SPE = 0.91

Screening	Raza et al.[155]	HR	N=172 (HR-ASD: 25; HR-AD: 30; HR-TD: 61; LR-AD: 13; LR-TD: 43)	N=95	36 months	DSM-IV criteria and clinical judgment based on all available developmental measures, including the ADOS, ADI-R, and MSEL.	Quantitative Checklist for Autism in Toddlers (Q-CHAT-10)	18 and 24 months	ROC	SEN=0.75; SPE=0.63; PPV=0.36 and NPV=0.90 (18 months); SEN=0.71, SPE=0.65, PPV=0.34, NPV=0.90 (24 months)
Screening	Zwaigenbaum et al.[10]	HR	N=225 (HR=150; LR=75)	Not clear	24 months	ICDS-10 or DSM-IV criteria and ADOS & ADI-R.	Autism Observation Scale for Infants (AOSI)	6–12 months	ROC	SEN=0.84; SPE=0.98
Screening	Ozonoff et al.[93]	HR	N=243 (HR=151, LR=92)	N=136	36 months	ADOS	Parent Concerns Questionnaire	6, 12, and 18 months	ROC	AUC=0.68; SEN=0.33; SPE=1.00; PPV=1.00; NPV=0.74 (6 months); 0.74, 0.83, 0.60, 0.58, 0.83 (12 months); 0.71, 0.57, 0.77, 0.60, 0.75 (18 months)

Continued

Table 1 List of studies using behavioral methods.—cont'd

Modalities	Study	Target group	N	Sex (male)	Age diagnosed	Method for diagnosis	Features for assessment	Age assessed	Analytical method	Accuracy
Screening	Sacrey et al.[79]	HR	$N=283$ (HR-ASD = 66; HR-N = 138; LR = 79)	$N=162$	36 months	Diagnoses were assigned using DSM-IV-TR or ICD-10 criteria, taking into account information from the ADI-R, ADOS, MSEL and VABS.	Autism Parent Screen for Infants (APSI)	6, 9, 12, 15, 18, and 24 months	ROC	SEN, SPE, PPV, and NPV for the total scale score were 0.67, 0.86, 0.47, and 0.83 (6 months); 0.59, 0.72, 0.65, and 0.76 (12 months); 0.65, 0.72, 0.68, and 0.77 (18 months)
Screening	Grodberg et al.[156]	HR	$N=45$ with suspected ASD or social communication concerns	$N=35$	Not clear	Best estimate diagnosis using DSM-5 criteria and ADOS-2 and ADI-R	Autism Mental Status Exam (AMSE)	18 months–5 years	ROC	SEN = 0.94; SPE = 1.00
Screening	Weitlauf et al.[91]	HR	$N=74$ HR	$N=42$	24–43 months	Clinical diagnosis based on DSM-IV-TR criteria, MSEL, VAB-II, and ADOS.	Modified Checklist for Autism in Toddlers Revised, with Follow-Up Interview (M-CHAT-R/F)	18 months	ROC	SEN = 0.78; SPE = 0.81; PPV = 0.75; NPV = 0.83
Screening	Capal et al.[157]	HR	$N=79$	Not clear	24 months	ADOS-2	AOSI	12 months	Logistic regression	SEN = 0.39; SPE = 0.89

Screening	Bussu et al.[94]	HR	N=232 (161 HR and 71 LR)	N=116	36 months	Clinical consensus best estimate diagnosis according to ICD-10 and DSM-5 criteria	AOSI, MSEL, VABS-II	8, 14, 24 and 36 months	Supervised classifier analysis using a least squares support vector machine	VABS daily living scores (14 months): AUC=0.713; SEN=0.796; SPE=0.522; ACC=0.575; PPV=0.285; NPV=0.915
Screening	Chiang et al.[95]	HR	Study1: 15 ASD, 15 DD/LI; Study2: 32 with Autism, 15 with PDD-NOS, and 30 with DD/LI.	N=10 (ASD); N=10 (DD)	25–36 months	Clinical evaluation based on DSM-IV and ADOS	Screening Tool for Autism in Two-Year-Olds - Taiwanese version (T-STAT)	24–35 months	ROC	SEN=0.93; SPE=0.74 (Study 1); EN=0.97; SPE=0.93 (Study 2)
Screening	Sacrey et al.[158]	HR; GP	N=188	HR-ASD=44; HR-N=67	36 months	Diagnoses were assigned using DSM-IV-TR or ICD-10 criteria, taking into account information from the ADI-R, ADOS, MSEL and VABS.	Autism Parent Screen for Infants (APSI); Infants (AOSI)	12 and 18 months	ROC	At 12 months, AUC=0.73 (APSI); AUC=0.73 (APSI and AOSI); At 18 months, AUC=0.69 (APSI); AUC=0.77 (APSI and AOSI)
Diagnostic tool	Guthrie et al.[81]	ASD	N=82	N=64	27–36 months	Clinical diagnosis based on review of all available information (e.g., ADOS/ADO-T scores, home observations, MSEL, VABS scores, etc) and DSM-IV criteria	ADOS-T	15–24 months	Percentage of confirmed diagnosis in Time 2	Stability of ASD diagnosis=100%

Continued

Table 1 List of studies using behavioral methods.—cont'd

Modalities	Study	Target group	*N*	Sex (male)	Age diagnosed	Method for diagnosis	Features for assessment	Age assessed	Analytical method	Accuracy
Diagnostic tool	Chawarska et al.[80]	ASD	$N=89$ (36 ASD; 28 PDD-NOS; 25 Non-ASD)	$N=66$	24–61 months	Time 1: Provisional diagnosis based on the DSM-IV criteria and ADOS-G; Time 2: confirmatory diagnosis based on developmental and health history, direct clinical impressions, ADI-R and ADOS-G.	ADOS-G	Time1: 13–27 months; Time2: 30–61 months	Percentage of confirmed diagnosis in Time 2	Stability of ASD diagnosis = 100%
Eye movements	Pierce et al.[104]	GP	$N=444$ including 115 ASD, 20 with ASD features, 57 DD, and 53 with other conditions (e.g., premature birth), 64 TD, and 25 unaffected with ASD siblings, and 110 previous their previous study.	ASD = 88; ASD Features = 15; TD = 35; DD = 45; TypSib ASD = 12; others = 26	Not clear	ADOS, MSEL, VABS	Eye movements during viewing a movie consisting of two geometric images (DGI) and dynamic social images (DSI) placed side by side.	10–49 months	ROC	SEN = 0.21; SPE = 0.98; PPV = 0.86

Eye movements	Pierce et al.[85]	GP	$N=110$ (ASD = 37, DD = 22, TD = 51)	ASD = 30, DD = 16, TD = 35	At least 24 months	ADOS-T, ADI-R	Visual fixation patterns in a preferential looking paradigm	14–42 months	ROC	AUC = 0.686, $p<0.001$; PPV = 1.00
Eye movements	Moore et al.[103]	GP	$N=227$ (76 ASD, 11 ASD features, 56 DD, 51 TD, 22 Other, and 11 TypSib)	ASD = 76; ASD Features = 10; TD = 30; DD = 36; TypSib ASD = 4; others = 11	3 years	DSM-IV or DSM-V criteria based on ADOS scores	Eye movements during viewing a 90-s video, the Complex Social GeoPref test	12–48 months	ROC	SEN = 0.35; SPE = 0.94; AUC = 0.75; PPV = 0.72; NPV = 0.78
Eye movements	Young et al.[105]	HR	$N=58$ (33 HR; 25 TD)	17 HR; 25 TD	24 months	Clinical diagnosis based on ADOS, M-CHAT, MSEL, behavioral observations, and DSM-IV criteria	1-min spontaneous face-to-face interaction between mother and infant; 1-min 'still-face' episode, and 1-min re-engagement.	6 months	Chi-square analysis	$\chi^2=1.51$, $p=0.47$
Diadic interaction	Wan et al.[108]	HR	$N=92$ (45 HR and 47 LR)	HR = 20; LR = 18	24 and 36 months	Expert clinical judgment based on ADOS, ADI-R using ICD-10 and DSM-5 criteria.	6-min videotaped episodes of parent-infant free play, AOSI, MSEL, Manchester Assessment of Caregiver-Infant Interaction (MACI)	6–10 months; 12–15 months	Logistic regression	Infant attentiveness: wald = 4.19; $p=0.04$; mutuality: wald = 4.92; $p=0.03$; Infant positive affect: wald = 6.77; $p=0.009$

Continued

Table 1 List of studies using behavioral methods.—cont'd

Modalities	Study	Target group	*N*	Sex (male)	Age diagnosed	Method for diagnosis	Features for assessment	Age assessed	Analytical method	Accuracy
Response to distress	Hutman et al.[111]	HR	HR = 103; LR = 55	ASD = 9; HR-N = 44; LR = 31	36 months	Clinical best estimate diagnosis using ADOS-G, ADI-R and DSM-IV criteria.	Response to another person's distress	12, 18, 24, and 36 months	Ordinal logistic random effects models	The best-fitting model yields a cross-validated error rate of 8.5%–9.9%
Response to name	Miller et al.[112]	HR	*N* = 156 (HR = 96 (20 ASD; 76 not ASD); LR = 60)	LR = 36 HR = 45 ASD = 15	36 months	ADOS and DSM-4 criteria	Orients to Name (adapted from the AOSI)	6, 9, 12, 15, 18, and 24 months	Generalized estimating equations with binomial variance function and logit link	SEN = 0.70; SPE = 0.70; PPV = 0.25
Motor skills	LeBarton et al.[113]	HR	*N* = 140 (LR = 51; HR = 89)	LR = 27; HR-N = 40; ASD = 12	24–36 months	Best clinical estimate based on diagnostic criteria in DSM-IV and the ADOS.	Peabody Developmental Motor Scales – 2 (PDMS-2) + MSEL	6 months	multinomial logistic regression analyses	At 6 month: $\chi^2 = 7.756$, $p = 0.021$
Social attention and communi cation	Barbaro et al.[159]	GP	*N* = 99 (autism = 37; ASD = 42; DD/ LD = 20)	AD = 28; ASD = 41; DD/LD = 14	24 months	ADOS, ADI-R, and in combination with clinical judgment.	Social Attention and Communication Study (SACS) Items	12, 18 and 24 months	Logistic regression	$\chi^2 = 72.32$, $p < 0.001$, predicting 96.2% of the AD/ASD group

Note: GP, general population; HR, high-risk; LR, low-risk; HR-ASD, high-risk developed ASD; HR-N, high risk with no ASD diagnosed; TD, typical development; DD, developmental delay; LD, language delay; ACC, accuracy; SEN, sensitivity; SPE, specificity; PPV, positive predictive value; NPV, negative predictive value.

is only 25%, suggesting a potential large number of false positive screening results for children who will not later be diagnosed with ASD. A study[84] known as The Social Attention and Communication Study (SACS), had a sample size of 20,770 children who were tested longitudinally at 8, 12, 18, and 24 months of age. It estimated a sensitivity ranging from 69% to 83.8%, and specificity from 99.8% to 99.9%, and positive predictive value (PPV) of 81%. The Communication and Symbolic Behavior Scales Developmental Profile Infant-Toddler Checklist (CSBS-DP-IT-Checklist) is also a parent-report questionnaire[85] for infants who were first assessed between 10 and 16 months and followed every 6 months until 36 months old. With a sample size of 10,479 infants, this screening tool is shown to have a PPV of 75%.

Checklist of Early Signs of Developmental Disorders (CESDD)[86] studied infants younger than 12 months including 3 months old. With a sample size of 6808, the CESDD reported an area under the curve (AUC) of the ROC of 0.90 (95% CI=0.84–0.97), overall sensitivity of 68%, specificity of 96%, PPV of 10%, and negative predictive value (NPV) of 99%. Nonetheless, the age of participants in this study ranged from 3 to 39 months and there is no clear information on their age of diagnosis. The Infant-Toddler Checklist (ITC)[87] assessed infants as young as 6 months. Its PPV and NPV were all above 70% for children 9–24 months of age, but not reliable for 6 months old. Other studies in this review such as The Baby and Infant Screen for Children with aUtIsm Traits (BISCUIT)[88] also demonstrated impressive screening results with a sample of 957 infants between 17 and 37 months that 13 items of BISCUIT-Part 1 were able to account for ~90% of the variance of either having an ASD or not from controls and 11 items of BISCUIT-Part 1 were able to classify 89% for autism and 88% for Pervasive Developmental Disorder-Not Otherwise Specified (PDD-NOS). The Brief Infant-Toddler Social Emotional Assessment (BITSEA)[89] reported AUC around 0.90 with a sample size of 3127 infants between 12 and 40 months old.

3.1.2 Screening tools for the high-risk population

It is estimated that the recurrence risk of ASD to be around 10.9%–18.7%.[31,90] Thus, infants at risk for ASD, including those with an older ASD sibling, are also target for ASD screening. The M-CHAT-R/F adapted from the CHAT, was also used to assess 74 high-risk infants at 18 months.[91] Results showed a sensitivity of 78%, specific of 81%, PPV of 75% and NPV of 83%, when it was used to classify ASD children from those with other developmental concerns and typically developing controls. The First Year Inventory (FYI) is a community-normed parent questionnaire designed to screen for signs of autism among 12-months-old infants. Using FYI (version 2.0), Lee et al.[92] studied 86 high-risk and 35 low-risk infants at 12 months old and conducted ASD clinical diagnosis at 36 months. Results suggested that the optimal balance of sensitivity and specificity criteria of this high-risk sample was 34% and 91%. The Parent Observation of Early Makrers Scale (POEMS) followed 108 high-risk infants from 3 months to 24 months. ASD diagnosis was confirmed at 36 months by ADI-R by phone interview. The overall sensitivity was 74% and specific was 73%. However, the PPV at 3, 6, 9, 12, 18, and 24 months of age was only 25%, 10%, 29%, 16%, 20%, 24%, respectively, suggesting potential referral of toddlers without ASD based on the screening results. Moreover, three screening tools also studied high-risk infants younger than 12 months, including the Parent Concerns Questionnaire (PCQ),[93] Autism Observation Scale for Infants (AOSI),[10,94] and Autism Parent Screen for Infants (APSI).[79] Scarey et al.[79] studied 79 infants from low-risk families and 204 from high-risk families when they were 6, 9, 12, 15, 18, and 24 months old. Diagnoses were assigned at 36 months old using DSM-IV-TR or ICD-10 criteria, taking into account information from the ADI-R, ADOS, MSEL and VABS. Results suggested that overall sensitivity, specificity, PPV, and NPV of the APSI were 67%, 86%, 47%, and 83%, respectively and 59%, 72%, 65%, and 76% even at 6 months. The AOSI[10] was also reported to have sensitivity and specificity of 84 and 98% for autism in siblings, respectively, after assessing 150 high-risk and 75 low-risk infants at 6–12 months and confirmed clinical diagnosis at 24 months. Using machine learning of data from AOSI, MSEL, VABS, Bussu et al.[94] examined 161 high-risk and 71 low-risk infants when they were 8, 14 and 24 months and assigned best estimate diagnosis according to ICD-10 criteria after reviewing all available information including ADOS, and ADI-R at 36 months. Results suggested that machine learning only managed to classify 8-month high-risk infants into ASD or non-ASD group at chance level.

Some of the screening tools have been translated and adapted to use in non-English speaking countries and their validity has been studied independently.[95–97] For example, CHAT and M-CHAT has been adapted into Chinese for Taiwan's general population.[97] It has shown to have a sensitivity over 83% and specific over 76% among 212 children between 18 and 24 months.

3.2 Observational and experimental studies

We focus on prospective studies that are designed to examine behavioral signs and markers of ASD with a priori selection of time points and measures based on theoretical-driven hypotheses. Compared with typically developing controls, individuals with ASD were found to demonstrate atypical behaviors, impairments and delays such as repetitive interests and behaviors, sensory-related behaviors, motor skills, social communication and emotional development, and eye gaze. Nonetheless, since we only include studies that report classification results of ASD diagnosis, only nine observational and experimental studies met our criteria out of 75 studies.

Among the nine studies included in this review, four studies used eye-tracking technique, which is a non-invasive method that provides objective measures for the distribution of eye gaze. Both adults and children with ASD were report to show atypical patterns of visual attention to objects and/or facial features compared to control groups.[98–102] Pierce and colleagues[85] used the preferential looking paradigm to evaluate toddlers' preference for dynamic geometric images (DGI) compared with dynamic social images (DSI). 110 toddlers between 14 and 42 months were studied from the general population, and results show that if they spent more than 69% of his or her time fixating on geometric patterns, the PPV for accurately classifying toddlers at 14 months as having ASD diagnoses was 100%. Using the Complex Social GeoPref eye-tracking test, Moor et al.[103] also studied toddlers from the community sample with a sample size of 227 toddlers between 12 and 48 months, and revealed a sensitivity of 35%, specificity of 94%, PPV of 72%, and NPV of 78%. With a larger sample size ($N=444$), Pierce et al.[104] used the preferential looking paradigm among infants between 10 and 49 months of age and 69% of fixation time on geometric images as the cutoff. However, the sensitivity for later ASD diagnosis was 21%, specificity was 98%, and PPV was 86%. Thus, the three eye-tracking studies revealed that the preference for geometric images of toddlers may be used as a potential sign of ASD, but the sensitivity was not optimal and may lead to a high false positive rate. Young et al.[105] studied 33 high-risk and 25 low-risk infants at 6 months and tracked their eye movements when they interacted with their face-to-face. They found that none of the infants previously identified as showing lower rates of eye contact had any signs of autism at outcome. In contrast, three infants who were diagnosed with autism demonstrated consistent gaze to the eye region and typical affective responses at 6 months.

Video analyses in the home and lab settings have also been employed in existing studies to classify ASD. The seminal study[106] using home videos provided the first opportunity of controlled, objective examination of everyday behaviors,[107] but these studies may be subject to biases related to sampling and the use of different coding schemes.[30] Wan et al.[108] studied 6-min videotaped episodes of parent-infant free play from 45 high-risk and 47 low-risk infants when they were 6–10 and 12–15 months old who received expert clinical judgment at 24 and 36 months. Results suggested that quality of interaction reflected by infants' positive affect and attentiveness to parents at 12 months but not 6 months predicted ASD diagnosis at 3 years using logistic regression models. Two studies reported the use machine learning of features from 162 2-min home videos[109] and 100 public videos[110] for the classification of ASD. Tariq et al.[109] reported the classifiers had sensitivity over 94.5% and Fusaro et al.[110] reported the classification accuracy was 96.8%, with 94.1% sensitivity and 100% specificity. Nonetheless, ASD participants in their studies have a mean age around 4 years and TD controls around 3 years old. It is not clear whether this kind of machine learning methods also demonstrate high sensitivity and specific among younger children. In addition, one study[86] did supervised classifier analysis using a least squares support vector machine based on behavioral data collected from HR participants when they were at 8, 14, 24 and 36 months. The model at 14 months demonstrated a sensitivity of 79.6% and specificity of 52.2%.

Hutman et al.[111] studied 156 infants (103 high-risk and 53 low-risk) longitudinally at 12, 18, 24, and 36 months and found that verbal mental age at 12 months and affective response to another person's distress at 18 months were the best predictors of diagnostic outcome at 3 years using stepwise logistic regression models and language skills and distress response was related at all time points, suggesting atypical response to negative emotion may interfere with social interaction and learning during early developmental stage. Miller et al.[112] focused on younger infants' response to name at 6, 9, 12, 15, 18, and 24 months. With a sample size of 156 infants (96 high-risk and 60 low-risk), they revealed that failure to give a response to name is an early sign of ASD that sensitivity and specificity for identifying ASD at 36 months based on at least 1 failure between 12 and 24 months were estimated at 0.70, and PPV at 0.25. LeBarton et al.[113] studied motor skills of 6 month-old infants using The Peabody Developmental Motor Scales—2 (PDMS-2) of 89 high--risk and 51 low-risk. They revealed that six-month PDMS-2 score distinguished between groups (low-risk or HR-neg or ASD) for the stationary subscale ($p = 0.021$) and Grasping subscale ($p = 0.043$).

3.3 Structural MRI

Table 2 includes all reviewed studies using neuroimaging. The first MRI study conducted prediction of ASD groups in 52 boys with a provisional diagnosis of autism and 15 typically developing children aged 1.7–5.2 years.[114] Diagnostic confirmation was obtained if the criteria of ADI-R, ADOS and DSM-IV were all met after 5 years of age. A discriminant function analysis was performed using the MRI brain measures including cerebellar white and gray matter volumes, cerebral white and gray matter volumes, and area of anterior and posterior vermis, which correctly classified 95.8% of the ASD cases and 92.3% of the control cases. These variables also correctly classified 85% of the ASD cases as lower functioning and 68% of the ASD cases as higher functioning. No cross-validation was conducted.

Shen and colleagues studied brain volume development on 55 infants (33 high-risk infants and 22 low-risk infants with no relatives with ASD) from 6 to 24 months of age.[115] Structural MRI data was scanned at three time points (6–9, 12–15, and 18–24 months of age). Ten infants were classified as meeting criteria for ASD, and the rest infants were classified as other developmental delays, or typical development at 24 months or later with a mean outcome age of 32.5 months. Using linear mixed model for repeated measures, the ASD group showed greater extra-axial fluid (an excessive amount of cerebrospinal fluid in the subarachnoid space surrounding the cortical surface of the brain) at 6–9 months compared with the other two groups. The excessive extra-axial fluid persisted and remained elevated at 12–15 and 18–24 months. A level of extra-axial fluid-to-brain volume, with the ratio of fluid to total volume = 0.14, yields 78% sensitivity and 79% specificity in predicting ASD cases at 12–15 months. The sample size is small, and there is no cross-validation for the predictive accuracy. Individual growth trajectories were not analyzed due to the small sample size.

The authors have replicated this study in a larger and independent cohort of 343 infants with 221 high-risk and 122 low-risk.[116] Forty-seven infants were diagnosed with ASD at 24 months and all of them are from high-risk group. A mixed effects model for repeated measures was employed. The HR-ASD had 18% greater extra-axial CSF volume at 6 months compared with low-risk and HR-neg groups. The volume remained elevated through 24 months of age. A cross-validated machine learning classification with a balance-boosted trees ensemble algorithm was performed. Results show extra-axial CSF volume at 6 months classified which high-risk infants would

Table 2 List of studies using neuroimaging.

Modalities	Study	Target group	*N*	Sex (male)	Age diagnosed	Method for diagnosis	Features for assessment	Age assessed	Analytical method	Accuracy
Structural MRI	Akshoomoff et al.[114]	Provisional diagnosed for autism	Controls = 15; provisional autism = 52; ASD = 42, PDD-NOS = 10	All males	5 years or later	ADI-R, ADOS, and DSM-IV	Cerebellar GM and WM volumes, and area of anterior and posterior vermis	1.7–5.2 y (3.6 ± 1.1 years)	Discriminant functional analysis	ASD: ACC = 0.958; Control: ACC = 0.923
Structural MRI	Shen et al.[115]	HR	HR = 33; LR = 22 (ASD: 10)	HR: 22; LR: 15	24–36 months	DSM-IV-TR and ADOS	Extra-axial CSF volume, brain volume	6–24 months	A ratio of extra-axial fluid to brain volume (0.14)	SEN = 0.78; SPE = 0.79
Structural MRI	Shen et al.[116]	HR	HR = 221 (HR-ASD = 47, HR-N = 174); LR = 122 (all non-ASD)	HR: 42 + 95; LR: 76	24 months	ADOS	Extra-axial CSF volume and cerebral volume	6, 12, and 24 months	RUSBoost trees	AUC = 0.69, SEN = 0.66, SPE = 0.68
Structural MRI	Shen et al.[117]	GP	ASD = 159; TD = 77	ASD:132; TD:49	2–4 years	ADOS and ADI-R	Extra-axial CSF volume and cerebral volume	2–4 years	RUSBoost trees	PPV = 0.83; ACC = 0.78; SEN = 0.84; SPE = 0.65
Structural MRI	Hazlett et al.[52]	HR	HR = 106 (HR-ASD = 15; HR-N = 91); LR = 42	HR:200; LR:69	24 months	(ADI-R)	Cortical surface area and thickness	6–24 months	Deep-learning network	PPV = 0.81, SEN = 0.88, SPE = 0.95
Structural MRI	Xiao et al.[119]	ASD and DD	ASD = 46; DD = 39	ASD: 41; DD: 34	18–37 months	DSM-IV-TR	cortical thickness	NA	SVM, random forest, naïve Bayes	ACC = 0.809, SEN = 0.813, SPE = 0.81, AUC = 0.88, MCC = 0.62

Diffusion MRI	Jin et al.[120,a]	HR	HR = 40; LR = 40	HR: 29; LR: 27	NA		White matter connectivities	6 months	Multi-kernel SVM	ACC = 0.76, AUC = 0.76
Functional MRI	Dinstein et al.[62]	ASD	ASD = 29; LD = 13; TD = 30	NA	12–46 months (mean = 29 months)	ADOS	Spontaneous cortical activity	12–46 months (mean = 29 months)	Use interhemispheric correlation values, ROC	SEN = 0.72; SPE = 0.84. for toddlers aged 12–24 months, SEN = 0.6; SPE = 0.8
Functional MRI	Lombardo et al.[63]	ASD and language delay	ASD = 60; LD = 19; TD = 24	NA	3–4 years	ADOS	fMRI speech-related activation and clinical intake measures	12–48 months	Partial least squares linear discriminant classifier with cross-validation	ACC = 0.68, AUC = 0.69, SEN =. 71, SPE = 0.67 with fMRI feature
Functional MRI	Emerson et al.[121]	HR	HR = 59; HR-ASD = 11	HR: 41	24 months	ADOS + ADI-R	Functional connectivity	6 months	SVM	PPV = 1, NPV = 0.96, SEN = 0.818, SPE = 1
EEG	Bosl et al.[122,a]	HR	HR = 46; control = 33	HR: 36; LR: 23	NA	NA	Multi-scale entropy of resting-state EEG	6–24 months	SVM, KNN, and naïve Bayesian	For boys: highest ACC = 1 at 9 months. For girls, highest ACC = 0.8 at age 6 months

Continued

Table 2 List of studies using neuroimaging.—cont'd

Modalities	Study	Target group	*N*	Sex (male)	Age diagnosed	Method for diagnosis	Features for assessment	Age assessed	Analytical method	Accuracy
EEG	Stahl et al.[123,a]	HR	HR = 19; control = 17	HR: 10; LR: 10	2–3 years	ICD-10 (infants' siblings)	ERP of direct gaze trials	9–10 months	Regularized discriminant function analysis, and SVM	LDA: ACC = 0.61, SEN = 0.65, SPE = 0.58, PPV = 0.58, NPV = 0.35. SVM: ACC = 0.64, SEN = 0.64, SPE = 0.64, PPV = 0.74, NPV = 0.47
EEG	Cohen et al.[127]	Risk for developmental problems	Abnormal initial ABR = 46 (ASD = 13); normal ABR = 28 (ASD = 1)	Abnormal initial ABR: 28; normal ABR: 13	2 and 3 years	6 diagnosed by PDDBI at 28 and 51 month; 8 diagnosed with ADI-R	ABR and AMA slope	4 months	Piece-wise linear regression, ROC	ACU = 0.85, SEN = 0.77, SPE = 0.88, PPV = 0.71, NPV = 0.91
EEG	Miron et al.[125]	ASD	ASD = 30; control = 30	24 in ASD; 24 in controls	NA	DSM-IV	ABR latencies	3 months	Optimal latency value based on ROC with leave-one-out procedure	SEN = 0.70, SPE = 0.80

EEG	Bosl et al.[129]	HR	HR = 99 (HR-ASD = 32; HR-N = 67); LR = 89 (LR-ASD = 3; LR-N = 86)	NA	18, 24, 36 months	ADOS	1026 nonlinear invariant EEG measures	3, 6, 9, 12, 24, 36 months	SVM	SEN = 0.82–1; SPE = 0.78–1; PPV = 0.78–1
EEG	Gabard-Durnam et al.[130]	HR	HR-ASD = 31; HR-N = 71; LR = 69	HR-ASD: 22; HR-neg: 34; LR:37	3 years	ADOS	Longitudinal EEG power parameters	3, 6, 9, 12, 24, 36 months	ROC of logistic regression models	HR-ASD vs HR-N: PPV = 0.72, SEN = 0.8182, SPE = 0.8235; HR-ASD vs LR: PPV = 0.6667, SEN = 0.8182, SPE = 0.8772
EEG	Dickinson et al.[131,b]	HR	HR = 36; LR = 29	HR: 23; LR: 18	18 months	ADOS-T	Functional connectivity with EEG	3 months	SVR	$r = 0.76$; $p = 0.02$

[a]Classification of high- and low-risk of ASD.
[b]Prediction of ASD symptoms measured by ADOS-T total score.
Note: GP, general population; HR, high-risk; LR: low-risk; HR-ASD: high-risk developed ASD; HR-N: high-risk with no ASD diagnosed; LR-ASD: low-risk developed ASD; LR-N: low-risk with no ASD diagnosed; TD: typical development; LD: language delay; ACC: accuracy; SEN: sensitivity; SPE: specificity; PPV: positive predictive value; NPV: negative predictive value.

be diagnosed with ASD at 24 months with an overall accuracy (area under the ROC curve) of 69% and corresponding 66% sensitivity and 68% specificity. This model was validated on the out of sample data published from their previous study,[115] yielding similar accuracy: overall accuracy of 72%, sensitivity of 80% and specificity of 67%.

Conducted by the same group of authors,[117] a follow-up work recruited ASD (including high- and low-risk) and typical developing children aged 2–4 years. MRI and behavioral measures were estimated including extra-axial CSF volume, brain volume, head circumference, sleep problems and familial risk status to classify the disease status using machine learning method. Prediction of the ASD diagnosis achieved a PPV of 0.83 with 95% CI of 76.2–88.3%, accuracy of 78%, sensitivity of 84% and specificity of 65%. The series studies suggest that extra-axial CSF volume is a reliable brain abnormal feature in three independent cohorts for ASD among both high-risk and non-high-risk young children.

The study performed by Hazlett et al.[52] measured cortical thickness and surface area, which are two components contributing to the brain volume but influenced by distinct genetic mechanisms.[118] Subjects to be predicted included 179 high-risk with 34 HR-ASD and 145 HR-neg infants. A three-stage deep learning network was used: sex, age-corrected intracranial volume, age-corrected surface area and cortical thickness measurements from 78 brain regions at 6 and 12 months. Tenfold cross-validation was used for performance evaluation. The prediction of the diagnosis outcome has a PPV of 81%, sensitivity of 88% and specificity of 95%. Further analysis shows that the prediction primarily relies on surface area information of 6–12-month-old individuals. Moreover, among 106 high-risk infants and 42 low-risk infants with all longitudinal data, cortical surface area hyper-expansion between 6 and 12 months of age precedes brain volume overgrowth observed between 12 and 24 months in 15 high-risk infants diagnosed with autism using ADI-R at 24 months.

A study identifying ASD from developmental delay children from 18 to 37 months of age used random forest, SVM, and Naive Bayes classifiers for prediction and cortical measures as features.[119] Random forest with thickness-based classification outperformed other classification using volume or surface area based features, which generated the average ACC of 80.9%, sensitivity of 81.3%, specificity of 81% and AUC of 0.88. Similar to the study by Shen et al. in 2018,[117] this study followed a cross-sectional design, and used machine learning methods to classify the ASD diagnosis from developmental delay children using the MRI scanned when the diagnosis was already made.

There are so far no studies using diffusion MRI on the prediction of the ASD diagnosis in infants or toddlers. Jin et al. worked on identifying infants at high-risk from low-risk with diffusion MRI data collected at 6 months of age.[120] Multi-kernel SVM classification was conducted using white matter connectivity features composed with multi-scale and multiple diffusion indices including fractional anisotropy, mean diffusivity, and average fiber length. The prediction reached an accuracy of 76% and AUC of 0.76.

3.4 Functional MRI

Dinstein et al.[62] studied 72 toddlers including 29 diagnosed with autism (average age: 29 months; range: 12–46 months), 13 with language delay (mean age: 19 months, range 13–27) and 30 typically developing controls (mean age: 28 months; range 13–46). The spontaneous cortical activity was obtained by regressing out the evoked auditory-evoked activations and the correlations between anatomically defined regions of interest were estimated. Toddlers with autism exhibited significantly weaker inter-hemispheric connectivity in putative language areas. The IFG and STG interhemispheric correlation values can identify the majority of autistic toddlers with a sensitivity of 72% and specificity of 84%. When considering only the young toddlers aged 12–24 months, it yielded a slightly lower sensitivity of 60% and specificity of 80%, which might be due to the higher percentage of language delay individuals in toddlers (11 out of 39) compared to the entire sample population (13 out of 72). The strength of synchronization was also positively correlated with verbal ability and negatively correlated with autism severity.

Another fMRI study conducted partial least squares linear discriminant classification with cross-validation to classify the poor and good language performance in 60 ASD children (36 good and 24 poor performance).[63] Both fMRI speech-related activation and clinical intake measures as features obtained before the ASD diagnosis. With the combined modalities including fMRI, ADOS, Mullen Scales of Early Learning, and Vineland Adaptive Behavior Scales, the prediction reached an accuracy of 80%, AUC of 0.81, sensitivity of 88% and specificity of 75%. With only fMRI derived features, the performance was lower with an ACC of 68%, AUC of 0.69, sensitivity of 71% and specificity of 67%.

Emerson and colleagues studied 59 high-risk infants and found that functional connectivity measured by fMRI at 6 months could predict which infant would receive an ASD diagnosis at 24 months using ADOS and ADI-R.[121] The study first identified functional connections at 6 months based on their association with ASD-related behaviors at 24 months (i.e., cores on social behavior, language, motor development, and repetitive behavior). Then machine learning using SVM models was conducted on those connections with cross-validation. The prediction achieved 100% PPV, 82% sensitivity (correctly predicting 9 of the 11 infants who received an ASD diagnosis at 24 months), and 100% specificity (correctly predicting 48 of 48 infants who were negative for ASD at 24 months of age).

3.5 EEG and ERP

Two early classification studies used EEG to classify the risk status for ASD in infants.[122,123] Bosl et al. calculated multi-scale entropy on the basis of resting-state EEG in infants aged 6–24 months.[122] This measure is considered as an estimation of physiological complexity that contains information about neural network dynamics. Machine learning algorithms including k-nearest neighbors (k-NN), SVM and naïve Bayesian classification (Bayes) were used with multi-scale entropy as a feature vector. Random permutations were run 100 times with 10-fold cross-validation for each classification calculation. Over 80% accuracies were obtained into control and high-risk at age 9 months. Prediction accuracy for boys was close to 100% at age 9 months and remains high (70%–90%) at ages 12 and 18 months. For girls, classification accuracy was highest at age 6 months, and declines afterwards. The results suggest that trajectories of EEG multi-scale entropy between 9 and 12 months can differentiated infant siblings at high- and low-risk for ASD. The other study[123] of identifying the risk status conducted classification on a set of previous published data including 19 high-risk infants and 17 low-risk controls aged around 10 months old.[124] The ERP recordings of the infant response to direct and averted gaze in static stimuli were analyzed. The classification accuracy for risk of ASD reached over 60% using the regularized discriminant function analysis (61%) and SVM (64%) methods with the direct and averted gaze data.

One potential early neurophysiological measure for detecting infants or toddlers who later develop ASD is the auditory brainstem response (ABR). ABR has been reported to exhibit prolonged latencies in young children with ASD.[125,126] Cohen and colleagues investigated the associations between the abnormal neonatal ABRs, 4-month arousal-modulated attention (AMA) visual preference and the later ASD behaviors in neonatal intensive care unit (NICU) graduates.[127] As 93% of ASD cases in the study were in the abnormal ABR group, ROC analysis was performed in the abnormal

ABR group using AMA slope as a predictor for the positive vs negative outcome of ASD diagnosis (6 using parent completion of PDDBI,[128] 8 with ADI-R) conducted at 3 years old. Using a cutoff AMA slope of 0.14, the AUC was 0.85, and maximum sensitivity, specificity, PPV and NPV were 77%, 88%, 71% and 91%, respectively. Another prediction study included 30 later diagnosed as ASD and 30 controls who had ABR tested between 0 and 3 months.[125] With a ROC analysis, the optimal criterion, ABR latency value, for separating the two groups was estimated. Classification accuracy was tested by leave-one-out procedure iterated 60 times. The classification sensitivity and specificity were 70% and 80% respectively.

A recent study collected EEG measurements from 99 high-risk for ASD, and 89 low-risk controls from 3 to 36 months longitudinally.[129] The features consisted of 9 nonlinear values computed on 6 frequency bands for each of 19 sensors, totaling 1026 measures. EEG recordings were collected at ages at 3, 6, 9, 12, 18, 24 and 36 months. Results show that the prediction of the ASD diagnostic outcome has specificity, sensitivity and PPV higher than 95% at age 6 months of testing for infants in low- and high- group. The prediction performance was higher of low-risk group compared to high-risk group for all ages of testing. There was a dip in predictive accuracy at 12 months of age. Prediction of ASD using ADOS calibrated severity scores can be taken as early as 3 months of age using only EEG data, and the EEG measures were strongly correlated with the actual measured scores.

With the same longitudinal cohort, a different approach has been conducted for differentiating the ASD outcomes, which analyzed the trajectories of EEG powers in infants.[130] Subjects included 31 HR-ASD (high-risk infants diagnosed ASD at 3 years old), 71HR-neg (high-risk infants not diagnosed ASD) and 69 low-risk control groups. The baseline EEG data recorded between 2 and 5 min was processed and the power of six frequency bands was calculated. Longitudinal EEG power parameters included the intercepts and slopes of the trajectories of EEG summed power across age windows. Binary logistic regression models across different developmental windows (3–12 months, 12–24 months, 3–36 months) and between groups (any two groups of HR-neg, HR-ASD and low-risk control groups) were used to estimate the predictability of EEG parameters adjusting for sex and parental education. Model selection was carried out with a hybrid stepwise selection approach. Frontal 3–12 month model provided the highest discriminant diagnostic outcomes. This measure produced higher sensitivity (81.82% ASD vs HR-neg; 82.35% for ASD vs low-risk) and specificity rates (86.27% ASD vs HR-neg; 87.72% for ASD vs low-risk) and higher PPV values (72% ASD vs HR-neg; 66.67% for ASD vs low-risk) compared to other models. Across models in this time window, low frequency bands were significant individual parameters differentiating ASD outcomes. Results suggest that delta and gamma frequency power dynamics during the first postnatal year best differentiate ASD diagnoses from others compared to other time window and frequency bands.

Functional connectivity measures of oscillatory phase coherence (6–12 Hz) using EEG were used for predicting ASD symptoms among high- and low-risk infants.[131] Multivariate pattern analysis was applied to examine functional patterns associated with later ASD symptoms. Support vector regression (SVR) algorithms was conducted to predict observed ASD symptoms using ADOS-T total score at 18 months using the EEG data collected at 3 months with $r=0.76$, and $p=0.02$. Most predictive features include lower frontal connectivity and higher right temporo-parietal connectivity for predicting higher ASD symptoms.

4. Discussion

In this review, we surveyed the literature on the classification and prediction studies of ASD in infants and toddlers. Some of the studies using screening tools have large sample sizes, for instance, eight out of 33 of them included greater than 1000 samples. The accuracy of classification using screening and behavior measures varies a lot, with the sensitivity values ranging from 20% to 100% and specificity values ranging from 48% to 100%. The higher accuracies were obtained when the screening or behavior assessments were conducted older than 1 year. Structural and functional MRI are two MRI modalities (no diffusion MRI) used in the classification of early diagnosis of ASD. The sensitivity and specificity with MRI derived features show ranges of 60%–95.8%, and 68%–100% respectively. With EEG signals, the neural physiological features produce similar accuracies compared to using MRI data, with sensitivity ranging 64%–100%, and specificity ranging 58%–100%. The sample sizes using neuroimaging features are relatively smaller compared to using non-imaging modalities, only one neuroimaging study included subjects over 100 ASD subjects.[117]

4.1 Summary of behavioral based prediction

We reviewed longitudinal studies that focused on using screening results and behavioral measures to classify or predict later ASD diagnosis. These studies provided classification outcomes using ROC or logistic regression.

Screening studies tend to involve large sample sizes when the general population is investigated, but this may not be true for all studies. The sample sizes of 20 studies targeted the general population range between 44 and 20,770. The 11 studies involving high-risk groups have sample sizes ranging between 30 and 283. When the general population was studied, the sensitivity ranges between 20% and 100% but the specificity ranges between 48% and 99%, suggesting the possibility of over referral children without ASD. Less than half of the children who were identified by the screening tests were later diagnosed with ASD. This is also the case for studies focusing on the high-risk group, which reported sensitivities between 33% and 94% and specificities between 73% and 100%, suggesting potential over referral for both groups. In addition, screening children before 12 months of age seems to be possible for predicting ASD, but more evidence is needed. The earliest age screened in high-risk infants is 3 months old,[42] with a sensitivity of 50% and specificity of 87%. One study demonstrated a sensitivity of 67% and a specificity of 87% among 6-month-old infants from high-risk families.[79] However, most of the studies we reviewed were not able to make reliable classification or prediction when participants were before 12 months old.

4.2 Summary of neuroimaging based prediction

The earliest MRI derived features for predicting ASD include cortical surface area measured at 6 and 12 months[52] and functional connectivities estimated as early as 6 months of age.[121] The prediction of the diagnosis for autism in high-risk infants has sensitivities above 80% and specificities above 90%. However it is worth to note that the number of diagnosed ASD toddlers is small (less than 20) in these studies. The extra-axial CSF volume has been reported to be predictive of ASD diagnosis in three independent datasets.[115–117] Although the findings seem to replicate in these independent datasets, the prediction metrics seem not yet strong enough to be clinically useful to predict individual outcome, where the sensitivities are 66%–78%, and specificities are 67%–79%. The cortical thickness has also been shown to be predictive of infants with ASD from developmental delayed toddlers with around 81% of sensitivity and specificity.[119]

Longitudinal EEG data, such as nonlinear invariant measures including multi-scale entropy as early as 3 months of age, and the trajectories of frontal lobe EEG power especially in low frequency bands from three to 12 months, have shown the ability to classify ASD diagnosed with ADOS at 3 years old.[129,130] The accuracy using these EEG measures seem to be higher compared to using other neuroimaging approaches such as MRI. The sensitivity and specificity were above 95% using EEG nonlinear measures obtained at age 6 months; and were above 80% using the trajectory indices, i.e., the intercepts and slopes of EEG power. Abnormal neonatal ABRs measured by EEG also showed predictability of ASD in infants with risks for developmental delays.[125,128]

Overall, neuroimaging based approaches appear to have higher accuracy for the early prediction of ASD compared to screening and behavioral methods. However, it is noteworthy to mention that all neuroimaging studies have included relatively small numbers of ASD subjects. It is possible that the classification models may be overfitting even though cross-validation procedure was conducted in the studies. The prediction performance needs to be externally validated with larger sample size to confirm the reliability of these imaging derived neural features.

4.3 Issues in existing prediction studies

It is noticed that there is large variance of the prediction performance in the prediction of ASD diagnosis. Possible factors contributing to this inconsistency may include etiologic heterogeneity of ASD, biased populations, age of the subjects when the features were collected, and different validation procedures. The concept of heterogeneity in autism covers three main domains: clinical heterogeneity, i.e., different clinical symptoms, biological heterogeneity, and environmental heterogeneity.[132] When the sample size is small, classification methods may produce low accuracy due to the difficulties of capturing features to predict ASD with different clinical symptoms, neurobiological mechanisms, or affected by different environmental factors. The reviewed studies targeted on different types of populations. Some studies included general population, whereas others targeted a specific population such as high risk with ASD, which may contribute to the mixed prediction performance. Among the screening studies we reviewed, there is a trend of higher accuracy if the predictive assessments were collected at an older age. Some of the reviewed studies did not include cross-validation process in the classification. Even for the ones with cross-validation included, the validation procedures may not be consistent among these studies. Overfitting of the predictive model would produce a high accuracy, which can hardly be obtained when generalizing the model to external samples. We also noticed that a few studies had either high sensitivity but low specificity, or low sensitivity but high specificity. The balance of the two indices needs further investigation.

The early prediction of ASD diagnosis using neuroimaging features is underexplored mainly due to the limited data for infants and toddlers. Optimized feature selection might be needed for improving the predictive accuracy. For example, the nonlinear invariant EEG based parameters from selected sensors show high sensitivity and specificity,[129] and the selection of different longitudinal models of EEG power under different frequency band can generate higher predictive accuracy.[130] The use of multi-modal features, e.g., combining neural imaging with demographic information[52] and behavioral features,[63] is also worth exploring, which has been shown to provide better prediction.

Many of the studies that we excluded from this review did not report classification or prediction results despite significant group differences of certain measures being reported between individuals with ASD and controls. Thus, it is difficult to generalize these findings to clinical settings. Some of the studies included in this review reported the ROC results for a cutoff score, or/and PPV (i.e., the proportion of children who were diagnosed with ASD among those who were identified as at risk of being diagnosed), NPV (the proportion of children who did not receive an ASD diagnosis and who were not identified as at risk). However, not all studies have taken into consideration the prevalence of ASD in the target population. For example, since the prevalence of ASD among children at risk is higher than from the general population, the PPV can also be higher than those from the general population sample. The optimal sampling schemes may involve highly unbalanced samples.[133]

Interaction between cultural contexts and genes may shape the functional organization of the human brain and behavior.[134] The avoidance of eye contact has been used as an important sign for ASD. However, the majority of these studies have been conducted in the west without taking cultural differences into consideration.[135] Existing studies report that individuals from Eastern and Western cultures may attend to and perceive the world differently.[136] Compared to individuals from Western cultures, individuals from Eastern countries tend to avoid direct eye contact during face-to-face interactions.[137] When viewing faces, individuals from Eastern cultures tend to fixate greatly on the nose (versus the eye) region of a face,[138,139] which was also reported among individuals with ASD in Western cultures.[101] Thus, future studies may need to consider cultural differences when early signs of ASD are examined.

4.4 Future directions

More advanced classification methods, such as deep learning using neuroimaging, are increasingly investigated in the identification of autism, but not in predicting the early diagnosis of ASD. This is largely due to the limited numbers of neuroimaging data and the lack of the diagnostic status of ASD in early years of age. Some deep learning methods have been applied using sharable large datasets containing children and adolescents. Heinsfeld and colleagues investigated the patterns of functional connectivity measured by fMRI to identify ASD of ABIDE (Autism Brain Imaging Data Exchange) database.[140] The results show 70% accuracy in classification of ASD versus control patients in the datasets of over 1000 subjects including 505 of ASD and 530 of typical developing children. One deep learning based approach using three-stage neural network has been shown to predict the diagnosis of ASD using cortical surface area measured at 6 an 12, producing a sensitivity of 88% and specificity of 95%.[52] Expending and sharing the neuroimaging data and the ASD diagnostic status of infants and toddlers are needed to further boost the research in predicting the early diagnosis. Increasing sample size will also enable the detection of more robust decision functions and features, and reliable evaluation of the classification performance.

Early brain development over time might be a promising approach for the investigation of biomarkers for predicting ASD. Early perturbations during basic circuit refinement may lead to widespread alterations, while later-occurring impacts on higher-order neural networks may contribute to more specific disruptions.[141] The temporal emergence of brain abnormalities, rather than a deviation at a certain time point, may rise to different subgroups of ASD and contribute to heterogeneous phenotypes.[130] Prospective and longitudinal studies examining early brain and behavior trajectories are critical for searching markers of the development of ASD and its subtypes.[52,116,130]

Genetic technologies have advanced our understanding of ASD etiology via identifying autism risk genes or loci. Over 100 genes and genomic regions have been shown to be associated with autism,[17,18] many of which target synaptic plasticity or protein synthesis.[142] However, it is still difficult to investigate which genetic variants induce the mechanisms,[143] and how the biological mechanisms may operate to give rise to ASD features. By bridging the gap between genetic risk variations and their effects on the brain, the imaging genetic approach provides promise for advancing our understanding of ASD etiology and the neurobiological pathways.[144] The combination of genetics and neuroimaging or behavior derived phenotypes may increase the power of predicting the diagnosis of ASD.

In summary, the neuroimaging features extracted under 1 year of age have shown promising predictive accuracies for the development of ASD in high-risk infants. However, the translation to clinical practice requires benchmark data for evaluation. The mixed predictive accuracies using behavioral and neuroimaging features might be partially due to the heterogeneity of ASD, the different targeted populations, the age when the features were collected, and various validation procedures. Further effort on systematic stratifications of ASD and external validation are needed for more accurate and robust early prediction of the diagnosis at an individual level.

References

1. American Psychiatric Association. *Diagnostic and Statistical Manual of Mental Disorders*. 5th ed. Washington, DC: Author; 2013.
2. Arora M, Reichenberg A, Willfors C, et al. Fetal and postnatal metal dysregulation in autism. *Nat Commun*. 2017;8:15493.
3. Bradshaw J, Steiner AM, Gengoux G, Koegel LK. Feasibility and effectiveness of very early intervention for infants at-risk for autism spectrum disorder: a systematic review. *J Autism Dev Disord*. 2015;45(3):778–794.
4. Dawson G. Early behavioral intervention, brain plasticity, and the prevention of autism spectrum disorder. *Dev Psychopathol*. 2008;20(3):775–803.
5. Green J, Pickles A, Pasco G, et al. Randomised trial of a parent-mediated intervention for infants at high risk for autism: longitudinal outcomes to age 3 years. *J Child Psychol Psychiatry*. 2017;58(12):1330–1340.
6. MacDonald R, Parry-Cruwys D, Dupere S, Ahearn W. Assessing progress and outcome of early intensive behavioral intervention for toddlers with autism. *Res Dev Disabil*. 2014;35(12):3632–3644.
7. Clark MLE, Vinen Z, Barbaro J, Dissanayake C. School age outcomes of children diagnosed early and later with autism Spectrum disorder. *J Autism Dev Disord*. 2018;48(1):92–102.
8. Zwaigenbaum L, Bauman ML, Choueiri R, et al. Early intervention for children with autism Spectrum disorder under 3 years of age: Recommendations for practice and research. *Pediatrics*. 2015;136(Suppl. 1):S60–S81.
9. Rotholz DA, Kinsman AM, Lacy KK, Charles J. Improving early identification and intervention for children at risk for autism spectrum disorder. *Pediatrics*. 2017;139(2), e20161061.
10. Zwaigenbaum L, Bryson S, Rogers T, Roberts W, Brian J, Szatmari P. Behavioral manifestations of autism in the first year of life. *Int J Dev Neurosci*. 2005;23(2–3):143–152.
11. Baio J, Wiggins L, Christensen DL, et al. Prevalence of autism spectrum disorder among children Aged 8 years—autism and developmental disabilities monitoring network, 11 sites, United States, 2014. *MMWR Surveill Summ*. 2018;67(6):1–23.
12. Daniels AM, Mandell DS. Explaining differences in age at autism spectrum disorder diagnosis: A critical review. *Autism*. 2014;18(5):583–597.
13. Zwaigenbaum L, Penner M. Autism spectrum disorder: advances in diagnosis and evaluation. *BMJ*. 2018;361, k1674.

14. Piven J, Elison JT, Zylka MJ. Toward a conceptual framework for early brain and behavior development in autism. *Mol Psychiatry*. 2017;22(10):1385–1394.
15. Shen MD, Piven J. Brain and behavior development in autism from birth through infancy. *Dialogues Clin Neurosci*. 2017;19(4):325–333.
16. Wolff JJ, Jacob S, Elison JT. The journey to autism: insights from neuroimaging studies of infants and toddlers. *Dev Psychopathol*. 2018;30(2):479–495.
17. Sanders SJ, He X, Willsey AJ, et al. Insights into autism spectrum disorder genomic Architecture and biology from 71 risk loci. *Neuron*. 2015;87(6):1215–\.
18. Satterstrom FK, Kosmicki JA, Wang J, et al. Large-scale exome sequencing study implicates both developmental and functional changes in the neurobiology of autism. *Cell*. 2020;180(3):568–584, [e523].
19. Elsabbagh M, Johnson MH. Getting answers from babies about autism. *Trends Cogn Sci*. 2010;14(2):81–87.
20. Maestro S, Muratori F, Cesari A, et al. Course of autism signs in the first year of life. *Psychopathology*. 2005;38(1):26–31.
21. Yirmiya N, Charman T. The prodrome of autism: early behavioral and biological signs, regression, peri- and post-natal development and genetics. *J Child Psychol Psychiatry*. 2010;51(4):432–458.
22. Ozonoff S, Iosif AM, Baguio F, et al. A prospective study of the emergence of early behavioral signs of autism. *J Am Acad Child Adolesc Psychiatry*. 2010;49(3):256–266. e251–252.
23. Landa RJ, Gross AL, Stuart EA, Faherty A. Developmental trajectories in children with and without autism spectrum disorders: the first 3 years. *Child Dev*. 2013;84(2): 429–442.
24. Short AB, Schopler E. Factors relating to age of onset in autism. *J Autism Dev Disord*. 1988;18(2):207–216.
25. Stone WL, Hoffman EL, Lewis SE, Ousley OY. Early recognition of autism. Parental reports vs clinical observation. *Arch Pediatr Adolesc Med*. 1994;148(2):174–179.
26. Baranek GT. Autism during infancy: a retrospective video analysis of sensory-motor and social behaviors at 9-12 months of age. *J Autism Dev Disord*. 1999;29(3):213–224.
27. Osterling J, Dawson G. Early recognition of children with autism: a study of first birthday home videotapes. *J Autism Dev Disord*. 1994;24(3):247–257.
28. Gillberg C, Ehlers S, Schaumann H, et al. Autism under age 3 years: a clinical study of 28 cases referred for autistic symptoms in infancy. *J Child Psychol Psychiatry*. 1990;31(6):921–934.
29. Rogers SJ. What are infant siblings teaching us about autism in infancy? *Autism Res*. 2009;2(3):125–137.
30. Zwaigenbaum L, Bryson S, Garon N. Early identification of autism spectrum disorders. *Behav Brain Res*. 2013;251:133–146.
31. Ozonoff S, Young GS, Carter A, et al. Recurrence risk for autism spectrum disorders: a baby siblings research consortium study. *Pediatrics*. 2011;128(3):e488–e495.
32. Sandin S, Lichtenstein P, Kuja-Halkola R, Larsson H, Hultman CM, Reichenberg A. The familial risk of autism. *JAMA*. 2014;311(17):1770–1777.
33. Elison JT, Wolff JJ, Reznick JS, et al. Repetitive behavior in 12-month-olds later classified with autism spectrum disorder. *J Am Acad Child Adolesc Psychiatry*. 2014;53(11):1216–1224.
34. Estes A, Zwaigenbaum L, Gu H, et al. Behavioral, cognitive, and adaptive development in infants with autism spectrum disorder in the first 2 years of life. *J Neurodev Disord*. 2015;7(1):24.
35. Swanson MR, Shen MD, Wolff JJ, et al. Subcortical brain and behavior phenotypes differentiate infants with autism versus language delay. *Biol Psychiatry Cogn Neurosci Neuroimaging*. 2017;2(8):664–672.

36. Shic F, Macari S, Chawarska K. Speech disturbs face scanning in 6-month-old infants who develop autism spectrum disorder. *Biol Psychiatry*. 2014;75(3):231–237.
37. Jones W, Klin A. Attention to eyes is present but in decline in 2–6-month-old infants later diagnosed with autism. *Nature*. 2013;504(7480):427–431.
38. Kim SH, Lord C. New autism diagnostic interview-revised algorithms for toddlers and young preschoolers from 12 to 47 months of age. *J Autism Dev Disord*. 2012;42(1):82–93.
39. Kim SH, Thurm A, Shumway S, Lord C. Multisite study of new autism diagnostic interview-revised (ADI-R) algorithms for toddlers and young preschoolers. *J Autism Dev Disord*. 2013;43(7):1527–1538.
40. Johnson CP, Myers SM. American Academy of Pediatrics Council on children with D. Identification and evaluation of children with autism spectrum disorders. *Pediatrics*. 2007;120(5):1183–1215.
41. Hazlett HC, Poe M, Gerig G, et al. Magnetic resonance imaging and head circumference study of brain size in autism: birth through age 2 years. *Arch Gen Psychiatry*. 2005;62(12):1366–1376.
42. Courchesne E, Karns CM, Davis HR, et al. Unusual brain growth patterns in early life in patients with autistic disorder: an MRI study. *Neurology*. 2001;57(2):245–254.
43. Hazlett HC, Poe MD, Gerig G, et al. Early brain overgrowth in autism associated with an increase in cortical surface area before age 2 years. *Arch Gen Psychiatry*. 2011;68(5):467–476.
44. Nordahl CW, Lange N, Li DD, et al. Brain enlargement is associated with regression in preschool-age boys with autism spectrum disorders. *Proc Natl Acad Sci U S A*. 2011;108(50):20195–20200.
45. Sparks BF, Friedman SD, Shaw DW, et al. Brain structural abnormalities in young children with autism spectrum disorder. *Neurology*. 2002;59(2):184–192.
46. Schumann CM, Bloss CS, Barnes CC, et al. Longitudinal magnetic resonance imaging study of cortical development through early childhood in autism. *J Neurosci*. 2010;30(12):4419–4427.
47. Carper RA, Moses P, Tigue ZD, Courchesne E. Cerebral lobes in autism: early hyperplasia and abnormal age effects. *Neuroimage*. 2002;16(4):1038–1051.
48. Schumann CM, Barnes CC, Lord C, Courchesne E. Amygdala enlargement in toddlers with autism related to severity of social and communication impairments. *Biol Psychiatry*. 2009;66(10):942–949.
49. Nordahl CW, Scholz R, Yang X, et al. Increased rate of amygdala growth in children aged 2 to 4 years with autism spectrum disorders: a longitudinal study. *Arch Gen Psychiatry*. 2012;69(1):53–61.
50. Estes A, Shaw DWW, Sparks BF, et al. Basal ganglia morphometry and repetitive behavior in Young children with autism Spectrum disorder. *Autism Res*. 2011;4(3):212–220.
51. Qiu T, Chang C, Li Y, et al. Two years changes in the development of caudate nucleus are involved in restricted repetitive behaviors in 2-5-year-old children with autism spectrum disorder. *Dev Cogn Neurosci*. 2016;19:137–143.
52. Hazlett HC, Gu HB, Munsell BC, et al. Early brain development in infants at high risk for autism spectrum disorder. *Nature*. 2017;542(7641):348.
53. Ben Bashat D, Kronfeld-Duenias V, Zachor DA, et al. Accelerated maturation of white matter in young children with autism: a high b value DWI study. *Neuroimage*. 2007;37(1):40–47.
54. Wolff JJ, Swanson MR, Elison JT, et al. Neural circuitry at age 6 months associated with later repetitive behavior and sensory responsiveness in autism. *Mol Autism*. 2017;8:8.
55. Solso S, Xu R, Proudfoot J, et al. Diffusion tensor imaging provides evidence of possible axonal overconnectivity in frontal lobes in autism spectrum disorder toddlers. *Biol Psychiatry*. 2016;79(8):676–684.

56. Conti E, Mitra J, Calderoni S, et al. Network over-connectivity differentiates autism spectrum disorder from other developmental disorders in toddlers: a diffusion MRI study. *Hum Brain Mapp.* 2017;38(5):2333–2344.
57. Xiao Z, Qiu T, Ke X, et al. Autism spectrum disorder as early neurodevelopmental disorder: evidence from the brain imaging abnormalities in 2-3 years old toddlers. *J Autism Dev Disord.* 2014;44(7):1633–1640.
58. Wolff JJ, Gu H, Gerig G, et al. Differences in white matter fiber tract development present from 6 to 24 months in infants with autism. *Am J Psychiatry.* 2012;169(6): 589–600.
59. Travers BG, Adluru N, Ennis C, et al. Diffusion tensor imaging in autism spectrum disorder: a review. *Autism Res.* 2012;5(5):289–313.
60. Prohl AK, Scherrer B, Tomas-Fernandez X, et al. Early white matter development is abnormal in tuberous sclerosis complex patients who develop autism spectrum disorder. *J Neurodev Disord.* 2019;11(1):36.
61. Redcay E, Courchesne E. Deviant functional magnetic resonance imaging patterns of brain activity to speech in 2–3-year-old children with autism spectrum disorder. *Biol Psychiatry.* 2008;64(7):589–598.
62. Dinstein I, Pierce K, Eyler L, et al. Disrupted neural synchronization in toddlers with autism. *Neuron.* 2011;70(6):1218–1225.
63. Lombardo MV, Pierce K, Eyler LT, et al. Different functional neural substrates for good and poor language outcome in autism. *Neuron.* 2015;86(2):567–577.
64. Eyler LT, Pierce K, Courchesne E. A failure of left temporal cortex to specialize for language is an early emerging and fundamental property of autism. *Brain.* 2012;135(Pt 3):949–960.
65. Shen MD, Li DD, Keown CL, et al. Functional connectivity of the amygdala is disrupted in preschool-aged children with autism Spectrum disorder. *J Am Acad Child Adolesc Psychiatry.* 2016;55(9):817–824.
66. Billeci L, Sicca F, Maharatna K, et al. On the application of quantitative EEG for characterizing autistic brain: a systematic review. *Front Hum Neurosci.* 2013;7:442.
67. Tierney AL, Gabard-Durnam L, Vogel-Farley V, Tager-Flusberg H, Nelson CA. Developmental trajectories of resting EEG power: an endophenotype of autism spectrum disorder. *PLoS One.* 2012;7(6), e39127.
68. Wilkinson CL, Levin AR, Gabard-Durnam LJ, Tager-Flusberg H, Nelson CA. Reduced frontal gamma power at 24 months is associated with better expressive language in toddlers at risk for autism. *Autism Res.* 2019;12(8):1211–1224.
69. Kolesnik A, Begum Ali J, Gliga T, Guiraud J, Charman T. Increased cortical reactivity to repeated tones at 8 months in infants with later ASD. *Transl Psychiatry.* 2019;9(1):46.
70. Orekhova EV, Elsabbagh M, Jones EJH, Dawson G, Charman T, Johnson MH. EEG hyper-connectivity in high-risk infants is associated with later autism. *J Neurodev Disord.* 2014;6:40.
71. Thut G, Nietzel A, Brandt SA, Pascual-Leone A. Alpha-band electroencephalographic activity over occipital cortex indexes visuospatial attention bias and predicts visual target detection. *J Neurosci.* 2006;26(37):9494–9502.
72. Toscani M, Marzi T, Righi S, Viggiano MP, Baldassi S. Alpha waves: a neural signature of visual suppression. *Exp Brain Res.* 2010;207(3–4):213–219.
73. Elsabbagh M, Mercure E, Hudry K, et al. Infant neural sensitivity to dynamic eye gaze is associated with later emerging autism. *Curr Biol.* 2012;22(4):338–342.
74. Jones EJ, Venema K, Earl R, et al. Reduced engagement with social stimuli in 6-month-old infants with later autism spectrum disorder: a longitudinal prospective study of infants at high familial risk. *J Neurodev Disord.* 2016;8:7.
75. Barbaro J, Dissanayake C. Autism spectrum disorders in infancy and toddlerhood: a review of the evidence on early signs, early identification tools, and early diagnosis. *J Dev Behav Pediatr.* 2009;30(5):447–459.

76. LeCun Y, Bengio Y, Hinton G. Deep learning. *Nature*. 2015;521(7553):436–444.
77. Siu AL, US Preventive Services Task Force (USPSTF), Bibbins-Domingo K, et al. Screening for autism spectrum disorder in young children: US preventive services task force recommendation statement. *JAMA*. 2016;315(7):691–696.
78. Baird G, Charman T, Baron-Cohen S, et al. A screening instrument for autism at 18 months of age: a 6-year follow-up study. *J Am Acad Child Adolesc Psychiatry*. 2000;39(6):694–702.
79. Sacrey LR, Bryson S, Zwaigenbaum L, et al. The autism parent screen for infants: predicting risk of autism spectrum disorder based on parent-reported behavior observed at 6-24 months of age. *Autism*. 2018;22(3):322–334.
80. Chawarska K, Klin A, Paul R, Macari S, Volkmar F. A prospective study of toddlers with ASD: short-term diagnostic and cognitive outcomes. *J Child Psychol Psychiatry*. 2009;50(10):1235–1245.
81. Guthrie W, Swineford LB, Nottke C, Wetherby AM. Early diagnosis of autism spectrum disorder: stability and change in clinical diagnosis and symptom presentation. *J Child Psychol Psychiatry*. 2013;54(5):582–590.
82. Robins DL, Casagrande K, Barton M, Chen CM, Dumont-Mathieu T, Fein D. Validation of the modified checklist for autism in toddlers, revised with follow-up (M-CHAT-R/F). *Pediatrics*. 2014;133(1):37–45.
83. Dietz C, Swinkels S, van Daalen E, van Engeland H, Buitelaar JK. Screening for autistic spectrum disorder in children aged 14-15 months. II: population screening with the early screening of autistic traits questionnaire (ESAT). Design and general findings. *J Autism Dev Disord*. 2006;36(6):713–722.
84. Barbaro J, Dissanayake C. Prospective identification of autism spectrum disorders in infancy and toddlerhood using developmental surveillance: the social attention and communication study. *J Dev Behav Pediatr*. 2010;31(5):376–385.
85. Pierce K, Conant D, Hazin R, Stoner R, Desmond J. Preference for geometric patterns early in life as a risk factor for autism. *Arch Gen Psychiatry*. 2011;68(1):101–109.
86. Dereu M, Warreyn P, Raymaekers R, et al. Screening for autism spectrum disorders in Flemish day-care centres with the checklist for early signs of developmental disorders. *J Autism Dev Disord*. 2010;40(10):1247–1258.
87. Wetherby AM, Brosnan-Maddox S, Peace V, Newton L. Validation of the infant-toddler checklist as a broadband screener for autism spectrum disorders from 9 to 24 months of age. *Autism*. 2008;12(5):487–511.
88. Matson JL, Fodstad JC, Dempsey T. What symptoms predict the diagnosis of autism or PDD-NOS in infants and toddlers with developmental delays using the baby and infant screen for aUtIsm traits. *Dev Neurorehabil*. 2009;12(6):381–388.
89. Kruizinga I, Visser JC, van Batenburg-Eddes T, Carter AS, Jansen W, Raat H. Screening for autism spectrum disorders with the brief infant-toddler social and emotional assessment. *PLoS One*. 2014;9(5), e97630.
90. Constantino JN, Zhang Y, Frazier T, Abbacchi AM, Law P. Sibling recurrence and the genetic epidemiology of autism. *Am J Psychiatry*. 2010;167(11):1349–1356.
91. Weitlauf AS, Vehorn AC, Stone WL, Fein D, Warren ZE. Using the M-CHAT-R/F to identify developmental concerns in a high-risk 18-month-old sibling sample. *J Dev Behav Pediatr*. 2015;36(7):497–502.
92. Lee HY, Vigen C, Zwaigenbaum L, et al. The performance of the first year inventory (FYI) screening on a sample of high-risk 12-month-olds diagnosed with autism Spectrum disorder (ASD) at 36 months. *J Autism Dev Disord*. 2019;49(12):4957–4973.
93. Ozonoff S, Young GS, Steinfeld MB, et al. How early do parent concerns predict later autism diagnosis? *J Dev Behav Pediatr*. 2009;30(5):367–375.
94. Bussu G, Jones EJH, Charman T, Johnson MH, Buitelaar JK, Team B. Prediction of autism at 3 years from Behavioural and developmental measures in high-risk infants: a longitudinal cross-domain classifier analysis. *J Autism Dev Disord*. 2018;48(7):2418–2433.

95. Chiang CH, Wu CC, Hou YM, Chu CL, Liu JH, Soong WT. Development of T-STAT for early autism screening. *J Autism Dev Disord*. 2013;43(5):1028–1037.
96. Fulceri F, Narzisi A, Apicella F, et al. Application of the repetitive behavior scale-revised–Italian version–in preschoolers with autism spectrum disorder. *Res Dev Disabil*. 2016;48:43–52.
97. Wong V, Hui LH, Lee WC, et al. A modified screening tool for autism (checklist for autism in toddlers [CHAT-23]) for Chinese children. *Pediatrics*. 2004;114(2): e166–e176.
98. Chawarska K, Macari S, Shic F. Context modulates attention to social scenes in toddlers with autism. *J Child Psychol Psychiatry*. 2012;53(8):903–913.
99. Klin A, Jones W. Altered face scanning and impaired recognition of biological motion in a 15-month-old infant with autism. *Dev Sci*. 2008;11(1):40–46.
100. Klin A, Lin DJ, Gorrindo P, Ramsay G, Jones W. Two-year-olds with autism orient to non-social contingencies rather than biological motion. *Nature*. 2009;459(7244): 257–261.
101. Dalton KM, Nacewicz BM, Johnstone T, et al. Gaze fixation and the neural circuitry of face processing in autism. *Nat Neurosci*. 2005;8(4):519–526.
102. Senju A, Southgate V, White S, Frith U. Mindblind eyes: an absence of spontaneous theory of mind in Asperger syndrome. *Science (New York, NY)*. 2009;325(5942): 883–885.
103. Moore A, Wozniak M, Yousef A, et al. The geometric preference subtype in ASD: identifying a consistent, early-emerging phenomenon through eye tracking. *Mol Autism*. 2018;9:19.
104. Pierce K, Marinero S, Hazin R, McKenna B, Barnes CC, Malige A. Eye tracking reveals abnormal visual preference for geometric images as an early Biomarker of an autism spectrum disorder subtype associated with increased symptom severity. *Biol Psychiatry*. 2016;79(8):657–666.
105. Young GS, Merin N, Rogers SJ, Ozonoff S. Gaze behavior and affect at 6 months: predicting clinical outcomes and language development in typically developing infants and infants at risk for autism. *Dev Sci*. 2009;12(5):798–814.
106. Adrien JL, Faure M, Perrot A, et al. Autism and family home movies: preliminary findings. *J Autism Dev Disord*. 1991;21(1):43–49.
107. Palomo R, Belinchon M, Ozonoff S. Autism and family home movies: a comprehensive review. *J Dev Behav Pediatr*. 2006;27(2 Suppl):S59–S68.
108. Wan MW, Green J, Elsabbagh M, et al. Quality of interaction between at-risk infants and caregiver at 12-15 months is associated with 3-year autism outcome. *J Child Psychol Psychiatry Allied Discip*. 2013;54(7):763–771.
109. Tariq Q, Daniels J, Schwartz JN, Washington P, Kalantarian H, Wall DP. Mobile detection of autism through machine learning on home video: a development and prospective validation study. *PLoS Med*. 2018;15(11), e1002705.
110. Fusaro VA, Daniels J, Duda M, et al. The potential of accelerating early detection of autism through content analysis of YouTube videos. *PLoS One*. 2014;9(4), e93533.
111. Hutman T, Rozga A, DeLaurentis AD, Barnwell JM, Sugar CA, Sigman M. Response to distress in infants at risk for autism: a prospective longitudinal study. *J Child Psychol Psychiatry*. 2010;51(9):1010–1020.
112. Miller M, Iosif AM, Hill M, Young GS, Schwichtenberg AJ, Ozonoff S. Response to name in infants developing autism spectrum disorder: a prospective study. *J Pediatr*. 2017;183:141–146, [e141].
113. LeBarton ES, Landa RJ. Infant motor skill predicts later expressive language and autism spectrum disorder diagnosis. *Infant Behav Dev*. 2019;54:37–47.
114. Akshoomoff N, Lord C, Lincoln AJ, et al. Outcome classification of preschool children with autism spectrum disorders using MRI brain measures. *J Am Acad Child Adolesc Psychiatry*. 2004;43(3):349–357.

115. Shen MD, Nordahl CW, Young GS, et al. Early brain enlargement and elevated extra-axial fluid in infants who develop autism spectrum disorder. *Brain.* 2013; 136(Pt 9):2825–2835.
116. Shen MD, Kim SH, McKinstry RC, et al. Increased extra-axial cerebrospinal fluid in high-risk infants who later develop autism. *Biol Psychiatry.* 2017;82(3):186–193.
117. Shen MD, Nordahl CW, Li DD, et al. Extra-axial cerebrospinal fluid in high-risk and normal-risk children with autism aged 2-4 years: a case-control study. *Lancet Psychiatry.* 2018;5(11):895–904.
118. Panizzon MS, Fennema-Notestine C, Eyler LT, et al. Distinct genetic influences on cortical surface area and cortical thickness. *Cerebral cortex (New York, NY: 1991).* 2009;19(11):2728–2735.
119. Xiao X, Fang H, Wu JS, et al. Diagnostic model generated by MRI-derived brain features in toddlers with autism spectrum disorder. *Autism Res.* 2017;10(4):620–630.
120. Jin Y, Wee CY, Shi F, et al. Identification of infants at high-risk for autism Spectrum disorder using multiparameter multiscale White matter connectivity networks. *Hum Brain Mapp.* 2015;36(12):4880–4896.
121. Emerson RW, Adams C, Nishino T, et al. Functional neuroimaging of high-risk 6-month-old infants predicts a diagnosis of autism at 24 months of age. *Sci Transl Med.* 2017;9(393), eaag2882.
122. Bosl W, Tierney A, Tager-Flusberg H, Nelson C. EEG complexity as a biomarker for autism spectrum disorder risk. *BMC Med.* 2011;9:18.
123. Stahl D, Pickles A, Elsabbagh M, Johnson MH. Novel machine learning methods for ERP analysis: a validation from research on infants at risk for autism. *Dev Neuropsychol.* 2012;37(3):274–298.
124. Elsabbagh M, Volein A, Csibra G, et al. Neural correlates of eye gaze processing in the infant broader autism phenotype. *Biol Psychiatry.* 2009;65(1):31–38.
125. Miron O, Ari-Even Roth D, Gabis LV, et al. Prolonged auditory brainstem responses in infants with autism. *Autism Res.* 2016;9(6):689–695.
126. Roth DA, Muchnik C, Shabtai E, Hildesheimer M, Henkin Y. Evidence for atypical auditory brainstem responses in young children with suspected autism spectrum disorders. *Dev Med Child Neurol.* 2012;54(1):23–29.
127. Cohen IL, Gardner JM, Karmel BZ, et al. Neonatal brainstem function and 4-month arousal-modulated attention are jointly associated with autism. *Autism Res.* 2013;6(1):11–22.
128. Cohen IL, Gomez TR, Gonzalez MG, Lennon EM, Karmel BZ, Gardner JM. Parent PDD behavior inventory profiles of young children classified according to autism diagnostic observation schedule-generic and autism diagnostic interview-revised criteria. *J Autism Dev Disord.* 2010;40(2):246–254.
129. Bosl WJ, Tager-Flusberg H, Nelson CA. EEG analytics for early detection of autism Spectrum disorder: a data-driven approach. *Sci Rep.* 2018;8:6828.
130. Gabard-Durnam LJ, Wilkinson C, Kapur K, Tager-Flusberg H, Levin AR, Nelson CA. Longitudinal EEG power in the first postnatal year differentiates autism outcomes. *Nat Commun.* 2019;10:4188.
131. Dickinson A, Varcin KJ. Early patterns of functional brain development associated with autism spectrum disorder in tuberous sclerosis complex. *Autism Res.* 2019;12(12): 1758–1773.
132. Wolfers T, Floris DL, Dinga R, et al. From pattern classification to stratification: towards conceptualizing the heterogeneity of autism spectrum disorder. *Neurosci Biobehav Rev.* 2019;104:240–254.
133. Steinberg DM, Fine J, Chappell R. Sample size for positive and negative predictive value in diagnostic research using case-control designs. *Biostatistics.* 2009;10(1):94–105.
134. Han S, Ma Y. A culture-behavior-brain loop model of human development. *Trends Cogn Sci.* 2015;19(11):666–676.

135. Henrich J, Heine SJ, Norenzayan A. The weirdest people in the world? *Behav Brain Sci.* 2010;33(2–3):61–83, [discussion 83–135].
136. Nisbett RE, Masuda T. Culture and point of view. *Proc Natl Acad Sci U S A.* 2003;100(19):11163–11170.
137. McCarthy A, Lee K, Itakura S, Muir DW. Display when thinking depends on culture and context. *J Cross Cult Psychol.* 2008;39(6):716–729.
138. Blais C, Jack RE, Scheepers C, Fiset D, Caldara R. Culture shapes how we look at faces. *PLoS One.* 2008;3(8), e3022.
139. Gobel MS, Chen A, Richardson DC. How different cultures look at faces depends on the interpersonal context. *Can J Exp Psychol.* 2017;71(3):258–264.
140. Heinsfeld AS, Franco AR, Craddock RC, Buchweitz A, Meneguzzi F. Identification of autism spectrum disorder using deep learning and the ABIDE dataset. *NeuroImage Clin.* 2018;17:16–23.
141. Thompson BL, Levitt P. The clinical-basic interface in defining pathogenesis in disorders of neurodevelopmental origin. *Neuron.* 2010;67(5):702–712.
142. Parikshak NN, Luo R, Zhang A, et al. Integrative functional genomic analyses implicate specific molecular pathways and circuits in autism. *Cell.* 2013;155(5):1008–1021.
143. Yuen RKC, Merico D, Bookman M, et al. Whole genome sequencing resource identifies 18 new candidate genes for autism spectrum disorder. *Nat Neurosci.* 2017;20(4): 602–611.
144. Fakhoury M. Imaging genetics in autism spectrum disorders: linking genetics and brain imaging in the pursuit of the underlying neurobiological mechanisms. *Prog Neuropsychopharmacol Biol Psychiatry.* 2018;80(Pt B):101–114.
145. Wu CC, Chu CL, Stewart L, Chiang CH, Hou YM, Liu JH. The utility of the screening tool for autism in 2-year-olds in detecting autism in Taiwanese toddlers who are less than 24 months of age: a longitudinal study. *J Autism Dev Disord.* 2020;50(4): 1172–1181.
146. Gray KM, Tonge BJ, Sweeney DJ, Einfeld SL. Screening for autism in young children with developmental delay: an evaluation of the developmental behaviour checklist: Early screen. *J Autism Dev Disord.* 2008;38(6):1003–1010.
147. Smith NJ, Sheldrick RC, Perrin EC. An abbreviated screening instrument for autism spectrum disorders. *Infant Ment Health J.* 2013;34(2):149–155.
148. Scambler DJ, Hepburn SL, Rogers SJ. A two-year follow-up on risk status identified by the checklist for autism in toddlers. *J Dev Behav Pediatr.* 2006;27(2 Suppl):S104–S110.
149. Giserman Kiss I, Feldman MS, Sheldrick RC, Carter AS. Developing autism screening criteria for the brief infant toddler social emotional assessment (BITSEA). *J Autism Dev Disord.* 2017;47(5):1269–1277.
150. Nah YH, Young RL, Brewer N. Using the autism detection in early childhood (ADEC) and childhood autism rating scales (CARS) to predict long term outcomes in children with autism spectrum disorders. *J Autism Dev Disord.* 2014;44(9): 2301–2310.
151. Choueiri R, Wagner S. A new interactive screening test for autism Spectrum disorders in toddlers. *J Pediatr.* 2015;167(2):460–466.
152. Turner-Brown LM, Baranek GT, Reznick JS, Watson LR, Crais ER. The first year inventory: a longitudinal follow-up of 12-month-old to 3-year-old children. *Autism.* 2013;17(5):527–540.
153. Stone WL, McMahon CR, Henderson LM. Use of the screening tool for autism in two-year-olds (STAT) for children under 24 months: an exploratory study. *Autism.* 2008;12(5):557–573.
154. Feldman MA, Ward RA, Savona D, et al. Development and initial validation of a parent report measure of the behavioral development of infants at risk for autism spectrum disorders. *J Autism Dev Disord.* 2012;42(1):13–22.

155. Raza S, Zwaigenbaum L, Sacrey LR, et al. Brief report: Evaluation of the Short quantitative checklist for autism in toddlers (Q-CHAT-10) as a brief screen for autism spectrum disorder in a high-risk sibling cohort. *J Autism Dev Disord*. 2019;49(5):2210–2218.
156. Grodberg D, Siper P, Jamison J, Buxbaum JD, Kolevzon A. A simplified diagnostic observational assessment of autism Spectrum disorder in early childhood. *Autism Res*. 2016;9(4):443–449.
157. Capal JK, Horn PS, Murray DS, et al. Utility of the autism observation scale for infants in early Identification of autism in tuberous sclerosis complex. *Pediatr Neurol*. 2017;75:80–86.
158. Sacrey LR, Zwaigenbaum L, Bryson S, et al. Parent and clinician agreement regarding early behavioral signs in 12- and 18-month-old infants at-risk of autism spectrum disorder. *Autism Res*. 2018;11(3):539–547.
159. Barbaro J, Dissanayake C. Early markers of autism spectrum disorders in infants and toddlers prospectively identified in the social attention and communication study. *Autism*. 2013;17(1):64–86.

CHAPTER FIVE

Resting-state abnormalities of posterior cingulate in autism spectrum disorder

Mei-Kei Leung, Way Kwok-Wai Lau*
Department of Special Education and Counselling, The Education University of Hong Kong, Hong Kong, China
*Corresponding author: e-mail address: waylau@eduhk.hk

Contents

Abstract

The posterior cingulate cortex (PCC) plays pivotal roles in cognitive, social and emotional processing, as well as early neural development that supports complex interactions among different neural networks. Alterations in its local and long-range connectivity during resting state are often implicated in neuropathology of neurodevelopmental disorders such as autism spectrum disorder (ASD). ASD is characterized by social and communication deficits, as well as restricted and repetitive behaviors and interests. Individuals with ASD demonstrate persistent disturbances in cognitive and social-emotional functioning, and their PCC exhibits both local and long-range resting state abnormalities compared to typically developing healthy controls. In terms of regional metrics, only the dorsal part of the PCC showed local underconnectivity. As to long-range connectivity measures, the most replicated finding in ASD studies is the

Progress in Molecular Biology and Translational Science, Volume 173
ISSN 1877-1173
https://doi.org/10.1016/bs.pmbts.2020.04.010

reduced functional coupling between the PCC and medial prefrontal cortex (MPFC), which may represent a core neuropathology of ASD unrelated to medication effects. Functional importance of these resting state abnormalities to ASD and directions of future study are discussed at the end of this chapter.

1. Anatomy of the PCC

The PCC is located at the posterior part of the cingulate gyrus. It also sits in the medial part of the inferior parietal lobe. The retrosplenial cortex (RSP), which is a thin layer of cortex on top of the posterior end of the corpus callosum, forms the inferior boundary of the PCC (Fig. 1). The precuneus forms the superior and posterior boundaries of the PCC. Other boundaries of the PCC include cingulate sulcus (superior boundaries), corpus callosum (inferior boundary), parieto-occipital sulcus (posterior boundaries), and mid-cingulate region (anterior boundary). It forms the posteromedial cortex with its adjacent brain regions, the RSP and precuneus. Cytoarchitectonically, the PCC comprises two distinct areas: Brodmann areas (BA) 23 and 31.

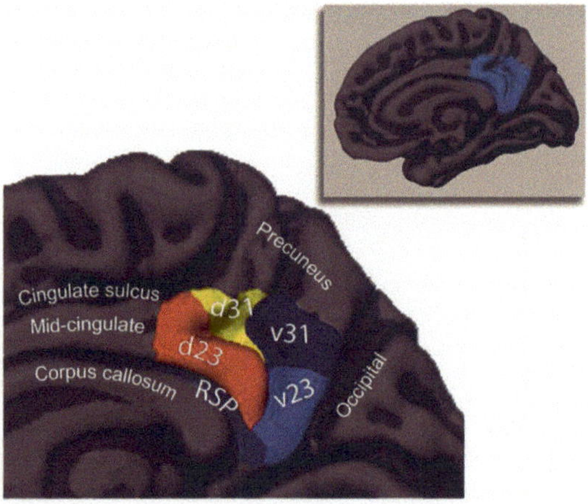

Fig. 1 Subdivisions of the posterior cingulate cortex (PCC) based on Brodmann area (BA). BA v23 and BA v31 correspond approximately to the ventral PCC (in cold colors). BA d23 and BA d31 correspond to the dorsal PCC (in warm colors). RSP, retrosplenial cortex. *Reproduced with permission from Leech R, Sharp DJ. The role of the posterior cingulate cortex in cognition and disease.* Brain. *2014;137(1):12–32, https://doi.org/10.1093/brain/awt162.*

According to the Vogt's model,[1,2] the PCC consists of the dorsal and ventral parts. The dorsal PCC is superior to the splenium of the corpus callosum while the ventral PCC is posterior to the splenium of the corpus callosum. The dorsal PCC is composed of BA d23 and d31 while the ventral PCC is composed of BA v23 and v31 (Fig. 1).

2. Structural connectivity of the PCC

Recent advances in magnetic resonance imaging (MRI) techniques and analytic methodology provide a comprehensive picture of the structural connectivity of PCC. Diffusion tensor imaging (DTI) can be used to reveal nerve fibers connection by tracking the diffusion of water molecules in white matter noninvasively. In white matter connections, the degree of diffusion is small, indicating the restriction of movement of water molecules inside intact nerve fibers. Conversely, the degree of diffusion of water molecules is large in brain structures outside nerve fibers, for instance, gray matter structure or cerebrospinal fluid.[3] It has been shown that the PCC has strong structural connections with core components of the default mode network (DMN) (the details of which will be discussed later). For instance, the long-range structural connection between the PCC and MPFC was consistently observed in a number of studies.[4–6] In particular, the dorsal PCC was structurally connected to the ventromedial prefrontal cortex via the cingulum bundle and the ventral PCC was structurally connected to the medial temporal lobe.[4] The computation of neural connections between the PCC and inferior parietal lobule (IPL) is, however, complicated by crossing fiber tracts and choice of tractography algorithm.[4] Nonetheless, a more recent report demonstrated the existence of white matter tracts between the PCC and bilateral IPL with different tractography approaches, particularly the probabilistic tractography, which is more effective on identifying the underlying connections between brain regions that experience considerable fiber crossing.[6]

3. Resting state fMRI: Intrinsic connectivity networks

Other than advancement in revealing the structural architecture of the brain, different functional connectivity networks have also been discovered with functional MRI (fMRI) in recent years. Through measuring the blood oxygenation level-dependent (BOLD) signal, fMRI can provide an estimate of neural activity in the brain with higher spatial resolution compared to electroencephalogram.[7] Most of the early fMRI studies mainly focused on measuring BOLD signal changes when a person is engaged in a specific task, to examine the neural correlates of the cognitive and/or affective processes associated with the task. These task-based fMRI studies provide valuable information about the underlying neural mechanisms of complex human behaviors. Numerous task-based fMRI studies were also conducted on people with ASD. As summarized in a previous systematic review and meta-analysis on ASD that involved different functions such as visual processing, motor function, auditory and language, and social cognition, functional alterations to brain regions that are responsible for social processing was the most consistent finding in ASD.[8] However, it was not feasible to compare the findings between different tasks because the tasks involved in different studies were very diverse. Furthermore, young children with ASD or infants at-risk of ASD are not able to follow instructions of difficult or complex tasks while being scanned. These render task-based fMRI methodology not an ideal candidate of producing biomarkers that feature ASD.[9]

Resting state fMRI is an alternative approach to examining the brain dynamics in both typically and atypically developing individuals.[10] Participants are instructed to lie in the MRI scanner, with their eyes closed or fixated on a crosshair while being scanned for around 6–10 min. They are asked to remain awake without thinking about anything specifically, thus the resting state is often referred to as a mind-wandering state that is associated with spontaneous or intrinsic (i.e. task free) activity.[11] Although it remains debatable whether this spontaneous, yet organized network activity reflects mind-wandering and/or other ongoing and largely unconscious processes (i.e. "intrinsic" functions),[11] resting state fMRI studies are more promising on producing replicable results compared to traditional task-based fMRI studies. Reproducible findings are important to research on identifying biomarker and individual differences. Furthermore, no task performance is required in resting state fMRI studies makes it particularly suitable for investigating abnormal brain maturation in pediatric and clinical participants with varying levels of functioning, such as people with ASD. Indeed, there are a lot of different kinds of resting state fMRI studies that look at differences in local activity and functional coupling of different brain regions between people with ASD and typically developing age-matched controls. When brain regions activate or deactivate together, they are said to be functionally connected. In the last two decades, different kinds of functional connectivity networks have been detected in both task-based and resting state fMRI studies. Specifically, through studying functional connectivity during resting state, intrinsic connectivity networks that share common spatio-temporal patterns of activity at rest were identified.[12] For instance, Yeo and colleagues[13] identified seven intrinsic connectivity networks including the DMN, visual, somatomotor, dorsal attention, ventral attention, limbic, and frontoparietal networks, using the 7-network estimate. Seeley and colleagues[14] defined a "salience network" [including the dorsal anterior cingulate cortex, orbitofrontal cortex (OFC) and insula] and "executive-control network" (including dorsolateral frontal and parietal cortices) using their resting state data. They found that the salience network activity correlated with anxiety while the executive-control network activity correlated with executive task performance. There could be different numbers of network parcellation, relying on the guidance from both computation estimation and expertise on neuroscience knowledge.

Among various parcellation methods, the DMN is one of the most studied intrinsic networks. Its coactivation patterns are very reproducible with high test-retest reproducibility in young healthy adults.[15] Its core components include the PCC, MPFC, medial temporal lobes, lateral and medial parietal cortex and precuneus.[4,5] The reason why the DMN is frequently studied is related to its special role in self-referential and introspective processing.[16] As summarized by Buckner and colleagues,[17] the main role of the DMN is on internally focused tasks such as remembering one's memory, planning for the future and understanding the perspectives of others via processing one's own thoughts and feelings. The authors proposed that the DMN is also implicated in navigating social interactions through the use of past experiences. Furthermore, it was observed that the DMN activity increased during resting state but decreased during task performance, which leads to its alternative name: the "task-negative" network.[16] When the task is more difficult, a suppression of DMN activity was detected, which indicated a reduction in introspective process.[18] Therefore, the activity of the DMN is opposed to other intrinsic connectivity networks (or "task-positive" networks) that are involved in attention, memory, motor functioning, or executive control, such as the salience and executive-control networks.[19]

4. The PCC as the central hub in DMN

The multiple interacting subsystems of the DMN are linked by hubs.[17] In graph theory, a hub is a node that has higher than average number of connections with other nodes (i.e. high degree).[20] The functional or structural connections of a brain region collected via fMRI or DTI methodologies can be used to study its network properties. The PCC is one of the prominent hubs in the adult brain.[20] It is also the first hub that exhibits the strongest connections of the DMN in the developmental trajectory,[21] which plausibly makes it one of the brain regions that has the highest metabolic rate.[22] As revealed by resting state fMRI and DTI data measured on the same human participants, there is a close correspondence between the functional connectivity of the PCC and its dense structural connectivity with the core components of the DMN,[4,5] providing evidence for its role as a core hub in the DMN.[23,24] Specifically, the PCC acts as an important node for different subsystems in the DMN to converge on for integrated processing.[17] For instance, the PCC forms the midline common core system of the DMN with the anterior MPFC to interact with two subsystems within the DMN: the medial temporal lobe subsystem and the dorsal MPFC subsystem.[25] While the medial temporal lobe subsystem is mainly responsible for mnemonic processes that provide building blocks of mental simulation, the dorsal MPFC subsystem responds to self-relevant mental simulations such as remembering the past, imagining the future, and conceiving the current viewpoint of others. Therefore, the interactions among the DMN support the construction of mental models of personally significant events, and contribute to making self-other distinctions.[25,26]

On the other hand, the PCC plays a special role in DMN to coordinate attentional control with other intrinsic networks in the brain. In general, it is thought that the PCC is involved in processes that require internally directed cognition, such as regulation and balancing of focus of attention to internal or external thoughts,[27] arousal and awareness[28] and self-referential thought.[29] However, Leech, Braga and Sharp[30] suggested that the PCC has a complex and dynamic pattern of functional connectivity, which supports its interaction with other intrinsic networks across the brain. For instance, subregions of the dorsal PCC showed strong but distinct patterns of connectivity with the frontoparietal networks during an attentionally demanding task while the ventral PCC maintained high connectivity to the remaining components of the DMN. These dynamic connectivity patterns are thought to be important for integrated processing of rapid environmental changes and internal needs, which provides evidence for the active role of the dorsal PCC on balancing between externally and internally directed cognition. Leech and Sharp[31] further proposed that the dorsal PCC is important for adjusting the stability of the brain network over time (i.e. "whole-brain metastability") to control attentional focus, whereas the ventral PCC is mainly for internally directed cognition. According to the authors, a high dorsal PCC activity supports rapid transition between different cognitive processes in a broad attentional state (i.e. high whole-brain metastability). In contrast, a low dorsal PCC activity allows stable cognitive processing for a specific task in a narrow attentional state (i.e. low whole-brain metastability). During resting state, a high dorsal PCC activity allows a broad attentional focus for the continuous flow of unconstrainted thoughts ("free-wheeling"), which is associated with rapid transitions between neural states, and rapid changes of activity in intrinsic connectivity networks. This account provides a theoretical framework to explain the complex and dynamic role of the PCC in interacting with different intrinsic networks in the brain for balancing between internal needs and external demands.[30]

In addition, the dorsal and ventral PCC showed dissociative functions and connectivity during task performance that requires cognitive control.[27] When the task became more difficult, the dorsal PCC showed higher integration within the DMN and more anticorrelation with the dorsal attentional network and other regions activated by the task (the task-positive network). Importantly, these patterns are associated with improved behavioral performance in attentional tasks.[32,33] Alternatively, the ventral PCC showed a totally opposite pattern (lower DMN integration and less anticorrelation with task-positive network) as the task difficulty increased.[27] The increases in functional connectivity between the dorsal PCC and task-positive network demonstrated that the dorsal PCC is necessary in modulating the interaction between the DMN and attentional networks for greater efficiency in allocating attention, which further provided evidence for its role in maintaining whole-brain metastability.

5. ASD

ASD is a complex neurodevelopmental disorder characterized by social and communication deficits, as well as restricted and repetitive behaviors and interests.[34] People with ASD often exhibit deficits in theory of mind (ToM), that is, difficulties in mentalizing or attributing mental states to self and others.[35] This may be related to their inability to develop an awareness of what is in their own and other individuals' minds (i.e. "mind-blindness"), which is likely to contribute to social communicative impairment in ASD.[36,37] In addition, compared to healthy controls, individuals with ASD suffer from difficulties in integrating information at both cognitive and neural levels, thus resulting in cognitive inflexibility[38] and poorer performance in tasks such as story comprehension that requires whole-brain integration.[39] Given the importance of PCC in balancing between externally and internally directed cognition and maintaining the whole-brain metastability, a question then follows: Is there any abnormality in neural activity of the PCC during resting state in people with ASD?

6. Local resting state abnormality of the PCC in ASD

A growing number of studies tried to investigate local resting state abnormality in different regions of the brain of subjects with ASD including the PCC. Some studies found local resting state underconnectivity of the PCC in children and adolescents with ASD compared to typically developing age-matched healthy controls,[40,41] however, such findings were not always consistent across studies.[42–44] Furthermore, in some other studies, significant local resting state underconnectivity was detected in brain regions other than the PCC such as the anterior cingulate cortex[44] or superior temporal region and insula[42] in subjects with ASD compared to typically developing controls.

On the other hand, widespread local resting state overconnectivity was reported in the frontal, temporal and/or occipital lobes in subjects with ASD,[40,41,44] whereas other studies did not find any local overconnectivity in the brain of subjects with ASD compared to controls.[43,45–47] It is noteworthy that one study did not even find any differences at all in local resting state activity in the entire brain between people with ASD and healthy controls.[48] Overall, these inconsistencies in findings could be driven by various factors such as small sample size, phenotypic heterogeneity of ASD subjects and/or the use of different analytical methods on processing and analyzing resting state data across studies.[49,50]

The use of online data-sharing initiatives that contains a large ASD dataset may address some of these problems. The team led by Dr. Adriana Di Martino introduced the Autism Brain Imaging Data Exchange-I (ABIDE-I)[51] that consists of more than 1000 resting state fMRI datasets along with other structural MRI and phenotypic information from more than 500 individuals with ASD. They later released the ABIDE-II that contains over 1000 additional datasets with more phenotypic details especially ASD symptoms.[52] With such large resting state datasets, it was found that the PCC has local reductions in terms of regional metrics such as regional homogeneity in subjects with ASD compared to typically developing controls.[51] Nair and colleagues[41] further separated the ABIDE-I dataset into eyes-open and eyes-closed subsamples and showed that the local underconnectivity in the PCC was only found in the eyes-open subsample.

However, the study conducted by Nair and colleagues[41] adopted the cluster-wise inference approach ($P<0.01$ uncorrected with a minimum cluster volume of 37 voxels), which was criticized for increasing false-positive rates, especially when a lenient cluster-defining threshold (e.g., uncorrected $P=0.01$) was used.[53] To address this problem, our team conducted a systematic and quantitative meta-analysis that included original studies of local resting state differences between subjects with ASD and typically developing healthy controls. One of the strengths of our meta-analysis is that we included studies no matter what kind of method(s) was used to compute local resting state activity or local connectivity (e.g. regional homogeneity, fractional amplitude of low frequency fluctuations, or independent component analysis).[54] Also, the characteristics of ASD subjects across studies differed in terms of their age, IQ, gender and disease severity. In other words, we aimed at identifying a consistent pattern of local resting state abnormalities that could truly and sensitively describe ASD beyond the concern of subject and/or method-wise differences. Furthermore, we deliberately excluded studies that only utilized online fMRI dataset(s) such as ABIDE to ensure that our meta-analysis results were not falsefully biased by similar results that were generated from the same data source(s). Our results showed that there was no consistent ASD-related resting state local overconnectivity,[54] which is in line with the recent conclusion drawn by Nair and colleagues[41(p174)]: " *'General local overconnectivity' in ASD is more myth than fact*". Instead, we detected a consistent pattern of local resting state underconnectivity in the dorsal PCC and right medial paracentral lobule in subjects with ASD compared to typically developing healthy controls.[54] In particular, the observed underconnectivity in the dorsal PCC may be associated with a weakened ability to interpret social signals in ASD, based on the behavioral profiling of our dorsal PCC result. The importance of local resting state abnormality of the dorsal PCC in ASD will be discussed in the final section of this chapter.

7. Long-range resting state abnormality of the PCC in ASD

Abnormal local connectivity in a brain region has two possible sides of influences: affecting the normal function of that specific brain region and/or its connections with other brain regions, hence altering brain function in relation to its connections. To better understand the role of a specific brain region in normal and pathological brain, it is therefore important to examine its local connectivity (how synchrony it is with the brain voxels right next to it) and its connectivity with other brain regions. Our meta-analysis finding of consistently lower resting state local connectivity in the dorsal PCC motivated the next question: Is there any abnormal resting state functional connection between the PCC and other brain regions in people with ASD?

In fact, various kinds of aberrant functional connectivity of the PCC in ASD have been reported in different studies. For instance, the PCC in subjects with ASD were found to have weaker connectivity with a number of brain regions in the frontal, parietal and/or temporal lobes, and at the same time stronger connectivity with some other brain regions in the frontal, parietal and/or temporal lobes, than healthy control subjects.[55–57] However, the findings could sometimes be in total opposite directions between different studies. For instance, Doyle-Thomas and colleagues[57] found that the resting state functional connectivity of PCC was stronger with the superior frontal gyrus and weaker with the inferior temporal gyrus in subjects with ASD relative to the control subjects. In contrast, Monk and colleagues[58] found that the resting state functional connectivity of PCC was weaker with the superior frontal gyrus and stronger with the temporal lobe in subjects with ASD relative to control subjects. In addition, it is noteworthy that some studies did not detect any hypoconnectivity of the PCC[55,59] while some others did not detect any hyperconnectivity of the PCC[60–62] in subjects with ASD compared to controls.

Similar to the local resting state abnormalities of the PCC in ASD, these inconsistent or opposite findings could be due to small sample size, age and heterogeneity of subjects, and the use of different analytical methods across studies. In addition, differences on how the PCC seed was defined may also contribute to the diversity in findings. Among some previous studies on resting state functional connectivity of the PCC in ASD, Yerys and colleagues[63] performed a well-controlled experiment that handled several methodological concerns arose from other earlier studies. They carefully selected participants with a relatively narrow age range (around 8–13 years old) that does not overlap with early childhood or emerging adulthood, which avoids confounding influences associated with developmental changes on brain connectivity (if any). Also, their sample size is moderate (22 ASD children vs 22 typically developing healthy controls), and they controlled for active psychotropic medication use on functional connectivity of the PCC.

As a result, they detected both hyperconnectivity and hypoconnectivity of the PCC in their ASD children compared to controls. The PCC was found to have increased resting state functional connectivity with the supramarginal gyrus, lingual gyrus, fusiform gyrus and occipital pole, and decreased resting state functional connectivity with the MPFC, OFC, nucleus accumbens and cerebellum in children with ASD. It is noteworthy that the reduction in functional connectivity between the PCC and MPFC is consistent with most previous reports[56,57,60] and is considered one of the best-replicated findings in ASD.[41] Since the PCC and MPFC are the two main hubs in the DMN,[64] decoupling between these two regions provides some evidence for the importance of the DMN in neuropathology of ASD.[19,65] In the study performed by Yerys and colleagues,[63] the use of active psychotropic medication was controlled for, it was therefore concluded that the reduced PCC functional connectivity with the MPFC was not related to medication effects. This conclusion is in line with a similar finding of lower functional connectivity between the PCC and MPFC in a large sample of children with ASD who were non-medicated.[57] In addition, Yerys and colleagues[63] reported a correlation between social symptoms of ASD and the PCC functional connectivity with the MPFC, which further strengthens the conclusion that this altered connection of PCC is unlikely to be caused by medication but potentially a core neuropathology of ASD.

Nonetheless, the findings of abnormal PCC connection in subjects with ASD are still too diverse to be concluded without empirical examinations. Other than conducting more carefully designed experiments with sufficient power, a systematic and quantitative meta-analysis will also shed light on the path of searching for a unified and consistent pattern of aberrant resting state functional connectivity of the PCC that is truly unique to ASD regardless of all methodological and subject differences.

8. Importance of local underconnectivity of the dorsal PCC in ASD

8.1 Cognitive inflexibility

As discussed earlier, the dorsal PCC is important for adjusting the stability of the brain network over time (i.e. "whole-brain metastability") to control attentional focus.[31] An increase in the dorsal PCC activity helps to maintain or boost whole-brain metastability and allows a broad attentional focus for the continuous flow of unconstrainted thoughts. Overall, this facilitates the rapid transition between different cognitive processes, as well as rapid changes of intrinsic connectivity networks, to adjust for the constantly changing external environment. Our meta-analysis findings demonstrated local underconnectivity of the dorsal PCC in ASD, which suggests the existence of a narrow attentional state in the mind of people with ASD when they are at rest. According to the metastability model,[31] rapid transition between different cognitive processes requires a broad attentional state. Instead, a narrow attentional state may hinder such transition, possibly resulting in cognitive inflexibility—one of the hallmark features of ASD.[66–68] It was found that a high symptom severity of repetitive behaviors in ASD, which means a high level of cognitive inflexibility,[69] was associated with a low discriminability between task-evoked and resting brain states in subjects with ASD.[67] In other words, it is possible that a weak modulation of brain states may underlie cognitive inflexibility in ASD, because it favors a narrow attentional state that does not support switching between brain states.

8.2 Impaired social-emotional processing

In addition, abnormal synchronization of the dorsal PCC may also be related to weakened ability on interpreting social affective signals in ASD. Our behavioral analysis showed that the locally underconnected dorsal PCC cluster in subjects with ASD fell onto a behavioral sub-domain of Emotion called "Other,"[54] that is not related to intensity of emotion, valence of emotion, positive emotion (e.g. joy or happiness) and negative emotion (e.g. fear, anxiety or sadness).[70] This suggests that the dorsal PCC cluster is not linked to evaluations of the strength or aversiveness of an emotional stimulus, nor the two opposite ends of emotion. Instead, it may be linked to the processing of social-emotional information, because the "Emotion: Other" sub-domain activation clusters overlap with several components of the so-called "social brain" such as the MPFC, PCC and amygdala, which are important for social-affective processing.[71] In line with this, it was recently showed that the delayed development in emotional understanding in ASD was associated with severity of social problems.[72] Indeed, according to the Diagnostic and Statistical Manual of Mental Disorders (DSM-V)[34] and International Statistical Classification of Diseases and Related Health Problems (ICD-10),[73] impaired social functioning and deficits in emotional understanding are persistent features of ASD. Since the PCC is implicated in both the DMN and ToM network, it is possible that the PCC plays an important and supportive role on social interaction by facilitating common-sense understanding in social behavior.[71] Therefore, the consistent finding of resting state underconnectivity in the dorsal PCC may be related to a weakened ability on interpreting social signals, especially those required affective processing, in people with ASD.

9. Importance of the PCC-MPFC hypoconnectivity in ASD

9.1 Disturbed large-scale cortical integration

Given that the PCC is highly interconnected to many other brain regions, long-range resting state abnormalities of the PCC in ASD may have some profound influences on multiple dimensions of functioning such as higher-order cognitive and complex social functions. In particular, the common observation of the PCC–MPFC hypoconnectivity, among various other PCC hyper- and hypoconnectivity reported in previous ASD studies,[56,57,60,63] has important influences on the initial structural and functional organization in the brain of individuals with ASD.[5,74] The PCC and MPFC overlap with components of the structural core of the brain, which consists of the posterior medial cortex and medial OFC. Structurally, the posterior medial cortex (including the PCC) had very rich and widespread interconnections with many other brain regions, and the medial OFC (overlapping with the MPFC) showed relatively less yet dense connections with some brain regions.[5] These structural connections are thought to be important for supporting communications across the entire brain by integrating information from functionally distinct brain modules.[5,74] Therefore, it is possible that the PCC–MPFC resting state hypoconnectivity commonly observed in people with ASD may hinder the initial neural development and then the subsequent interactions among different brain systems in the course of development. Indeed, it was found that the connection between the PCC and MPFC emerged as early as in 2-week-old neonates and steadily increased from 2 weeks to 2 years old,[21] which is well within the critical period of early brain development. If this main connection of the structural core undergoes abnormal development at an early stage, the whole-brain communication phenomenon that is supposed to emerge gradually among different cortices could be compromised. This corroborates with the notion put forward by Just and colleagues[39 (p1817)]: "*...autism entails preservation and possibly enhancement of the function of individual cortical centres, but at the same time entails poorer integration of information at higher levels of processing that require more coordination among cortical centres.*" Such dissociation can be found in multiple domains in autism, for instance, language, memory, and sensory and motor functioning, suggesting the reduced anterior-posterior functional synchrony may affect large-scale integration of cortical function and thereby possibly underlie all higher-order cognitive deficits in ASD.[39]

9.2 Impaired cognitive performance

Other than affecting the whole-brain communication, the PCC–MPFC hypoconnectivity may also bring about specific influences on cognitive function in ASD. According to Fair and colleagues,[75–77] this kind of long-range functional connectivity is supposed to be gradually increased within the same network—a phenomenon that is called "functional integration"—at an early stage of development. At the same time, functional connectivity between brain regions of different brain networks is supposed to be gradually decreased, which is called "functional segregation." Therefore, the commonly observed long-range PCC–MPFC hypoconnectivity in the DMN in previous ASD studies[56,57,60,63] suggests that the DMN may be less integrated with itself in people with ASD. This reduced integration of the DMN was found to be associated with lower connectivity between networks anticorrelated with the DMN, i.e. the typical task-positive networks such as the prefrontal-based motor control network and the parietal–extrastriate network.[64] Furthermore, the influences exerted by the PCC and MPFC on their respective anticorrelated networks appeared to be stronger than the other way around.[64] Taken together, it appears that the PCC–MPFC hypoconnectivity during resting state may play an active role on weakening the coordination of their anticorrelated networks that are linked with allocation of attentional resources and hence affecting cognitive performance in ASD.

9.3 Impaired social cognition

The reduced PCC-MPFC coupling may also be related to abnormal social functioning in ASD. According to Raichle,[78] the MPFC is anatomically connected to several regions in the brain (e.g., the hypothalamus, amygdala, and periaqueductal gray matter of the midbrain), which form a circuit to process sensory and visceral information in the external world and body received by the OFC.[79] This sensory-visceromotor circuitry is potentially involved in higher-order processing that affects social behavior, mood control, and motivational drive.[78] Damage to the MPFC has been shown to cause drastic changes in social-related behavior and cognition, such as empathy, moral judgment, and ToM.[79–81] Specifically, damage to the OFC, of which the medial portion partially overlaps with the MPFC,[79] may cause impairment in interpersonal behavior.[82] Patients with OFC damage failed to maintain expected and comfortable distances with both strangers and close others, versus patients with damage in the dorsolateral prefrontal cortex and healthy controls. These findings suggested the importance of the OFC in controlling the preference on interpersonal distance and inhibiting inappropriate social behavior.[82] The authors emphasized the way that the OFC evaluates and regulates how one's behavior in the present moment matches one's higher order goals is not necessarily a self-aware process. This corroborates the idea of lack of self-insight (or mindblindness) in patients with ASD, which may precipitate social communicative impairment in ASD.[36,37] Our default instinct is supposed to operate or manipulate objects that are within reach, and individuals with OFC damage may experience difficulties in resisting such an impulse.[82] The medial OFC partially overlaps with the MPFC.[79] Therefore, it is possible that patients with ASD and abnormal PCC–MPFC functional connectivity may experience similar difficulties due to the lack of an inhibitory process that monitors social appropriateness.

In addition, reduced interactions between the PCC and MPFC may also explain why it is difficult for individuals with ASD to build connections with others.[83] The PCC is responsible for processing information related to others (other-referential processing) while the MPFC is supposed to process information related to the self (self-referential processing), decoupling of the PCC and MPFC might therefore affect the processing of both types of information. The ability of linking oneself with others is thought to be a fundamental component of social cognition and competence. However, deficits in ToM are persistent features of individuals with ASD.[37] The degree of ToM deficits was associated with social communication symptoms and restricted and repetitive behaviors.[84] A recent research study on a group of healthy subjects revealed that the dorsal PCC and anterior part of MPFC are more associated with interpersonal subjective well-being, while the ventral PCC and posterior part of MPFC are more for intrapersonal subjective well-being.[85] Future study could explore whether the resting state connectivity between the dorsal PCC and anterior MPFC is exclusively damaged in individuals with ASD, in order to investigate the degree of neural alterations on the self-to-other-concerned processing in ASD.

10. Future directions

In this chapter, we reviewed up-to-date literatures that provided evidence to support the involvement of the PCC abnormalities in ASD pathology. Interestingly, different types of the PCC abnormalities have been identified across studies, indicating extensive involvement and plausibly differential roles of the PCC in contributing to cognitive, social and behavioral symptoms of ASD. As the core hubs within the DMN, reduced resting state functional connectivity between the PCC and MPFC is likely to be an important and consistent neural imaging feature that is responsible for the impaired integrative processing in ASD at whole-brain level. However, it remains unclear if the PCC-MPFC hypoconnectivity is the only consistent finding for long-range resting state connectivity, given that there are other reported findings of hyper- and hypoconnectivity of the PCC. Systematical review and meta-analyses would be helpful to comprehensively search for a consistent pattern of abnormal resting state functional connectivity of the PCC in ASD. Future studies with strong methodological design such as large sample size and well-controlled parameters (especially those that are sensitive to processing of resting state data) are warranted, which also help provide more data for more powerful meta-analyses. In addition, most studies that found local or long-range resting state abnormalities of the PCC were performed on adolescents or young adults with ASD. It is possible that the developments of the PCC and its neural connectivity happen at a very early stage of life. For instance, it was found that the PCC-MPFC connectivity existed as early as at around 2 weeks of age in healthy individuals.[21] Indeed, it was recently found that the resting state functional connectivity of DMN regions including the PCC was lower in children with ASD (1–8 years old) compared to typically developing children.[86] This suggests that the PCC may be involved in the early neuropathology of ASD. It remains to be investigated whether the local and long-range resting state abnormalities of the PCC reviewed in this chapter would also be consistently observed in children with ASD, or even younger individuals at risk of ASD. These specific abnormalities could serve as cues on the path of discovering early neural markers that can reliably differentiate children with ASD from typically developing children. Also, it would be interesting to examine the timing of onset of these abnormalities of the PCC. In individuals with ASD, is the PCC born with abnormalities in the first place? Or, does it develop normally at the beginning but somehow its development becomes undesirable at a certain age? Identifying the core neural abnormality that truly signifies ASD helps narrow down the investigation to the most important neurobiological event relevant to the onset of ASD.

Future studies should further explore whether the reduced local resting state connectivity of the dorsal PCC is associated with the behavioral symptom of cognitive inflexibility in ASD, which may help explain the inability of people with ASD to flexibly adapt to changes. Also, it is important to differentiate whether the dorsal and ventral PCC undergo different alterations in terms of connectivity with other brain regions including the MPFC. Given that the dorsal and ventral PCC have distinct functions and connectivity patterns,[27] and only the dorsal PCC demonstrated consistent local resting state differences in people with ASD,[54] future studies are warranted to find out if the PCC-MPFC hypoconnectivity that is consistently observed in previous ASD studies is specific to the dorsal and/or ventral PCC.

Last but not least, it remains unclear why the PCC is also involved in different neuropsychiatric disorders at different developmental stages (e.g. mild cognitive impairment,[87] attention deficit hyperactivity disorder,[88] schizophrenia,[89] depression[90] and so on). There is still a long way on investigating how and when the PCC is involved in the neuropathology of ASD. Nonetheless, consistent patterns of the dorsal PCC and PCC-MPFC abnormalities provide some promises to further investigations focusing on the PCC for prevention and treatment of this heterogeneous neurodevelopmental disorder.

References

1. Vogt BA. *Cingulate Neurobiology and Disease*. Oxford: Oxford University Press; 2009.
2. Vogt BA, Vogt L, Laureys S. Cytology and functionally correlated circuits of human posterior cingulate areas. *Neuroimage*. 2006;29(2):452–466.
3. Sato T, Hasan K, Alexander AL, Minato K. Structural connectivity in white matter using the projected diffusion-tensor distance. *Stud Health Technol Inform*. 2001; 84(pt 2):929–932.
4. Greicius MD, Supekar K, Menon V, Dougherty RF. Resting-state functional connectivity reflects structural connectivity in the default mode network. *Cereb Cortex*. 2009;19(1):72–78.
5. Hagmann P, Cammoun L, Gigandet X, et al. Mapping the structural core of human cerebral cortex. *PLoS Biol*. 2008;6(7):e159.
6. Khalsa S, Mayhew SD, Chechlacz M, Bagary M, Bagshaw AP. The structural and functional connectivity of the posterior cingulate cortex: comparison between deterministic and probabilistic tractography for the investigation of structure-function relationships. *Neuroimage*. 2014;102(pt 1):118–127.
7. Steinbrink J, Villringer A, Kempf F, Haux D, Boden S, Obrig H. Illuminating the BOLD signal: combined fMRI-fNIRS studies. *Magn Reson Imaging*. 2006;24(4): 495–505.

8. Philip RC, Dauvermann MR, Whalley HC, Baynham K, Lawrie SM, Stanfield AC. A systematic review and meta-analysis of the fMRI investigation of autism spectrum disorders. *Neurosci Biobehav Rev*. 2012;36(2):901–942.
9. Uddin LQ, Dajani DR, Voorhies W, Bednarz H, Kana RK. Progress and roadblocks in the search for brain-based biomarkers of autism and attention-deficit/hyperactivity disorder. *Transl Psychiatry*. 2017;7(8), e1218.
10. Uddin LQ, Supekar K, Menon V. Typical and atypical development of functional human brain networks: insights from resting-state FMRI. *Front Syst Neurosci*. 2010;4:21.
11. Kucyi A. Just a thought: how mind-wandering is represented in dynamic brain connectivity. *Neuroimage*. 2018;180(pt B):505–514.
12. Smith SM, Fox PT, Miller KL, et al. Correspondence of the brain's functional architecture during activation and rest. *Proc Natl Acad Sci USA*. 2009;106(31):13040–13045.
13. Yeo BT, Krienen FM, Sepulcre J, et al. The organization of the human cerebral cortex estimated by intrinsic functional connectivity. *J Neurophysiol*. 2011;106(3): 1125–1165.
14. Seeley WW, Menon V, Schatzberg AF, et al. Dissociable intrinsic connectivity networks for salience processing and executive control. *J Neurosci*. 2007;27(9):2349–2356.
15. Meindl T, Teipel S, Elmouden R, et al. Test-retest reproducibility of the default-mode network in healthy individuals. *Hum Brain Mapp*. 2010;31(2):237–246.
16. Mak LE, Minuzzi L, MacQueen G, Hall G, Kennedy SH, Milev R. The default mode network in healthy individuals: a systematic review and meta-analysis. *Brain Connect*. 2017;7(1):25–33.
17. Buckner RL, Andrews-Hanna JR, Schacter DL. The brain's default network: anatomy, function, and relevance to disease. *Ann N Y Acad Sci*. 2008;1124:1–38.
18. Singh KD, Fawcett IP. Transient and linearly graded deactivation of the human default-mode network by a visual detection task. *Neuroimage*. 2008;41(1):100–112.
19. Menon V. Large-scale brain networks and psychopathology: a unifying triple network model. *Trends Cogn Sci*. 2011;15(10):483–506.
20. Cao M, Huang H, Peng Y, Dong Q, He Y. Toward developmental connectomics of the human brain. *Front Neuroanat*. 2016;10:25.
21. Gao W, Zhu H, Giovanello KS, et al. Evidence on the emergence of the brain's default network from 2-week-old to 2-year-old healthy pediatric subjects. *Proc Natl Acad Sci USA*. 2009;106(16):6790–6795.
22. Raichle ME, MacLeod AM, Snyder AZ, Powers WJ, Gusnard DA, Shulman GL. A default mode of brain function. *Proc Natl Acad Sci USA*. 2001;98(2):676–682.
23. de Pasquale F, Corbetta M, Betti V, Della PS. Cortical cores in network dynamics. *Neuroimage*. 2018;180:370–382.
24. Busler JN, Yanes JA, Bird RT, Reid MA, Robinson JL. Differential functional patterns of the human posterior cingulate cortex during activation and deactivation: a meta-analytic connectivity model. *Exp Brain Res*. 2019;237(9):2367–2385.
25. Andrews-Hanna JR, Reidler JS, Sepulcre J, Poulin R, Buckner RL. Functional-anatomic fractionation of the brain's default network. *Neuron*. 2010;65(4):550–562.
26. Li W, Mai X, Liu C. The default mode network and social understanding of others: what do brain connectivity studies tell us. *Front Hum Neurosci*. 2014;8:74.
27. Leech R, Kamourieh S, Beckmann CF, Sharp DJ. Fractionating the default mode network: distinct contributions of the ventral and dorsal posterior cingulate cortex to cognitive control. *J Neurosci*. 2011;31(9):3217–3224.
28. Vogt BA, Laureys S. Posterior cingulate, precuneal and retrosplenial cortices: cytology and components of the neural network correlates of consciousness. *Prog Brain Res*. 2005;150:205–217.
29. Brewer J, Garrison K, Whitfield-Gabrieli S. What about the "self" is processed in the posterior cingulate cortex? *Front Hum Neurosci*. 2013;7:647.

30. Leech R, Braga R, Sharp DJ. Echoes of the brain within the posterior cingulate cortex. *J Neurosci.* 2012;32(1):215–222.
31. Leech R, Sharp DJ. The role of the posterior cingulate cortex in cognition and disease. *Brain.* 2014;137(pt 1):12–32.
32. Hampson M, Driesen NR, Skudlarski P, Gore JC, Constable RT. Brain connectivity related to working memory performance. *J Neurosci.* 2006;26(51):13338–13343.
33. Kelly AM, Uddin LQ, Biswal BB, Castellanos FX, Milham MP. Competition between functional brain networks mediates behavioral variability. *Neuroimage.* 2008;39(1): 527–537.
34. American Psychiatric Association. *Diagnostic and Statistical Manual of Mental Disorders.* 5th ed. American Psychiatric Publishing; 2013.
35. Cantio C, White S, Madsen GF, Bilenberg N, Jepsen JRM. Do cognitive deficits persist into adolescence in autism? *Autism Res.* 2018;11(9):1229–1238.
36. Lombardo MV, Baron-Cohen S. The role of the self in mindblindness in autism. *Conscious Cogn.* 2011;20(1):130–140.
37. Akechi H, Kikuchi Y, Tojo Y, Hakarino K, Hasegawa T. Mind perception and moral judgment in autism. *Autism Res.* 2018;11(9):1239–1244.
38. Yasuda Y, Hashimoto R, Ohi K, et al. Cognitive inflexibility in Japanese adolescents and adults with autism spectrum disorders. *World J Psychiatr.* 2014;4(2):42–48.
39. Just MA, Cherkassky VL, Keller TA, Minshew NJ. Cortical activation and synchronization during sentence comprehension in high-functioning autism: evidence of underconnectivity. *Brain.* 2004;127(Pt 8):1811–1821.
40. Maximo JO, Keown CL, Nair A, Müller R-A. Approaches to local connectivity in autism using resting state functional connectivity MRI. *Front Hum Neurosci.* 2013;7:605.
41. Nair S, Jao Keehn RJ, Berkebile MM, Maximo JO, Witkowska N, Müller R-A. Local resting state functional connectivity in autism: site and cohort variability and the effect of eye status. *Brain Imaging Behav.* 2018;12(1):168–179.
42. Paakki J-J, Rahko J, Long X, et al. Alterations in regional homogeneity of resting-state brain activity in autism spectrum disorders. *Brain Res.* 2010;1321:169–179.
43. Mueller S, Keeser D, Samson AC, et al. Convergent findings of altered functional and structural brain connectivity in individuals with high functioning autism: a multimodal MRI study. *PLoS One.* 2013;8(6), e67329.
44. Jann K, Hernandez LM, Beck-Pancer D, et al. Altered resting perfusion and functional connectivity of default mode network in youth with autism spectrum disorder. *Brain Behav.* 2015;5(9), e00358.
45. von dem Hagen EAH, Stoyanova RS, Baron-Cohen S, Calder AJ. Reduced functional connectivity within and between 'social' resting state networks in autism spectrum conditions. *Soc Cogn Affect Neurosci.* 2013;8(6):694–701.
46. Bos DJ, van Raalten TR, Oranje B, et al. Developmental differences in higher-order resting-state networks in autism spectrum disorder. *Neuroimage Clin.* 2014;4:820–827.
47. Itahashi T, Yamada T, Watanabe H, et al. Alterations of local spontaneous brain activity and connectivity in adults with high-functioning autism spectrum disorder. *Mol Autism.* 2015;6:30.
48. Bernas A, Aldenkamp AP, Zinger S. Wavelet coherence-based classifier: a resting-state functional MRI study on neurodynamics in adolescents with high-functioning autism. *Comput Methods Programs Biomed.* 2018;154:143–151.
49. Saad ZS, Gotts SJ, Murphy K, et al. Trouble at rest: how correlation patterns and group differences become distorted after global signal regression. *Brain Connect.* 2012; 2(1):25–32.
50. Smitha KA, Akhil Raja K, Arun KM, et al. Resting state fMRI: a review on methods in resting state connectivity analysis and resting state networks. *Neuroradiol J.* 2017;30(4):305–317.

51. Di Martino A, Yan C-G, Li Q, et al. The autism brain imaging data exchange: towards a large-scale evaluation of the intrinsic brain architecture in autism. *Mol Psychiatry.* 2014;19(6):659–667.
52. Di Martino A, O'Connor D, Chen B, et al. Enhancing studies of the connectome in autism using the autism brain imaging data exchange II. *Sci Data.* 2017;4:170010.
53. Eklund A, Nichols TE, Knutsson H. Cluster failure: why fMRI inferences for spatial extent have inflated false-positive rates. *Proc Natl Acad Sci USA.* 2016;113(28): 7900–7905.
54. Lau WKW, Leung M-K, Lau BWM. Resting-state abnormalities in autism spectrum disorders: a meta-analysis. *Sci Rep.* 2019;9(1):3892.
55. Lynch CJ, Uddin LQ, Supekar K, Khouzam A, Phillips J, Menon V. Default mode network in childhood autism: posteromedial cortex heterogeneity and relationship with social deficits. *Biol Psychiatry.* 2013;74(3):212–219.
56. Eilam-Stock T, Xu P, Cao M, et al. Abnormal autonomic and associated brain activities during rest in autism spectrum disorder. *Brain.* 2014;137(pt 1):153–171.
57. Doyle-Thomas KAR, Lee W, Foster NEV, et al. Atypical functional brain connectivity during rest in autism spectrum disorders. *Ann Neurol.* 2015;77(5):866–876.
58. Monk CS, Peltier SJ, Wiggins JL, et al. Abnormalities of intrinsic functional connectivity in autism spectrum disorders. *Neuroimage.* 2009;47(2):764–772.
59. Fishman I, Keown CL, Lincoln AJ, Pineda JA, Müller R-A. Atypical cross talk between mentalizing and mirror neuron networks in autism spectrum disorder. *JAMA Psychiat.* 2014;71(7):751–760.
60. Weng S-J, Wiggins JL, Peltier SJ, et al. Alterations of resting state functional connectivity in the default network in adolescents with autism spectrum disorders. *Brain Res.* 2010;1313:202–214.
61. Rudie JD, Hernandez LM, Brown JA, et al. Autism-associated promoter variant in MET impacts functional and structural brain networks. *Neuron.* 2012;75(5):904–915.
62. Jung M, Kosaka H, Saito DN, et al. Default mode network in young male adults with autism spectrum disorder: relationship with autism spectrum traits. *Mol Autism.* 2014;5:35.
63. Yerys BE, Gordon EM, Abrams DN, et al. Default mode network segregation and social deficits in autism spectrum disorder: evidence from non-medicated children. *Neuroimage Clin.* 2015;9:223–232.
64. Uddin LQ, Kelly AM, Biswal BB, Castellanos FX, Milham MP. Functional connectivity of default mode network components: correlation, anticorrelation, and causality. *Hum Brain Mapp.* 2009;30(2):625–637.
65. Joshi G, Arnold Anteraper S, Patil KR, et al. Integration and segregation of default mode network resting-state functional connectivity in transition-age males with high-functioning autism spectrum disorder: a proof-of-concept study. *Brain Connect.* 2017;7(9):558–573.
66. Turner M. Annotation: repetitive behaviour in autism: a review of psychological research. *J Child Psychol Psychiatry.* 1999;40(6):839–849.
67. Uddin LQ, Supekar K, Lynch CJ, et al. Brain state differentiation and behavioral inflexibility in autism. *Cereb Cortex.* 2015;25(12):4740–4747.
68. Watanabe T, Lawson RP, Wallden YSE, Rees G. A neuroanatomical substrate linking perceptual stability to cognitive rigidity in autism. *J Neurosci.* 2019;39(33):6540–6554.
69. Lopez BR, Lincoln AJ, Ozonoff S, Lai Z. Examining the relationship between executive functions and restricted, repetitive symptoms of autistic disorder. *J Autism Dev Disord.* 2005;35(4):445–460.
70. Lancaster JL, Laird AR, Eickhoff SB, Martinez MJ, Fox PM, Fox PT. Automated regional behavioral analysis for human brain images. *Front Neuroinform.* 2012;6:23.

71. Subbaraju V, Sundaram S, Narasimhan S. Identification of lateralized compensatory neural activities within the social brain due to autism spectrum disorder in adolescent males. *Eur J Neurosci*. 2018;47(6):631–642.
72. Høyland AL, Nærland T, Engstrøm M, Lydersen S, Andreassen OA. The relation between face-emotion recognition and social function in adolescents with autism spectrum disorders: a case control study. *PLoS One*. 2017;12(10), e0186124.
73. World Health Organization. *ICD-10: International Statistical Classification of Diseases and Related Health Problems*. 10th ed. World Health Organization; 2004.
74. van den Heuvel MP, Sporns O. Rich-club organization of the human connectome. *J Neurosci*. 2011;31(44):15775–15786.
75. Fair DA, Cohen AL, Dosenbach NU, et al. The maturing architecture of the brain's default network. *Proc Natl Acad Sci USA*. 2008;105(10):4028–4032.
76. Fair DA, Cohen AL, Power JD, et al. Functional brain networks develop from a "local to distributed" organization. *PLoS Comput Biol*. 2009;5(5), e1000381.
77. Fair DA, Dosenbach NU, Church JA, et al. Development of distinct control networks through segregation and integration. *Proc Natl Acad Sci USA*. 2007;104(33): 13507–13512.
78. Raichle ME. The brain's default mode network. *Annu Rev Neurosci*. 2015;38:433–447.
79. Zald DH, Andreotti C. Neuropsychological assessment of the orbital and ventromedial prefrontal cortex. *Neuropsychologia*. 2010;48(12):3377–3391.
80. Beadle JN, Paradiso S, Tranel D. Ventromedial prefrontal cortex is critical for helping others who are suffering. *Front Neurol*. 2018;9:288.
81. Schneider B, Koenigs M. Human lesion studies of ventromedial prefrontal cortex. *Neuropsychologia*. 2017;107:84–93.
82. Perry A, Lwi SJ, Verstaen A, Dewar C, Levenson RW, Knight RT. The role of the orbitofrontal cortex in regulation of interpersonal space: evidence from frontal lesion and frontotemporal dementia patients. *Soc Cogn Affect Neurosci*. 2016;11(12):1894–1901.
83. Burrows CA, Laird AR, Uddin LQ. Functional connectivity of brain regions for self- and other-evaluation in children, adolescents and adults with autism. *Dev Sci*. 2016;19(4):564–580.
84. Jones CRG, Simonoff E, Baird G, et al. The association between theory of mind, executive function, and the symptoms of autism spectrum disorder. *Autism Res*. 2018;11(1):95–109.
85. Jo H, Ou YY, Kung CC. The neural substrate of self- and other-concerned wellbeing: an fMRI study. *PLoS One*. 2019;14(10), e0203974.
86. Funakoshi Y, Harada M, Otsuka H, Mori K, Ito H, Iwanaga T. Default mode network abnormalities in children with autism spectrum disorder detected by resting-state functional magnetic resonance imaging. *J Med Invest*. 2016;63(3.4):204–208.
87. Lau WK, Leung MK, Lee TM, Law AC. Resting-state abnormalities in amnestic mild cognitive impairment: a meta-analysis. *Transl Psychiatry*. 2016;6, e790.
88. Jiang K, Yi Y, Li L, et al. Functional network connectivity changes in children with attention-deficit hyperactivity disorder: a resting-state fMRI study. *Int J Dev Neurosci*. 2019;78:1–6.
89. Zhang S, Yang G, Ou Y, et al. Abnormal default-mode network homogeneity and its correlations with neurocognitive deficits in drug-naive first-episode adolescent-onset schizophrenia. *Schizophr Res*. 2020;215:140–147.
90. Zhong X, Pu W, Yao S. Functional alterations of fronto-limbic circuit and default mode network systems in first-episode, drug-naïve patients with major depressive disorder: a meta-analysis of resting-state fMRI data. *J Affect Disord*. 2016;206:280–286.

CHAPTER SIX

Neurobiology of sensory processing in autism spectrum disorder

Phoebe Pui Pui Cheung*, Benson Wui Man Lau
Department of Rehabilitation Sciences, Hong Kong Polytechnic University, Hung Hom, Hong Kong
*Corresponding author: e-mail address: pp.cheung@polyu.edu.hk

Contents

Abstract

Altered sensory processing and perception has been one of the characteristics of autism spectrum disorder (ASD). In this chapter, we review the neural underpinnings of sensory abnormalities of ASD by examining the literature on clinical, behavioral and neurobiological evidence that underlies the main patterns of sensory integration function and dysfunction. Furthermore, neural differences in anatomy, function and connectivity of different regions underlying sensory processing are also discussed. We conclude that sensory integration intervention is built on the premise of neuroplasticity to improve function and behavior for individuals with ASD.

1. Introduction

Autism spectrum disorder (ASD) is defined clinically by impairment in communication, social interaction, and behavioral flexibility.[1] There is strong evidence to support that the "autistic brain" is functionally and

Progress in Molecular Biology and Translational Science, Volume 173
ISSN 1877-1173
https://doi.org/10.1016/bs.pmbts.2020.04.020

structurally different from the brain of typically developing (TD) children. Moreover, there exists a wide range of phenotypic variations involving cognitive ability, the pace of language development and the co-morbidity of other diagnoses like epilepsy or attention deficit and hyperactivity disorders.[2]

ASD is a complex neurodevelopmental disorder characterized by heterogeneity of skills and functioning. A variety of models have been developed to describe the disorders which included the temporal binding hypothesis, the intense world theory and the sensory integration theory.[3] One of the most common features to individuals with ASD across the spectrum are the atypical behavioral responses to sensory stimuli. Literature showed that over 96% of children with ASD report hyper and hypo-sensitivities in multiple domains, including the vestibular, somatosensory, visual, auditory, olfactory and gustatory system.[4,5] Atypical sensory processing in ASD is postulated to be one of the cause core features of ASD and can interfere with development and affect the interaction with the environment.[6] However, the neurobiological underpinnings of sensory abnormalities in ASD have been less addressed in the literature. The sensory behavioral differences range from mild to severe and the functional impairment can endure throughout the adulthood.[4,7] Early abnormalities in sensory processing can be compounded over time, creating a maladaptive developmental trajectory of functional impairment.

Sensory integration theory was proposed by Dr. A. Jean Ayres who was an occupational therapist and neuropsychologist. She developed theory and intervention to help and treat children with learning and behavioral challenges.[8,9] In this theory, different sensory system laid an important foundation for adaptive behavior. It involves perceiving, organizing, and interpreting information received through sensory systems which included taste, touch, smell, sight, auditory, vestibular that support movement and development.[8,9] Ayres hypothesized that poor sensory stimulus registration and modulation impairments were commonly seen in individuals with ASD.[8]

This chapter reviews the abnormalities in each affected sensory system of ASD by integrating clinical, behavioral and neurobiological evidence to explore the sensory impairments in ASD. To begin with, we will examine the existing construct of sensory registration and sensory modulation in Ayres Sensory Integration (ASI) hypothesis in ASD. Then we will discuss the possible mechanisms by which atypical sensory processing across the core sensory systems including vestibular, proprioceptive, tactile, visual, auditory, olfactory and gustatory systems. While several primary sensory and association cortex areas may be involved in sensory processing and integration of related information, we will briefly discuss different neural regions in midbrain and cerebral cortex and address their roles in processing sensory stimuli in individuals with ASD.

2. Atypical sensory processing in ASD

According to the American Psychiatric Association's Diagnostic Manual for Mental Disorders-Fifth Edition (DSM-5), hyper- and hypo-sensory reactivity have been included as diagnostic criteria of ASD.[2] These criteria, however, manifest differently across individuals with ASD. For examples, some individuals seem to be unaware of certain auditory, visual or tactile stimuli (classified as hypo-responsivity), while others may avoid the same stimuli altogether (classified as hyper-responsivity).[1] These modulation deficits can contribute to and/or overlap with the core characteristics of ASD and hinder their participation in everyday activities.

Ayres proposed that children with ASD not only fail to register sensory input properly but also has trouble modulating input that they register. Sensory registration is "the detection of sensory sensations within the central nervous system"[10] (p. 3). For example, children with ASD may not detect the presence of novel stimulus, e.g., the sensation of a puff of air on their necks, the way TD children do. Ayres hypothesized that registration problems are related to poor central nervous system (CNS) registration of sensory stimuli.

Impairments in emotion-related brain regions like limbic system not only has impact on registration, but also on sensory modulation.[8] Limbic system comprised several subcortical nuclei and cortical structures including the insula, hypothalamus, hippocampus, para-hippocampal gyrus, amygdala, fornix, mammillary body, septal nuclei and thalamus.[11,12] Structural abnormalities in the forebrain limbic system were postulated to dysfunction in sensory modulation, which could lead to emotional problems.[13] Studies show that adults with ASD have reduced amygdala and hippocampus volume as compared to typical individuals.[14] A few white matter studies have reported variation in tracts connecting emotion-related brain regions to motor, auditory and cognitive processes.[3,15]

Ayres defined modulation as the "brain's regulation of its own activity"[10] (p. 182). Based on her clinical observation, she suggested that hypo- or hyper-reactivity to vestibular and tactile sensations, may manifest gravitational insecurity (i.e., fear of movement) and tactile defensiveness

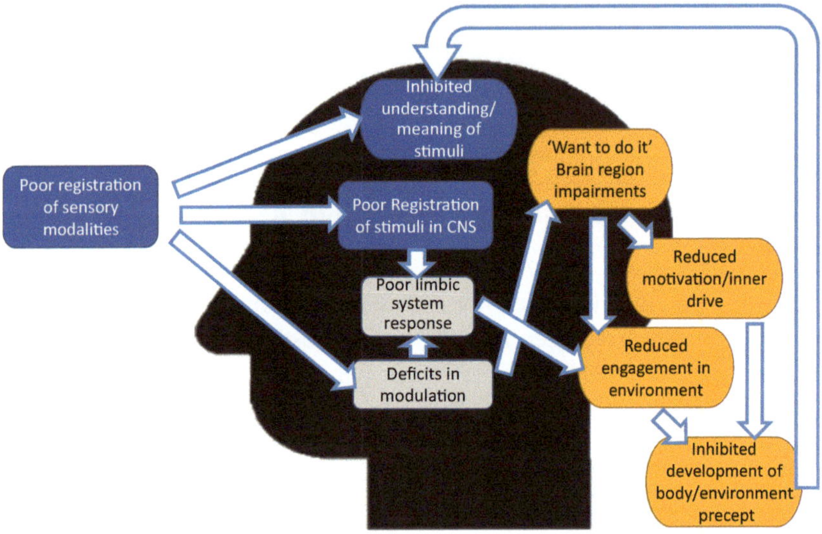

Fig. 1 A proposed figure of Ayres Sensory Integration hypothesis in autism spectrum disorder. *Adapted from Kilroy E, Aziz-Zadeh L, Cermak S. Ayres theories of autism and sensory integration revisited: what contemporary neuroscience has to say.* Brain Sci. *2019;9 (3):68. doi: 10.3390/brainsci9030068.*

(fight, fright or flight reaction to light touch that other children consider normal sensation). Ayres' postulation about abnormal modulation of sensory input in ASD would have implications for therapy. Multisensory experiences are embedded in our daily life and are crucial to our life participation. In TD children, multisensory inputs allow processing and interpretation of sensory events in the environment. However, individual with ASD of poor modulation show weaker evoked potentials to auditory and somatosensory stimuli and atypical auditory-somatosensory integration, which would overwhelm them and may hinder their functioning in their own environment.[11] Fig. 1 summarizes the proposed visualization of sensory integration hypothesis in ASD by Kilroy and her colleagues.[10]

3. Vestibular system

Vestibular system has a significant role on brain and behavior function.[8–10] Research shows that vestibular connection to many brain regions that serve as arousal regulation, static and dynamic postural control, balance and equilibrium responses, bilateral coordination, maintenance of stable visual field and spatial perception of body movement.[11,12]

One example is the rapid or rotary movement of the body through space is associated with increased alertness and autonomic responses that activate the arousal system via the brainstem reticular formation.[12] On the other hand, slow and rhythmic movement, such as gentle rocking, decreased arousal experienced as calming or drowsiness. Vestibular information travels via the vestibulospinal tracts activates neck and trunk muscle to uphold the postural and head control. Vestibular information also carried in the medial longitudinal fasciculus to the cranial nerves control the ocular motor muscles to support coordinated eye and head movements. Furthermore, since vestibular system is a bilateral system that affects muscle activation of left and right sides, it contributes to bilateral motor coordination. With the anatomical connection with vision and proprioception, the vestibular system contributes to anticipatory motor actions and sequences.[3,16] In fact, decreased vestibular response in post-rotary nystagmus test were postulated to deficits in bilateral integration, decreased extension against gravity, poor equilibrium reactions and decreased coordination of eye and head movement.[6]

It is still being controversial whether the vestibular system is affected in ASD. Studies have shown that functions related to the vestibular system, including maintaining balance[17] and responsivity to vestibular stimuli,[18] are found to be significantly different between ASD and TD individuals. While the vestibulo-ocular reflex (vof) of keeping fixation on an object is found in ASD individuals and the accuracy of the reflex is not significantly different from TD, the latency to start saccade was shown to be significantly increased in ASD.[19] Furthermore, the gain (angle of eye movement divided by angle of head movement) and ratio of eye velocity to head velocity of vof were also found to increase in ASD subjects.[20] The difference between ASD and TD subjects was attributed to delayed visual information processing or lack of cerebellar inhibitory input to the vestibular nuclei, rather than a dysfunctional vestibular system.[19] Though vestibular input is usually involved in the sensory-based therapy (e.g., sensory integration therapy) for children affected by ASD, the aim of the therapy is to organize the sensory system via different sensory modalities and not necessarily pinpointing a dysfunction in the vestibular system.[21] As also suggested by an early study, the abnormal responsivity toward vestibular stimuli in ASD may be due to the altered sensory processing, and the signs of hyper-responsivity and hypo-responsivity may not be homogenous in ASD individuals.[18]

Recently, neuroscientists have shown that temporo-parietal junction, anterior parietal lobe, posterior parietal and medial superior temporal cortices, cingulate gyrus and retrosplenial cortex, and hippocampal and para-hippocampal cortices all receives vestibular information.[22] With the widespread projection to different neural regions, vestibular system may play an important role in learning and behavioral difficulties of children.

4. Somatosensory system

Somatosensory system is consisted of touch and proprioception. Touch is considered one of the most basic ways in which individuals interact with the world around them.[3] Proprioception is the sense in which we perceive the position and movement of our body.[23] Individuals with ASD were reported to have hypo- and hyper-reactivity to tactile stimuli, with these responses varying according to stimuli and context. Specifically, abnormal detection of tactile stimuli and a lack of habituation to tactile stimuli were reported by them. Literature showed that tactile sensations project to the primary sensory cortex and the secondary somatosensory cortex, a region responsible for object manipulation, tactile discrimination and grasp and release.[24] Structural differences were noted in several primary sensory cortical areas,[25] for instance, the planum temporale (PT) and posterior superior temporal gyrus (pSTG) were found to be significantly larger in ASD, which is in contrast to TD individuals. Also, EEG and MEG studies have shown significant differences in response latencies in the auditory, somatosensory and visual cortices.[26,27] Literature also supports that somatosensory-vestibular-visual integration occurs in many central nervous system (CNS) levels including the vestibular nuclei, the thalamus, and the cortex. For example, vestibular activation increases tactile discrimination and may decrease pain. Tactile input project to the posterior parietal cortex would integrate with visual information and motor signals and to the orbitofrontal cortex contributes to affect.[11]

Another study reported the changes in blood oxygenation level-dependent (BOLD) signal of ASD participants exhibited diminished responses in the posterior cingulate cortex and the insula, particularly for pleasant and neutral tactile stimuli, compared to the typical developing group.[28] For the unpleasant tactile stimuli, ASD participants exhibited greater BOLD signal than the control group in affective somatosensory processing area, including the posterior cingulate cortex and the insula. There is evidence to show that dysfunction in touch and tactile perception are closely tied to social dysfunction and core symptoms of ASD. For example, hyper-responsivity to tactile stimuli (i.e., tactile defensiveness) of children with ASD have been found to relate to social impairment (i.e., avoiding unintentional touch and move away from crowd), and tactile hypo-responsivity was associated with poorer social functioning and non-verbal communication skills.[3] Differences in tactile processing and atypical tactile behaviors in ASD are observed in early infancy. Atypical touch during infancy can develop into critical social deficits in later life, including forming a secure attachment to caregivers. Therefore, dysfunction in tactile processing early in development can have crucial social and interpersonal implications.

5. Visual system

Similar to the somatosensory system, individuals with ASD were reported to have atypical visual behavior that include avoid visual stimuli (e.g., covering eyes at bright light) or seek out intense visual stimuli (e.g., staring at the sunlight). The atypical visual behavior is likely to be caused by atypical visual processing and perception, including contrast sensitivity, boundary detection and color perception.[29–31] The visual perceptual dysfunctions are likely to be found at dynamic situations rather than in static spatial tasks.[32] While the ability of ASD individuals to identify orientation of simple objects might be even superior than TD controls, increase in information in an experimental stimulus, which required much input from other cortical areas to process, would decrease the ability to correctly identify the orientations. The detection of moving stimulus with high contrast to the background was found worse in ASD individuals.[33] Furthermore, individuals with ASD were also found to perform worse than age-matched controls in detection of coherent motion of target stimuli.[34] These different findings converge on a common atypical property of ASD: motion detection of objects at different levels of processing, possibly include both subcortical and cortical visual pathways.

After receiving light signal by photoreceptors (rods and cones) in the retina, the signal will be transduced to the interneurons termed retinal ganglion cells, and subsequently to the lateral geniculate nucleus (LGN) in the thalamus, and further transduced to the cortical regions for visual perception. In the context of perception of moving objects, the visual pathway of primate and human can be classified according to function of detecting static or moving objects, namely, the parvocellular or magnocellular pathways, respectively.[35] Parvocellular pathways represent the visual pathway which involves the parvocellular geniculate subdivisions of the LGN, and the retinal ganglion cells which provide input to this division is called P-type retinal ganglion cells. The other LGN subdivision is the magnocellular division, and the ganglion cells that transduce signals to this subdivision are termed

M-type ganglion cells. Parvocellular pathway transmit information on color, shape of objects and spatial relationship. In contrast, magnocellular pathway transmit information related to motion. In other words, parvocellular pathway is specialized in detecting high contrast, spatial while magnocellular pathway is specialized in low contrast, temporal resolution. With the deficit in detecting moving objects, it is possible that the magnocellular pathway is affected in ASD.[36]

Apart from subcortical detection of moving objects, it is hypothesized that the cortical processing of visual stimuli is also affected in ASD. This is supported by the observation that second-order motion processing is worse in ASD children compared to TD controls.[37] First-order motion processing is the perception of movement of an object, which has relatively high contrast to the background, by changing the contrast between the object and background. Compared to first-order, the second-order motion processing is a more complex process, in which the object stands out from the background according to the texture. The second-order motion processing requires additional neural computation and integration of information. This points out to the possible dysfunction in cortical regions related to visual processing: the "visual streams" which consist of the occipital lobe and particular regions on dorsal and ventral surface of the cerebrum.[38] It is hypothesized that there are two visual streams at the cortical level of human, namely the dorsal and ventral streams. The ventral stream originates from the visual/occipital cortex, extends to the ventral part of cerebrum including temporal lobe and limbic system. This stream receives input from the parvocellular pathway and is believed to be responsible for perceiving shapes of objects. The dorsal stream extends from the visual cortex to the parietal lobe and interconnected to the ventral stream. This stream is associated with visually guided action and manipulation of objects. There are reports support that both streams may be involved in the motion-detection deficit in ASD,[39,40] but there is no conclusive finding whether both systems are involved, and whether the two streams are associated differentially to different etiologies of ASD. Nevertheless the hypothesis of dysfunctional neuro-integration is a plausible explanation of the complex motion detection malfunction in ASD, and emerging studies are providing evidence for this hypothesis.

Poor facial recognition, facial memory, facial discrimination and deficits in facial emotion recognition of ASD have been well documented in the literature and this is another recognized dysfunction related to vision. When being presented with static face drawings, there was hypo-activation in the areas related to face recognition in ASD subjects, including fusiform gyrus, superior temporal sulcus and occipital face area.[41] Since fusiform cortex is a part of the ventral visual stream and essential for facial processing, the dysfunction may be associated with alteration in activation and function in this region.[42] Thus the visual stream hypothesis may be applicable on interpreting the poor facial processing. Difficulties in processing visual information in a timely manner are seen in ASD, leading to reduced attention to faces. These differences in early life may affect learning and social participation, as perception of social cues drives visual attention, and as a result, impact on social development which built on joint attention and observation.

6. Auditory system

Auditory hypo- (i.e., lack of orientation of auditory stimuli) and hyper-responsivity (i.e., over-sensitivity to sudden loud noise) are commonly seen in individuals with ASD. Atypical auditory processing with enhanced pitch perception, impaired perception of prosody, and diminished auditory stream segregation[22] were reported. One way to measure the flow of auditory information is through traditional auditory brainstem response (ABR) in which the electrical activity evoked from a series of clicks or tones is recorded using surface electrodes. Auditory information from the vestibulocochlear nerve (cranial nerve VIII) project to cochlear nuclei and the superior olivary complex in the brain stem and the inferior colliculus in the midbrain.[2] Literature supported a significantly longer CN III–V interpeak latency in ASD compared to the TD group. There was also a prolonged latency in child and adolescent ASD cohort.[43]

Besides brainstem studies, cortical auditory processing using event-related potentials (ERPs) with electroencephalography (EEG) and magnetoencephalography (MEG) was another area for investigation. Both EEG and MEG studies show atypical latencies in the early peaks that project to the primary auditory cortex located within Heschl's gyrus in the superior temporal cortex and association auditory cortices.[2] Structural imaging studies of adults with ASD reveal increased cortical thickness in Heschl's gyrus.[44] Other studies also found underrating of less emotionally intense stimuli reported by participants of ASD which was associated with the lower signals project to right caudate nucleus compared to TD controls.[45] Although individuals with ASD did not have a basic auditory impairment, they were reported to have difficulty filtering auditory information which interferes with the ability to perceive other information in the presence of competing auditory input. Thus, atypical auditory processing in ASD occurs early in development including altered auditory perception and lack of preferential attention to auditory stimuli, impacts speech processing, social engagement, and language development.

7. Olfactory and gustatory systems

There are relatively few established hypotheses and theories on the abnormalities of olfaction and gustation in ASD, but it is generally agreed that individuals with ASD have hypo- and hyper-responsiveness (i.e., sensitivity to smell) with chemosensation (the senses which receive environmental chemical signals). The findings of olfactory dysfunction in ASD are heterogeneous, partly due to the complex etiology of ASD, the use of different olfactory tests and plenty of confounding factors like intelligence and age.[46]

Alteration in olfactory perception of individuals with ASD was documented as early as in 1990s,[47] in which the affected children were shown to response much negatively to unpleasant smells than TD controls. The affected individuals even rate pleasant odors (cinnamon, pineapple and clove) to be less pleasant than TD ones.[48] Later it was shown in a study with parental rating Short Sensory Profile that the affect individuals may avoid certain smells or odors from foods.[49] The ability for the affected children to identify odors in Sniffing Stick test was also found to be worsen in correctly identifying the odors and to have a higher than normal detection threshold, though children with high functioning autism may not show worse performance.[50,51] Interestingly, the olfactory function in adults with ASD is similar to healthy individuals.[52]

The availability of animal models of autism provides an appropriate tool for studying the neurobiology of the olfactory deficit in ASD. Due to the fact that ASD pathology is heterogeneous, the types of animal models which simulate the disorders are hugely diverse. Since olfaction is a major form of sensation for rodents to demonstrate social and communicating behaviors, the disrupted olfactory system may subsequently affects the findings from social ability test of the animals, which is considered similar to the social dysfunction of human affected by ASD and usually tested in animal models of ASD. If there is a dysfunction in the olfactory system, both olfaction and social functions are likely to be affected.

In utero exposure to substance known to cause autism in human is a typical way to establish an autistic model. One of the models is created by exposing rat fetus to valproic acid (VPA) in utero, which is comparable to exposure to VPA in human prenatal stage.[53] In this model with highly similar etiology to human, the affected offspring showed delayed nest-seeking behavior regulated by the olfactory cues. Injection of polyinosinic-polycytidylic acid (Poly l:C) at gestation day 9.5 is another model, in which Poly l:C acts like a double-stranded viral RNA to mimic maternal infection and activate the immune system.[54] The affected offspring demonstrate reduced neurogenesis (generation of new neurons) in the olfactory bulb. As olfactory bulb is the first anatomical region along the olfactory pathway for integration of olfactory signals, the reduction of neuronal proliferation may affect the ability of odor discrimination. The affected offspring showed compromised olfactory discrimination, this suggests that reduction in olfactory bulb neurogenesis may be a possible mechanism underlying the olfactory dysfunction in ASD.

Transgenic animal is another typical form of animal modeling. Collapsing response mediator protein (CRMP) is a family of proteins suggested to involved in neuronal development, dendritic growth and axonal guidance.[55] The mutation of CRMP was reported to be associated with autism.[56] Transgenic mice without Crmp4 expression (i.e., Crmp4-knockout mice) showed impaired olfactory ability and hyperactivity in the olfactory bulb, and this supports the perspective that the altered olfaction in ASD may be attributed to abnormality of the olfactory bulb. Another line of transgenic mice model of autism, point mutations in neuroligin-3 (NL-3), also showed simultaneous dysfunction in olfaction and lack of social novelty preference.[57] Neuroligins are family of postsynaptic proteins which form a trans-synaptic bridge to facilitate development of synapse; mutation of neuroligins was found to associate with autistic phenotype.[58] Since neuroligins are essential for the formation of both excitatory glutamatergic and inhibitory GABAnergic synapses, this family of proteins virtually affect the neurodevelopment of various brain regions including somatosensory cortex, hippocampus and prefrontal cortex. In particular, NL-3 was found to express in all neurons in the olfactory bulb and in olfactory ensheathing cells,[59,60] which suggest the potential involvement of NL-3 in olfactory dysfunction of ASD. Due to the wide expression of neuroligin in the brain, cortical regions which involve olfactory perception including piriform, entorhinal, thalamocortical and orbitofrontal cortex[61] may possibly be involved in the olfactory dysfunction. The altered cytoarchitecture of the primary olfactory cortex (i.e., piriform cortex) was reported in a postmortem morphological study of ASD patients, in which an increased glial cell density in layer II of the piriform cortex was found.[62] No significant differences in neuronal cell densities, including pyramidal and non-pyramidal cells, were found between ASD and TD subjects. In addition, in layer I of the piriform cortex from ASD subjects, the amount of corpora amylacea (small masses which may plays a role in debris clearance) was significantly higher than TD subjects. The cause and consequences of the increased glial dentistry and corpora amylacea remain to be determined, but it was hypothesized to be caused by neuro-inflammation. Furthermore, the activity of piriform cortex in adults with ASD was found to be decreased by functional imaging.[63] Collectively these suggest the olfactory dysfunction in ASD is associated with morphological and functional alteration, which is supported by evidence from both animal and human studies.

Apart from the perspective of abnormal changes in the olfactory system, the alteration in olfactory function was also suggested to be caused by resistance to change/repetitive behavior in ASD in another animal study, which demonstrated the inability to shift to favorable olfactory cue in BTBR T + *tf*/J mice (an autism animal model showing behavioral profile of low social preference, selective deficit in reversal learning and lacking anxiety-like behavior[64]). While other mice breeds showed flexible adaptation to receive an appetitive outcome in an exploratory hole-board task, the transgenic mice failed to change its preference in hole-picking. In addition to rodent, a drosophila model of Fragile X syndrome, a known leading genetic cause of ASD in human, also demonstrates impaired olfactory behaviors.[65]

However, the disruption in olfaction and the social function may not be ubiquitous in all animal models of ASD: the olfactory function and social preference of novelty were found to be normal in a mice model with oxytocin knockout (National Institute of Mental Health line in Bethesda and the Baylor/Emory line[66]). Shank1 knockout mice, with deletion of SHANK gene which encodes synaptic scaffolding proteins in postsynaptic regions, also showed no alteration in olfactory learning to discriminate social odors.[67] However, optogenetically induced oxytocin expression was shown to enhance social odor differentiation of same sex and olfactory exploration.[68] These findings suggest olfactory dysfunction may not be found in all individuals with ASD, but intervention for ASD may be possible to improve the olfaction.[69]

Though the observation of food selectivity in ASD has been recognized by clinical practitioners for several decades, the first clinical case report on food selectivity in autism was reported relatively recently in 2003.[70] In this single subject design study, a male individual with autism and profound intellectual disabilities showed a consistent refusal to vegetables but not other food types including protein, starch and fruit. The acceptance of vegetable was increased after simultaneous addition of condiment like ketchup and mustard. The acceptance, however, could not be maintained by classical conditioning after the simultaneous presentation. It was suggested that the flavor and texture of vegetable elicited the refusal. Later, a study on the parental report of sensory reactivity in children with autism found that the affected children had significantly more sensory symptoms, in particular abnormality in responses to taste and smell.[49] Short Sensory Profile was used in the study and the abnormality observed including avoidance to certain tastes/textures of food, specific preference of certain food taste, texture and temperatures. The finding was echoed by self-reported food selectivity by adults and adolescents with high functioning autism, in which the participants reported preference on foods which are familiar (i.e., food neophobia) and avoided foods with particular textures or strong flavors.[71] Worse taste identification was shown in adults with autism, with more frequent misidentification of a taste as salty or no taste, which led to an inability to correctly identify sour, bitter and sweet tastes.[72] The worsen identification was further supported by electro-gustometry findings.[50] It is suggested that food selectivity or neophobia in ASD individuals is associated with the magnitude of autistic traits, but unlikely to be modulated by olfactory sensitivity.[73]

Similar to olfaction, it remains elusive on the biological causes of the alteration of olfactory and taste sensations in ASD. One of the proposed mechanisms is the altered arginine-vasopressin (AVP, or anti-diuretic hormone) signaling. AVP is a neuropeptide/hormone which has a range of functions, including reducing volume of urine produced, mediation of social behavior, including pair-bond formation and communication.[74] Since deficits in social behavior is a core symptom of ASD, AVP and its receptors (AVPR1a and AVPR1b) have long been the target genes to be studied in related research, and postulated to be the causes of social and behavior deficits in ASD.[75] Specifically, the lengths of two microsatellite regions (RS1 and RS3) in the promotor regions of AVPR1a are associated with altered expression of mRNA of AVPR1a, and altered social behavior in individuals with ASD.[76] Single nucleotide polymorphism in AVPR1b (rs28373064) was linked to sociability and empathy,[77] while cerebrospinal fluid level of AVP is lower in children with ASD. From clinical studies, it is shown that blocking AVPR1a and increasing available AVP may improve social behaviors in ASD individuals.[78,79]

The involvement of AVP in regulating taste can be shown by an ex vivo study using frog gustatory organs[80]: after perfusion of AVP to the basolateral membrane side of taste cells, the sensitivity of taste cells toward salty and sour was increased. AVP may modulate taste perception via modulating the activity of epithelial sodium ion channels (ENaCs), which are found to selectively permit sodium to get across the cell membrane. With the presence of AVP, the threshold of stimulating the host cells by sodium influx is lowered, which in turns increase the sensitivity to salty or sour tastes.[81] On the other hand, AVP administered in the CNS stimulates salt intake, and the action is likely to be modulated by ENaCs expressed on magnocellular neurons of paraventricular nucleus and suprachiasmatic nucleus.[82] This is coherent with the function of AVP which was discovered early: to retain water from urine production by increasing reabsorption of water from the renal collecting duct through increasing salt content in the blood stream. Combining with the observation that CSF level of AVP is lower in autistic individuals, the sensitivity to salty and sour tastes may be lower than the other tastes, hence prompts the affected individuals to choose salty and sour food. Though the current understanding of AVP on taste may not be able to explain the complex food selectivity variations of ASD individuals, the AVP signaling may still be useful to explain food selectivity from the perspective of altered taste perception.

The preference and avoidance of particular tastes were found to link to the differential expression of taste-related genes. Engrailed-2 (En2) is a transcriptional factor essential for normal neural development, in which transgenic mice lacking En2 was shown to display autistic traits.[83] The knockout mice showed a preference to sodium chloride solution and decreased lick response to fructose solution. Interestingly, the expression of taste receptor genes including Scnn1a, T2R140, T1R3, Trpm5 and Pkd1I3 are altered in the En2 mice, which is likely to induce a different taste perception in the transgenic mice. T2R38, which encodes bitter taste receptor, was also hypothesized to be affected in ASD and causes the hypersensitivity to bitter taste.[84] Genetic association studies showed that polymorphism of EN2 is associated with autism.[85] ASD-associated haplotype of ASD (rs1861972–rs1861973 A–C genotype) was found to be increased in postmortem cerebellar samples of ASD individuals, which further confirms the involvement of EN2 in ASD.[86] It is still elusive in human that whether the SNPs in EN2 is the cause of altered gustation in ASD, and this is a possible research gap for future studies.

Interestingly, some clinical researchers made use of the food selectivity in autism for behavioral modification treatment.[87] In a single case study, a child with autism showed pica (ingestion of objects that are inedible) and food refusal to tapioca pudding. The researchers dipped the inedible objects in the pudding and this significantly reduced the frequency of pica, and the conditioning was shown to maintain for at least a year. When compared to the first case report, pairing of stimuli may be more effective when a naturally aversive stimulus is used.

Though olfactory and gustation are traditionally thought to be less important than vision and hearing in human being, the manifestations of alteration in the chemosensations may be potential biomarkers of autism.[88] As the chemosensory systems may have potential association with the pathophysiology of neuropsychiatric disorders, the sensory dysfunctions may provide clues to explore the underlying mechanisms of the disorder, including dysfunctional immune system and neurogenesis.[89] A study showed that olfactory bulb dysgenesis which results in reduced vasopressin and oxytocin receptor binding could be one of the neural foundations of ASD.[90] Another study also found functional connectivity of the amygdala with the orbitofrontal cortex which has impact on the socio-emotional recognition and behavioral regulation in individuals with ASD.[91] These findings suggest that future research is needed to further examine the relationship between olfaction and social cognition and behavioral regulation of ASD.

8. Neuroplasticity and sensory integration

Remodeling of the human brain in response to experience has been supported by neuroscience research. Experience-dependent learning is the ongoing process of creation and organization of neuronal connections that occur as a result of experiences. Changing the input by providing novel experiences and enriched learning opportunities leads to brain changes which include increased gray matter density, angiogenesis, glial volume and neurogenesis.[92] Other animal model studies also showed that enriched environment with novel sensory, motor and cognitive stimulation can initiate functional changes in the brain.[93] These investigators concluded that the enriched environment principles are aligned with the premise of sensory integration intervention. The hallmark of sensory integration intervention is based on the concept of neuroplasticity, through individually tailed sensorimotor activities in the context of play in creating changes of function and behavior of individuals with ASD.[6,10]

9. Conclusion

Abnormal sensory registration and sensitivity in ASD have significant clinical and social implications. With the advancement of neuroimaging and other innovative technologies, tremendous gains have been made over the past few decades to guide our understanding of the neurobiology of sensory processing. It is hoped that these advancements will provide insight and tailor appropriate and effective intervention in sensory integration and sensory processing to improve participation in people with ASD.

Acknowledgment

This work is supported by the University Grant Council (HK) fund support to Lau BW (Gerneral Research Fund, 15164216).

References

1. American Psychiatric Association. *Diagnostic and Statistical Manual of Mental Disorders*. American Psychiatric Association; 2013. https://doi.org/10.1176/appi.books.9780890425596.
2. Marco E, Hinkley L, Hill S, Nagarajan SS. Sensory processing in autism: a review of neurophysiologic findings. *Pediatr Res*. 2011;69:48–54. Undefined. https://www.nature.com/articles/pr9201193. Accessed February 28, 2020.
3. Thye MD, Bednarz HM, Herringshaw AJ, Sartin EB, Kana RK. The impact of atypical sensory processing on social impairments in autism spectrum disorder. *Dev Cogn Neurosci*. 2018;29:151–167. https://doi.org/10.1016/j.dcn.2017.04.010.
4. Leekam SR, Nieto C, Libby SJ, Wing L, Gould J. Describing the sensory abnormalities of children and adults with autism. *J Autism Dev Disord*. 2007;37(5):894–910. https://doi.org/10.1007/s10803-006-0218-7.
5. Tomchek SD, Dunn W. Sensory processing in children with and without autism: a comparative study using the short sensory profile. *Am J Occup Ther*. 2007;61(2):190–200. https://doi.org/10.5014/ajot.61.2.190.
6. Mailloux Z, Mulligan S, Roley SS, et al. Verification and clarification of patterns of sensory integrative dysfunction. *Am J Occup Ther*. 2011;65(2):143–151. https://doi.org/10.5014/ajot.2011.000752.
7. Crane L, Goddard L, Pring L. Sensory processing in adults with autism spectrum disorders. *Autism*. 2009;13(3):215–228. https://doi.org/10.1177/1362361309103794.
8. Traboulsi EI, Maurhenee IH. Ophthalmologic findings in mucolipidosis III (pseudo-Hurler polydystrophy). *Am J Ophthalmol*. 1986;102(5):592–597. https://doi.org/10.1016/0002-9394(86)90529-5.
9. Ayres AJ. *Sensory Integration and Learning Disorders*. Western Psychological Services; 1972.
10. Kilroy E, Aziz-Zadeh L, Cermak S. Ayres theories of autism and sensory integration revisited: what contemporary neuroscience has to say. *Brain Sci*. 2019;9(3):68. https://doi.org/10.3390/brainsci9030068.
11. Lane SJ, Mailloux Z, Schoen S, et al. Neural foundations of Ayres sensory integration®. *Brain Sci*. 2019;9(7):153. https://doi.org/10.3390/brainsci9070153.
12. Kandel ER, Schwartz JH, Jessell TM. *Principles of Neural Science*. McGraw-Hill, Health Professions Division; 2000. http://www.sidalc.net/cgi-bin/wxis.exe/?IsisScript=FCL.xis&method=post&formato=2&cantidad=1&expresion=mfn=006218. Accessed February 28, 2020.
13. Kemper TL, Bauman M. Neuropathology of infantile autism. *J Neuropathol Exp Neurol*. 1998;57(7):645–652. https://doi.org/10.1097/00005072-199807000-00001.
14. Aylward EH, Minshew NJ, Goldstein G, et al. MRI volumes of amygdala and hippocampus in non-mentally retarded autistic adolescents and adults. *Neurology*. 1999;53(9):2145–2150. https://doi.org/10.1212/wnl.53.9.2145.

15. Pryweller JR, Schauder KB, Anderson AW, et al. White matter correlates of sensory processing in autism spectrum disorders. *Neuroimage Clin.* 2014;6:379–387. https://doi.org/10.1016/j.nicl.2014.09.018.
16. Ayres AJ. Cluster analyses of measures of sensory integration. *Am J Occup Ther.* 1977;31(6):362–366. http://www.ncbi.nlm.nih.gov/pubmed/879252. Accessed February 28, 2020.
17. Molloy CA, Dietrich KN, Bhattacharya A. Postural stability in children with autism spectrum disorder. *J Autism Dev Disord.* 2003;33(6):643–652. https://doi.org/10.1023/b:jadd.0000006001.00667.4c.
18. Ayres AJ, Tickle LS. Hyper-responsivity to touch and vestibular stimuli as a predictor of positive response to sensory integration procedures by autistic children. *Am J Occup Ther.* 1980;34(6):375–381. https://doi.org/10.5014/ajot.34.6.375.
19. Furman JM, Osorio MJ, Minshew NJ. Visual and vestibular induced eye movements in verbal children and adults with autism. *Autism Res.* 2015;8(6):658–667. https://doi.org/10.1002/aur.1481.
20. Carson TB, Wilkes BJ, Patel K, et al. Vestibulo-ocular reflex function in children with high-functioning autism spectrum disorders. *Autism Res.* 2017;10(2):251–266. https://doi.org/10.1002/aur.1642.
21. Section On Complementary And Integrative Medicine, Council on Children with Disabilities, American Academy of Pediatrics, Zimmer M, Desch L. Sensory integration therapies for children with developmental and behavioral disorders. *Pediatrics.* 2012;129(6):1186–1189. https://doi.org/10.1542/peds.2012-0876.
22. Hitier M, Besnard S, Smith PF. Vestibular pathways involved in cognition. *Front Integr Neurosci.* 2014;8:59. https://doi.org/10.3389/fnint.2014.00059.
23. Ackerley R, Kavounoudias A. The role of tactile afference in shaping motor behaviour and implications for prosthetic innovation. *Neuropsychologia.* 2015;79(Pt B):192–205. https://doi.org/10.1016/j.neuropsychologia.2015.06.024.
24. Wolff P, Shepard J. Causation, touch, and the perception of force. *Psychol Learn Motiv.* 2013;58:167–202. https://doi.org/10.1016/B978-0-12-407237-4.00005-0.
25. Gage NM, Juranek J, Filipek PA, et al. Rightward hemispheric asymmetries in auditory language cortex in children with autistic disorder: an MRI investigation. *J Neurodev Disord.* 2009;1(3):205–214. https://doi.org/10.1007/s11689-009-9010-2.
26. Edgar JC, Khan SY, Blaskey L, et al. Neuromagnetic oscillations predict evoked-response latency delays and core language deficits in autism Spectrum disorders. *J Autism Dev Disord.* 2015;45(2):395–405. https://doi.org/10.1007/s10803-013-1904-x.
27. Miyazaki M, Fujii E, Saijo T, et al. Short-latency somatosensory evoked potentials in infantile autism: evidence of hyperactivity in the right primary somatosensory area. *Dev Med Child Neurol.* 2006;49(1):13–17. https://doi.org/10.1017/S0012162207000059.x.
28. Ackerley R, Hassan E, Curran A, Wessberg J, Olausson H, McGlone F. An fMRI study on cortical responses during active self-touch and passive touch from others. *Front Behav Neurosci.* 2012;6:51. https://doi.org/10.3389/fnbeh.2012.00051.
29. Behrmann M, Thomas C, Humphreys K. Seeing it differently: visual processing in autism. *Trends Cogn Sci.* 2006;10(6):258–264. https://doi.org/10.1016/j.tics.2006.05.001.
30. Franklin A, Sowden P, Burley R, Notman L, Alder E. Color perception in children with autism. *J Autism Dev Disord.* 2008;38(10):1837–1847. https://doi.org/10.1007/s10803-008-0574-6.
31. Vandenbroucke MWG, Scholte HS, van Engeland H, Lamme VAF, Kemner C. A neural substrate for atypical low-level visual processing in autism spectrum disorder. *Brain.* 2008;131(4):1013–1024. https://doi.org/10.1093/brain/awm321.
32. Bertone A, Mottron L, Jelenic P, Faubert J. Enhanced and diminished visuo-spatial information processing in autism depends on stimulus complexity. *Brain.* 2005; 128(10):2430–2441. https://doi.org/10.1093/brain/awh561.

33. Koh HC, Milne E, Dobkins K. Contrast sensitivity for motion detection and direction discrimination in adolescents with autism spectrum disorders and their siblings. *Neuropsychologia*. 2010;48(14):4046–4056. https://doi.org/10.1016/j.neuropsychologia.2010.10.008.
34. Spencer J, O'Brien J, Riggs K, Braddick O, Atkinson J, Wattam-Bell J. Motion processing in autism: evidence for a dorsal stream deficiency. *Neuroreport*. 2000;11(12):2765–2767. https://doi.org/10.1097/00001756-200008210-00031.
35. Laycock R, Crewther SG, Crewther DP. A role for the "magnocellular advantage" in visual impairments in neurodevelopmental and psychiatric disorders. *Neurosci Biobehav Rev*. 2007;31(3):363–376. https://doi.org/10.1016/j.neubiorev.2006.10.003.
36. Weinger PM, Zemon V, Soorya L, Gordon J. Low-contrast response deficits and increased neural noise in children with autism spectrum disorder. *Neuropsychologia*. 2014:63. https://doi.org/10.1016/j.neuropsychologia.2014.07.031.
37. Bertone A, Mottron L, Jelenic P, Faubert J. Motion perception in autism: a "complex" issue. *J Cogn Neurosci*. 2003;15(2):218–225. https://doi.org/10.1162/089892903321208150.
38. Hebart MN, Hesselmann G. What visual information is processed in the human dorsal stream? *J Neurosci*. 2012;32(24):8107–8109. https://doi.org/10.1523/JNEUROSCI.1462-12.2012.
39. Pellicano E, Gibson LY. Investigating the functional integrity of the dorsal visual pathway in autism and dyslexia. *Neuropsychologia*. 2008;46(10):2593–2596. https://doi.org/10.1016/j.neuropsychologia.2008.04.008.
40. Ronconi L, Gori S, Federici A, et al. Weak surround suppression of the attentional focus characterizes visual selection in the ventral stream in autism. *Neuroimage Clin*. 2018;18:912–922. https://doi.org/10.1016/j.nicl.2018.02.014.
41. Humphreys K, Hasson U, Avidan G, Minshew N, Behrmann M. Cortical patterns of category-selective activation for faces, places and objects in adults with autism. *Autism Res*. 2008;1(1):52–63. https://doi.org/10.1002/aur.1.
42. Adamson K, Troiani V. Distinct and overlapping fusiform activation to faces and food. *Neuroimage*. 2018;174:393–406. https://doi.org/10.1016/j.neuroimage.2018.02.064.
43. Kwon S, Kim J, Choe BH, Ko C, Park S. Electrophysiologic assessment of central auditory processing by auditory brainstem responses in children with autism spectrum disorders. *J Korean Med Sci*. 2007;22(4):656–659. https://doi.org/10.3346/jkms.2007.22.4.656.
44. Hyde KL, Samson F, Evans AC, Mottron L. Neuroanatomical differences in brain areas implicated in perceptual and other core features of autism revealed by cortical thickness analysis and voxel-based morphometry. *Hum Brain Mapp*. 2009;31(4):556–566. https://doi.org/10.1002/hbm.20887.
45. Gebauer L, Skewes J, Hørlyck L, Vuust P. Atypical perception of affective prosody in autism spectrum disorder. *Neuroimage Clin*. 2014;6:370–378. https://doi.org/10.1016/j.nicl.2014.08.025.
46. Schecklmann M, Schwenck C, Taurines R, et al. A systematic review on olfaction in child and adolescent psychiatric disorders. *J Neural Transm*. 2013;120(1):121–130. https://doi.org/10.1007/s00702-012-0855-2.
47. Soussignan R, Schaal B, Schmit G, Nadel J. Facial responsiveness to odours in normal and pervasively developmentally disordered children. *Chem Senses*. 1995;20(1):47–59. https://doi.org/10.1093/chemse/20.1.47.
48. Hrdlicka M, Vodicka J, Havlovicova M, Urbanek T, Blatny M, Dudova I. Brief report: significant differences in perceived odor pleasantness found in children with ASD. *J Autism Dev Disord*. 2011;41(4):524–527. https://doi.org/10.1007/s10803-010-1084-x.
49. Rogers SJ, Hepburn S, Wehner E. Parent reports of sensory symptoms in toddlers with autism and those with other developmental disorders. *J Autism Dev Disord*. 2003;33(6):631–642. http://www.ncbi.nlm.nih.gov/pubmed/14714932. Accessed February 28, 2020.

50. Bennetto L, Kuschner ES, Hyman SL. Olfaction and taste processing in autism. *Biol Psychiatry*. 2007;62(9):1015–1021. https://doi.org/10.1016/j.biopsych. 2007.04.019.
51. Dudova I, Vodicka J, Havlovicova M, Sedlacek Z, Urbanek T, Hrdlicka M. Odor detection threshold, but not odor identification, is impaired in children with autism. *Eur Child Adolesc Psychiatry*. 2011;20(7):333–340. https://doi.org/10.1007/s00787-011-0177-1.
52. Tavassoli T, Baron-Cohen S. Olfactory detection thresholds and adaptation in adults with autism spectrum condition. *J Autism Dev Disord*. 2012;42(6):905–909. https://doi.org/10.1007/s10803-011-1321-y.
53. Schneider T, Przewłocki R. Behavioral alterations in rats prenatally exposed to valproic acid: animal model of autism. *Neuropsychopharmacology*. 2005;30(1):80–89. https://doi.org/10.1038/sj.npp.1300518.
54. Liu Y-H, Lai W-S, Tsay H-J, Wang T-W, Yu J-Y. Effects of maternal immune activation on adult neurogenesis in the subventricular zone-olfactory bulb pathway and olfactory discrimination. *Schizophr Res*. 2013;151(1–3):1–11. https://doi.org/10.1016/j.schres.2013.09.007.
55. Tsutiya A, Nishihara M, Goshima Y, Ohtani-Kaneko R. Mouse pups lacking collapsin response mediator protein 4 manifest impaired olfactory function and hyperactivity in the olfactory bulb. *Eur J Neurosci*. 2015;42(6):2335–2345. https://doi.org/10.1111/ejn.12999.
56. Ohtani-Kaneko R. Crmp4-KO mice as an animal model for investigating certain phenotypes of autism spectrum disorders. *Int J Mol Sci*. 2019;20(10):2485. https://doi.org/10.3390/ijms20102485.
57. Radyushkin K, Hammerschmidt K, Boretius S, et al. Neuroligin-3-deficient mice: model of a monogenic heritable form of autism with an olfactory deficit. *Genes Brain Behav*. 2009;8(4):416–425. https://doi.org/10.1111/j.1601-183X.2009.00487.x.
58. Biswas S, Reinhard J, Oakeshott J, Russell R, Srinivasan MV, Claudianos C. Sensory regulation of neuroligins and neurexin I in the honeybee brain. Giurfa M, ed. *PLoS One*. 2010;5(2):e9133. https://doi.org/10.1371/journal.pone.0009133.
59. Varoqueaux F, Aramuni G, Rawson RL, et al. Neuroligins determine synapse maturation and function. *Neuron*. 2006;51(6):741–754. https://doi.org/10.1016/j.neuron.2006.09.003.
60. Gilbert M, Smith J, Roskams AJ, Auld VJ. Neuroligin 3 is a vertebrate gliotactin expressed in the olfactory ensheathing glia, a growth-promoting class of macroglia. *Glia*. 2001;34(3):151–164. https://doi.org/10.1002/glia.1050.
61. Wilson DA, Xu W, Sadrian B, Courtiol E, Cohen Y, Barnes DC. Cortical odor processing in health and disease. *Prog Brain Res*. 2014;208:275–305. https://doi.org/10.1016/B978-0-444-63350-7.00011-5.
62. Menassa DA, Sloan C, Chance SA. Primary olfactory cortex in autism and epilepsy: increased glial cells in autism. *Brain Pathol*. 2017;27(4):437–448. https://doi.org/10.1111/bpa.12415.
63. Koehler L, Fournel A, Albertowski K, et al. Impaired odor perception in autism spectrum disorder is associated with decreased activity in olfactory cortex. *Chem Senses*. 2018;43(8):627–634. https://doi.org/10.1093/chemse/bjy051.
64. Moy SS, Nadler JJ, Poe MD, et al. Development of a mouse test for repetitive, restricted behaviors: relevance to autism. *Behav Brain Res*. 2008;188(1):178–194. https://doi.org/10.1016/j.bbr.2007.10.029.
65. Franco LM, Okray Z, Linneweber GA, Hassan BA, Yaksi E. Reduced lateral inhibition impairs olfactory computations and behaviors in a drosophila model of fragile X syndrome. *Curr Biol*. 2017;27(8):1111–1123. https://doi.org/10.1016/j.cub.2017.02.065.
66. Crawley JN, Chen T, Puri A, et al. Social approach behaviors in oxytocin knockout mice: comparison of two independent lines tested in different laboratory environments. *Neuropeptides*. 2007;41(3):145–163. https://doi.org/10.1016/j.npep.2007.02.002.

67. Silverman JL, Turner SM, Barkan CL, et al. Sociability and motor functions in Shank 1 mutant mice. *Brain Res.* 2011;1380:120–137. https://doi.org/10.1016/j.brainres.2010.09.026.
68. Oettl L-L, Ravi N, Schneider M, et al. Oxytocin enhances social recognition by modulating cortical control of early olfactory processing. *Neuron.* 2016;90(3):609–621. https://doi.org/10.1016/j.neuron.2016.03.033.
69. Du L, Zhao G, Duan Z, Li F. Behavioral improvements in a valproic acid rat model of autism following vitamin D supplementation. *Psychiatry Res.* 2017;253:28–32. https://doi.org/10.1016/j.psychres.2017.03.003.
70. Ahearn WH. Using simultaneous presentation to increase vegetable consumption in a mildly selective child with autism. *J Appl Behav Anal.* 2003;36(3):361–365. https://doi.org/10.1901/jaba.2003.36-361.
71. Kuschner ES, Eisenberg IW, Orionzi B, et al. A preliminary study of self-reported food selectivity in adolescents and young adults with autism spectrum disorder. *Res Autism Spectr Disord.* 2015;15–16:53–59. https://doi.org/10.1016/j.rasd.2015.04.005.
72. Tavassoli T, Baron-Cohen S. Taste identification in adults with autism spectrum conditions. *J Autism Dev Disord.* 2012;42(7):1419–1424. https://doi.org/10.1007/s10803-011-1377-8.
73. Stafford LD, Tsang I, López B, Severini M, Iacomini S. Autistic traits associated with food neophobia but not olfactory sensitivity. *Appetite.* 2017;116:584–588. https://doi.org/10.1016/j.appet.2017.05.054.
74. Heinrichs M, von Dawans B, Domes G. Oxytocin, vasopressin, and human social behavior. *Front Neuroendocrinol.* 2009;30(4):548–557. https://doi.org/10.1016/j.yfrne.2009.05.005.
75. Cataldo I, Azhari A, Esposito G. A review of oxytocin and arginine-vasopressin receptors and their modulation of autism spectrum disorder. *Front Mol Neurosci.* 2018;11:27. https://doi.org/10.3389/fnmol.2018.00027.
76. Israel S, Lerer E, Shalev I, et al. Molecular genetic studies of the arginine vasopressin 1a receptor (AVPR1a) and the oxytocin receptor (OXTR) in human behaviour: from autism to altruism with some notes in between. *Prog Brain Res.* 2008;170:435–449. https://doi.org/10.1016/S0079-6123(08)00434-2.
77. Wu N, Shang S, Su Y. The arginine vasopressin V1b receptor gene and prosociality: mediation role of emotional empathy. *Psych J.* 2015;4(3):160–165. https://doi.org/10.1002/pchj.102.
78. Umbricht D, Del Valle RM, Hollander E, et al. A single dose, randomized, controlled proof-of-mechanism study of a novel vasopressin 1a receptor antagonist (RG7713) in high-functioning adults with autism spectrum disorder. *Neuropsychopharmacology.* 2017;42(9):1914–1923. https://doi.org/10.1038/npp.2016.232.
79. Parker KJ, Oztan O, Libove RA, et al. A randomized placebo-controlled pilot trial shows that intranasal vasopressin improves social deficits in children with autism. *Sci Transl Med.* 2019;11(491):eaau7356. https://doi.org/10.1126/scitranslmed.aau7356.
80. Okada Y, Miyamoto T, Sato T. Vasopressin increases frog gustatory neural responses elicited by NaCl and HCl. *Comp Biochem Physiol A Comp Physiol.* 1991;100(3):693–696. https://doi.org/10.1016/0300-9629(91)90391-o.
81. Gilbertson TA, Roper SD, Kinnamon SC. Proton currents through amiloride-sensitive Na+ channels in isolated hamster taste cells: enhancement by vasopressin and cAMP. *Neuron.* 1993;10(5):931–942. https://doi.org/10.1016/0896-6273(93)90208-9.
82. Bennett JB. Repairing cuts and lacerations of the hand. *Postgrad Med.* 1988;83(8):157–161. https://doi.org/10.1080/00325481.1988.11700312.
83. Gupta A, Li X, Di Cicco-Bloom E, Bello NT. Altered salt taste response and increased tongue epithelium Scnna1 expression in adult Engrailed-2 null mice. *Physiol Behav.* 2018;194:410–419. https://doi.org/10.1016/j.physbeh.2018.06.030.

84. Riccio MP, Franco C, Negri R, et al. Is food refusal in autistic children related to TAS2R38 genotype? *Autism Res*. 2018;11(3):531–538. https://doi.org/10.1002/aur.1912.
85. Sen B, Singh AS, Sinha S, et al. Family-based studies indicate association of Engrailed 2 gene with autism in an Indian population. *Genes Brain Behav*. 2010;9(2):248–255. https://doi.org/10.1111/j.1601-183X.2009.00556.x.
86. Choi J, Ababon MR, Soliman M, et al. Autism associated gene, engrailed2, and flanking gene levels are altered in post-mortem cerebellum. *PLoS One*. 2014;9(2):e87208. https://doi.org/10.1371/journal.pone.0087208.
87. Ferreri SJ, Tamm L, Wier KG. Using food aversion to decrease severe pica by a child with autism. *Behav Modif*. 2006;30(4):456–471. https://doi.org/10.1177/0145445504272970.
88. Small DM, Pelphrey KA. Autism spectrum disorder: sniffing out a new biomarker. *Curr Biol*. 2015;25(15):R674–R676. https://doi.org/10.1016/j.cub.2015.06.050.
89. Moscavitch S-D, Szyper-Kravitz M, Shoenfeld Y. Autoimmune pathology accounts for common manifestations in a wide range of neuro-psychiatric disorders: the olfactory and immune system interrelationship. *Clin Immunol*. 2009;130(3):235–243. https://doi.org/10.1016/j.clim.2008.10.010.
90. Brang D, Ramachandran VSS. Olfactory bulb dysgenesis, mirror neuron system dysfunction, and autonomic dysregulation as the neural basis for autism. 74(5):919–921. https://doi.org/10.1016/j.mehy.2008.11.048.
91. Bachevalier J, Loveland KA. The orbitofrontal-amygdala circuit and self-regulation of social-emotional behavior in autism. *Neurosci Biobehav Rev*. 2006;30(1):97–117. https://doi.org/10.1016/j.neubiorev.2005.07.002.
92. Thomas AG, Dennis A, Bandettini PA, Johansen-Berg H. The effects of aerobic activity on brain structure. *Front Psychol*. 2012;3:86. https://doi.org/10.3389/fpsyg.2012.00086.
93. Reynolds S, Lane SJ, Richards L. Using animal models of enriched environments to inform research on sensory integration intervention for the rehabilitation of neurodevelopmental disorders. *J Neurodev Disord*. 2010;2(3):120–132. https://doi.org/10.1007/s11689-010-9053-4.

CHAPTER SEVEN

The role of the endocannabinoid system in autism spectrum disorders: Evidence from mouse studies

Susanna Pietropaolo[a,b,*], Luigi Bellocchio[b,c], Inés Bouzón-Arnáiz[d,e], Benjamin K. Yee[f]

[a]University of Bordeaux, Bordeaux Cedex, France
[b]CNRS, INCIA, UMR 5287, Bat B2, Pessac Cedex, France
[c]INSERM, U1215 NeuroCentre Magendie, Bordeaux Cedex, France
[d]Institute for Bioengineering of Catalonia (IBEC), The Barcelona Institute of Science and Technology, Barcelona, Spain
[e]Barcelona Institute for Global Health (ISGlobal, Hospital Clínic—Universitat de Barcelona), Barcelona, Spain
[f]Department of Rehabilitation Sciences, Faculty of Health & Social Sciences, The Hong Kong Polytechnic University, Hong Kong, China
*Corresponding author: e-mail address: susanna.pietropaolo@u-bordeaux.fr

Contents

Abstract

A substantive volume of research on autism spectrum disorder (ASD) has emerged in recent years adding to our understanding of the etiopathological process. Preclinical models in mice and rats have been highly instrumental in modeling and dissecting the contributions of a multitude of known genetic and environmental risk factors. However, the translation of preclinical data into suitable drug targets must overcome three critical hurdles: (i) ASD comprises a highly heterogeneous group of conditions that can markedly differ in terms of their clinical presentation and symptoms, (ii) the plethora of genetic and environmental risk factors suggests a complex, non-unitary, etio-pathology, and (iii) the lack of consensus over the myriad of preclinical models, with respect to both construct validity and face validity. Against this backdrop, this

Progress in Molecular Biology and Translational Science, Volume 173
ISSN 1877-1173
https://doi.org/10.1016/bs.pmbts.2020.04.016

Chapter traces how the endocannabinoid system (ECS) has emerged as a promising target for intervention with predictive validity. Recent supportive preclinical evidence is summarized, especially studies in mice demonstrating the emergence of ASD-like behaviors following diverse genetic or pharmacological manipulations targeting the ECS. The critical relevance of ECS to the complex pathogenesis of ASD is underscored by its multiple roles in modulating neuronal functions and shaping brain development. Finally, we argue that important lessons have been learned from the novel mouse models of ASD, which not only stimulate game-changing innovative treatments but also foster a consensual framework to integrate the diverse approaches applied in the search of novel treatments for ASD.

1. Animal models of autism spectrum disorder (ASD)

Autism spectrum disorder (ASD) is probably one of the central nervous system pathological conditions whose diagnostic criteria have undergone the greatest number of revisions in recent decades. Autism—as the disorder was originally named—was first proposed as a nosological entity by Leo Kanner.[1] His initial criteria and definitions were refined by Rutter,[2] and subsequently underwent several reformulations through successive editions of the Diagnostic and Statistical Manual of Mental Disorders (through DSM-III to the latest DSM-5). According to the original criteria, a child may be diagnosed with autism when the characteristic symptoms, namely, pervasive social impairment and deficits in language and/or communication, emerged before 30 months of age, in the absence of the delusions and hallucinations typically found in schizophrenic patients.[3] Subsequent criteria in DSM-III (1980) have conceptualized autism as a pervasive developmental disorder, and recognized that autistic symptomatology is not confined to early childhood by dropping the requirement for onset of symptoms before 30 months.[4] Nowadays, the diagnostic criteria of ASD are defined by symptoms grouped under three distinct domains: social dysfunction, qualitative deficits in verbal/nonverbal communication and a restricted range of interests associated with the occurrence of repetitive behaviors. These triad of symptoms have been central and maintained throughout the most recent editions of DSM (from DSM-III to DSM-5), although the boundary between social dysfunction and communication deficits may be blurred in practice and may be merged under a more general domain of social/communication deficits.

The introduction of the concept of *spectrum disorder* has fundamentally shifted our understanding of autism from a collection of phenomenologically related developmental disorders to a set of conditions with wide-ranging degrees of impairments thought to stem from the same underlying mechanism. Hence, ASD is now considered as a spectrum of conditions comprising diverse psychopathological patterns that were previously classified as separate entities, namely, autism, Asperger's syndrome, childhood disintegrative disorder and pervasive developmental disorder not otherwise specified.[5] The concept of "spectrum" accommodates the wide range of symptoms and the varying degrees of severity between ASD patients, including the expression of both the core triad of symptoms and additional symptoms.[6] Language deficits and delays, for instance, may be completely absent in Asperger's syndrome, as is mental retardation. Apparent divergent pathological expression may also be incorporated such that ASD patients may either underreact or overreact to sensory stimuli. Similarly, some patients may show some degrees of intellectual impairment while some can display very unevenly developed cognitive skills, including specific savant behavior.

Hence, the heterogeneity of ASD is now understood as intrinsic to the disorder. The complex and heterogeneous manifestations of ASD as well as its definition on social/communication behavior are obvious challenge to attempts in designing valid and reliable animal models of ASD. The task is already daunting given the precise etiology of the disorder remains unknown, and the range of risk markers includes complex, multiple genetic and environmental factors. The relative contributions and underlying mechanisms of specific factors still remain undefined, let alone their interaction from a developmental perspective. The absence of diagnostic biomarkers of autistic pathology implies that the construction and evaluation of animal models essentially have to focus on the behavioral levels—namely, face validity. However, assessing the face validity of animal models of ASD is highly challenging, because some of the core behavioral symptoms are not easily translated to rodents, especially to mice. Hence, a large body of recent research has concentrated on developing more appropriate and reliable methods to assess the core ASD-like behaviors in mice. Faced with the difficulty of developing animal models based on any hypothesized unitary explanation of ASD encompassing across constellations of symptoms with varying severity, one favored alternative approach is to model individual symptoms.[7] This approach essentially abandons a top-down approach that builds on the conception of a spectrum disorder with an assumed common singular etiological process. Instead, it remains open to the possibility that the current ASD classification is an arbitrary collection of distinct disorders as their heterogeneous clinical presentations would suggest.

To evaluate social interest in juvenile and adult mice, various paradigms have been developed based on some methods to quantify social preference. These include the automated three compartment test,[8] and various forms of direct interaction test for evaluating juvenile play, affiliative behaviors or aggression. On the other hand, communication is typically studied in laboratory mice through the emittance of ultrasound vocalizations (USVs) in diverse social contexts across distinct development stages.[9,10] These include infant-mother communication in early post-natal days,[11,12] and social communication between adults that may be recorded during social encounters between male and female mice, or between female mice.[13,14] Attempts to measure the third core symptom—repetitive behaviors and insistence on the sameness—in mice are often made indirectly by assessing behavioral flexibility or perseveration,[15] such as reversal learning in mazes or operant chambers, or the quantification of home-cage or novelty-induced behavioral stereotypies, such as excessive self-grooming. Additional ASD-relevant symptoms such as anxiety, sensory abnormalities, hyperactivity, and cognitive abnormalities can be readily evaluated in mice using well-established and standardized behavioral protocols.

There is no agreement yet as to what constitutes a valid animal model of ASD, nor the set of necessary criteria that must be met. It remains a matter of intense debate.[16] The most intuitive position may initially emphasize that an ideal model should mimic faithfully *all* core autistic features as well as some of the notable secondary symptoms. The most widely accepted criteria for a valid model would require the presence of the core triad of symptoms (poor social interaction, deficits in communication, and inflexible or repetitive behaviors).[10,16,17] The presence of additional behavioral abnormalities, such as hyper-reactivity, memory deficits and anxiety, may boost the general validity of the model, but they are usually not considered as essential.[18] However, a more pragmatic position may relax the criteria of a *useful* model as one that should capture behavioral features resembling at least one of the autistic core symptoms. The pragmatic value of such models of ASD has also been emphasized in other disease models.[19]

Over the last decades, we have seen the emergence of an impressive variety of mouse models based on very different strategies [for several recent reviews, see Refs. 18, 20–22]. One strategy is to identify specific mouse strains with inherent phenotypes resembling some major ASD-like symptoms, such as poor sociability, communicative impairment, and behavioral inflexibility. The best-known example is the inbred BTBR T^+ *tf*/J mouse strain.[23,24] It is widely employed in ASD research, and it has been proposed as a tool to identify novel genes involved in the control of social behaviors. This specific model, or approach in general, is however not without difficulties. First, the choice of control strain is problematic. Is C57BL/6 strain the most appropriate control simply because it is currently the most common strain used in behavioral research? Second, there is the danger that the observed ASD-like behavior may stem from strain-specific traits of emotionality as such, that may not constitute an intrinsic pathological state.[25]

Other mouse models rely on exposure to adverse environmental factors that are often performed prenatally by subjecting the pregnant dams to defined stressors or agents suspected to trigger etiological processes underlying ASD. These are often based on risk factors suggested by epidemiological data such as teratogens, immune activation or viral infections.[26,27] The epidemiological link of such models readily lends them some degrees of "construct validity"—even if the causative mechanism is far from clear at first. The valproic acid model is a well-known preparation in this category. It was initially developed in rats[28,29] and has since been extended to mice.[30,31] Nonetheless, the focus on environmental factors alone may inadvertently overlook the contributions of genetic (as opposed to environmental) factors.

By contrast, genomic manipulations designed to duplicate known genetic alterations associated with ASD, including highly specific genetic mutations, deletion, or chromosomal abnormalities, are ideally suited to study the genetics of ASD. The extent to which one may learn about the etiological mechanism of ASD from such models, however, is limited by the known heterogeneity of the genetic alterations in ASD that include both heritable and de novo genetic variations. These essentially encompass almost all susceptibility genes implicated in the regulation of neuronal function and development.[32] Some of the more notable examples include molecular mechanism implicated in chromatin remodeling and regulation of transcription (e.g., the *MECP2* and *FMR1* genes), actin cytoskeleton dynamics (the *TSC1*, *TSC2* and *NF1* genes), synaptic scaffolding proteins (e.g., the protein encoded by the *SHANK3* gene), receptors and transporters (e.g., the *GRIN2A*, *GRIK2*, *GABAR*, *SLC6A4*, *SLC25A13*, *OXTR*, and *AVPR1 genes*), second-messenger systems (the *PRKCB1*,

CACNA1C, and *NBEA* genes), post-synaptic cell-adhesion such as neuroligin (encoded by the genes *NLGN3* and *NLGN4*), and secreted proteins (e.g., those encoded by genes *RELN* and *LAMB1*) (for a review see Ref. 33).

An exhaustive account of all implicated genes is supposed to be manyfold greater than this short list[34] and it is beyond the scope of this Chapter. It is however important to emphasize that the identified roles of these genes in normal brain development have provided the basis of the so-called "many genes, common pathways" hypothesis of ASD etiology.[35] It emphasizes the contribution of defective synaptic functioning and abnormal brain connectivity in ASD as key pathological processes leading ultimately to altered information processing that underlies the emergence of multiple ASD symptoms. While this may suggest that a holistic approach is necessary to fully appreciate the full contribution of concomitant genetic mutations in multiple loci to the development of ASD, it also suggests a somewhat paradoxical conclusion. Often mutation or deletion of any one gene taken from this growing list is often sufficient to trigger the whole range of autistic phenotypes through disruption of a handful of common mechanistic pathways. If so, mutant mouse model based on gene mutations may indeed be instrumental in isolating such critical pathways and unraveling the underlying mechanism.[36]

Finally, one seemingly circuitous strategy is to utilize other developmental pathologies featuring ASD-like traits with a known and clearly defined single genetic cause that can be faithfully reproduced in the mouse. One such example is the *Fmr1*-KO mouse line[37]—a monogenic model of Fragile X syndrome, a X-linked disorder attributable to the lack of the protein encoded by the *FMR1* gene, known as FMRP.[38] The *Fmr1*-KO mouse is the most common monogenic cause of ASD. *Fmr1*-KO mice showed ASD-like deficits, namely, attenuated social interaction, poor social memory, deficits in reversal learning and altered USVs (for a review, see Ref. 39). Criticisms of this approach include concerns over its generalizability to ASD patients who do not carry the *FMR1* mutation, such that its relevance may be limited to a subset of patients.[40] Nonetheless, as mentioned above, there is increasing evidence suggesting that murine models of single gene disruption are valuable tools for understanding the more general etiopathological mechanism of ASD.[9,10] Our laboratory has been developing this line of investigation in a major part of our recent research.[25]

The research field of ASD drug discovery transcends conventional boundaries between disciplines as defined by methodology, levels of explanations, which is necessary to dissect a disorder as complex as ASD. Admittedly, all of the mouse models of ASD developed by the diverse approaches outlined above have been utilized for the discovery of new therapeutic targets for ASD or at least a subset of relevant conditions or symptoms. In what follows, we will summarize the experimental evidence in support of the roles of the endocannabinoid system (ECS) in the etiopathology of ASD. In the process, we shall make extensive use of the diverse strategies behind mouse models of ASD outlined above. To this end, we will first review studies showing abnormalities in the functionality of ECS in specific mouse models of ASD, including models based on strain differences (the BTBR T^{+} *tf*/J mouse), prenatal exposure to environmental agents (e.g., the valproic acid model), candidate genes (e.g., the *Neuroligin-3*(R451C) mutation), or monogenic developmental disorder with ASD characteristics (e.g., the *Fmr1*-KO mouse). We will then summarize the evidence for the reversal of ASD-like behaviors in such mouse models following pharmacological modulation of the ECS, which substantial our claim that ECS is a promising new drug target for ASD. Last, we will evaluate the possibility of inducing ASD-like behaviors in mice through pharmacological or genetic manipulations of the ECS, and thus their potential as novel etiopathological models for ASD.

2. Understanding the etiopathology of ASD: The role of the endocannabinoid system

Understanding the etiology of ASD is an extremely challenging task, as this disorder is attributable to the combination of multiple genetic and environmental factors, acting through a myriad of complex disease mechanisms, including imbalances between neuronal excitation and inhibition as well as hypo- and or hyper-connectivity.[41] As already mentioned, the genetics of ASD are highly heterogeneous, with hundreds of ASD susceptibility genes.[32] The complexity seems to defy any simple definition of pathogenic mechanisms that can fully account for the interaction and permutative effects of the polygenic mutations, as well as the contribution of environmental influence.[42] Indeed, investigation into the combination of genetic risk factors with environmental adversity has generated increasing interest in ASD research,[7] and it seems to be a key to understanding the marked heterogeneity in clinical presentation among ASD patients.[42,43] Recent studies have even suggested that all this multitude of causal factors may functionally converge on to a small subset of functional pathways that are more amenable to drug discovery and theorizing of ASD.[44] The functional convergence onto particular signaling pathways and the synaptopathology shared across related but genetically distinct ASD conditions raised hope that a common therapeutic strategy may be applicable to a wide spectrum of conditions in spite of divergent causal processes.[44] This reductionistic framework is in keeping with the "many genes, common pathways" hypothesis.[35] As a general theory, it posits that the multiple causal factors converge to altered synaptic functioning and ultimately defective information processing underlies the symptoms production in ASD. The convergence is taken as evidence for a common path to toward etiopathogenesis.

Successfully dissecting the etiopathological process of ASD can lead directly to the identification of new therapeutic targets. Available pharmacological treatments of ASD are only limited for relatively non-core symptoms such as anxiety, irritability, aggressiveness and epilepsy. They are typically controlled by anxiolytics, antipsychotics and anticonvulsants, respective. Such symptom-oriented treatments did not address the cause of ASD, and they are associated with heavy side effects that are especially undesirable for children. To address the urgent need for better medications, the impetus in animal research therefore is to identify new molecular targets for powerful interventions through the dissection of common ASD etiopathological mechanisms.

A case in point is the identification of dysregulated activity-dependent protein synthesis in the pathogenesis of ASD. It represents a potential common causal pathway or convergent effect of most ASD susceptibility genes, which invariably impair synaptic activity. This predicts that compounds designed to ameliorate this convergent dysregulation via complementary molecular pathways may be efficacious, e.g., inhibition of mTOR pathway by rapamycin and selective inhibition of mGluR1/mGluR5 receptors to boost synaptic activity, e.g.[45,46]

In our search for common causal pathways for ASD, the ECS has demonstrated etiological relevance and potential as a novel drug target. The ECS encompasses a range of regulatory systems that utilize endocannabinoids—endogenous lipid-based retrograde messengers that bind to cannabinoid receptors. Cannabinoid receptors are G-protein-coupled receptors comprising two subtypes: CB1 receptors (CBR1) and CB2 receptors (CBR2), encoded by genes *CNR1* and *CNR2*, respectively. They are ubiquitously found in the central and peripheral nervous systems of vertebrates.

First, CB1Rs are abundantly expressed in the brain.[47] They are particularly enriched in the neocortex, hippocampus, amygdala, and cerebellum—all closely involved in the development of ASD. Second, the ECS has emerged as an important modulator of neuronal functions. Endocannabinoid signaling regulates synaptogenesis and neuronal interconnectivity during development.[48] Disruption of these processes underlies the ASD pathology.[41] As illustrated in Fig. 1, CB1 receptors are located pre-synaptically in both GABAergic and glutamatergic neurons,[49] where they are activated by endogenous ligands known as endocannabinoids, consisting of anandamide (AEA) and 2-arachidonoyl-glycerol (2-AG). These endogenous cannabinoids are produced at post-synaptic site following membrane depolarization or activation of metabotropic receptors.[49] Calcium elevation generated by these two types of events is able to trigger plasma membrane lipid remodeling which generates as final products AEA, by the biosynthetic enzyme N-acyl phosphatidylethanolamine-specific phospholipase D (NAPE-PLD) and 2-AG, by diacylglycerol lipase alpha (DAGL-a).[49] The retrograde binding of AEA and 2-AG to pre-synaptic CB1 receptors can trigger several

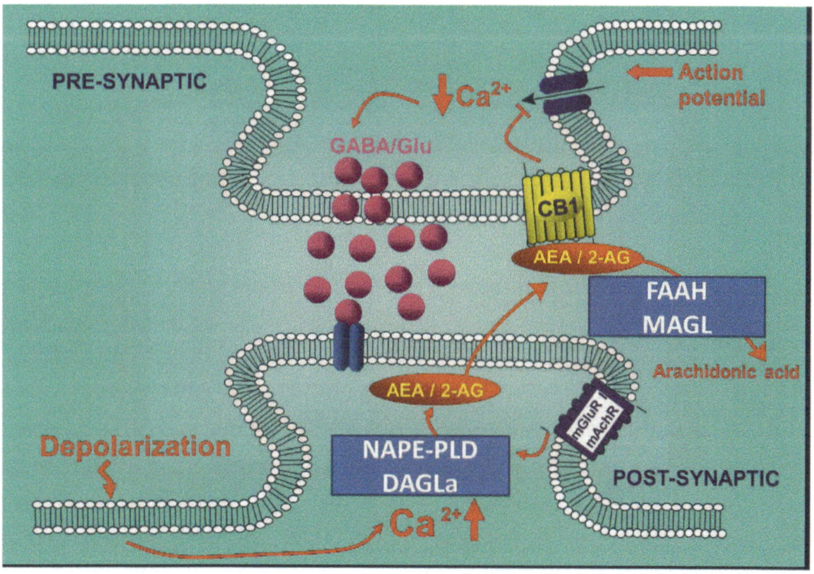

Fig. 1 Schematic representation of ECS functionality. AEA: anandamide, 2-AG: 2-arachidonoyl-glycerol. NAPE-PLD: *N*-acyl phosphatidylethanolamine-specific phospholipase D, DAGL-a: Diacylglycerol lipase alpha. FAAH: fatty acid amide hydrolase, MAGL: monoacyl-glycerol lipase.

intracellular pathways to reduce intracellular calcium and inhibit neurotransmitter release.[49] Endocannabinoids are terminated by reuptake mechanism. AEA is degraded by fatty acid amide hydrolase (FAAH), and 2-AG by monoacyl-glycerol lipase (MAGL).

The ECS differs from classical neurotransmitter systems as there is no constitutive tone. It is characterized by on-demand production of endocannabinoids, CB1 receptor activation, followed by the immediate degradation of the endocannabinoids.[49] Several pharmacological tools are available for studying the ECS. It can be activated by CB1R agonist, e.g., cannabinoid Δ(9)-tetrahydrocannabinol (THC) or indirectly by pharmacological inhibition of AEA and 2-AG reuptake and degradation. The latter is useful for activity-dependent potentiation of endocannabinoid signals and would not have an effect in the absence of endocannabinoids production.[50] The ECS can be blocked by blockade of CB1 receptor (e.g., rimonabant or SR141716A) or inhibition of endocannabinoids biosynthesis.[49] In addition, transgenic mouse models with constitutive and cell type specific deletion of the gene encoding CB1R[51,52] as well as the genes of associated metabolic proteins (NAPE-PLD, DAGL-a, FAAH and MAGL)[53] are available. With these genetical modified engineered mice, a plethora of ECS-dependent regulation of brain and peripheral functions have been identified.[54]

The ECS is therefore a neuromodulatory network of lipid signaling pathways, and includes the endocannabinoids, together with their receptors and the associated metabolic enzymes.[55] Although endocannabinoids exert their central effects mainly through the CB1R, they also interact with CB2R which is involved in the regulation of inflammatory processes and neuromodulation of immune system.[56,57] In additional, they can also interact with targets in addition to CB1R and CB2R, including the transient receptor potential vanilloid type 1 channel, peroxisome proliferator-activated receptor-alpha, peroxisome proliferator-activated receptor-gamma and the G-protein-coupled receptor GPR55.[58] The etiopathological role of alterations in ECS functionality has been already suggested for several psychiatric[59] and neurological disorders,[60] e.g., schizophrenia, anxiety, depression, neurodegenerative diseases, epilepsy.[54] This is not surprising, considering the major impact of ECS in the modulation of neuronal and synaptic function as well as brain development.

A range of pathological behaviors implicated in ASD (that are also found in other mental disorders, such as schizophrenia and anxiety) are modulated by the ECS. The impairments span across cognitive functions, emotionality, circadian rhythms, epileptic seizures and social interaction.[52,61–63] A direct link between ECS function and ASD has been obtained with human neuroimaging. It has been demonstrated that variations of the cannabinoid receptor gene are involved in the modulation of social reward responsivity, specifically in the striatal response to happy faces,[64] a response specifically altered in ASD patients. Post-mortem analyses have revealed a reduction of CB1R expression in several areas of ASD brains.[65] Plasma AEA, but not 2-AG, was found to be significantly lower in ASD children relative to controls.[66.67]

It is still not clear whether the reduced AEA levels could be due to underactivity of NAPE-PLD (the synthetic enzyme of AEA) or increased activity of FAAH (the catabolic enzyme of AEA). Furthermore, rare genetic variants to the core protein of the ECS (genes *CNR1*, *CNR2*, *DAGLA*, *MGLL* and *FAAH*) have been identified in molecular testing data from a large sample of patients with a broad spectrum of neurological disorders. The heterozygous rare variants in the *DAGLA* gene were significantly associated with seizures and ASD.[68] In addition, higher expression of CB2R in blood monocytes has been reported in ASD patients,[69] whereby ECS mechanism may lead to immune pathology in ASD, because CB2R is known to modulate the development, migration, proliferation, and effector functions of immune cells.[70] It is conceivable that CB2R-dependent mechanism is implicated in immune dysfunction reported in ASD patients.[71,72] Encouraged by the large body of evidence from ASD patients, attempts have been made to dissect the ECS-dependent etiopathology of ASD with animal models.

3. Alterations of ECS in animal models of ASD: Toward novel therapeutic approaches

Alterations in ECS functionality have been described in several animal models of ASD[73] based on various approaches described in Section 1. Here, we will review studies investigating ECS in ASD rodent models, from approaches based on monogenic diseases (e.g., the *Fmr1*-KO mouse), strain differences (e.g., the BTBR T^+ *tf*/J mouse strain), other candidate genes (e.g., the transgenic mice) and environmental toxins exposure (e.g., the valproic acid model).

The ASD model that has been the most extensively studied in relation to the ECS is the *Fmr1*-KO mouse model of fragile-X syndrome—a monogenic developmental disorder associated with ASD.[39] As fragile-X syndrome also has no therapeutics, research effort has also been made to examine the ECS for potential treatments. Alterations and imbalance in EC signaling have been identified in *Fmr1*-KO mice, suggesting that their normalization via 2-AG, AEA, cannabinoid CB1 and CB2 receptors may yield therapeutic potential. First, the coupling between hippocampal mGluR activation and eCB mobilization was enhanced in *Fmr1*-KO mice, without alterations to the expression of CB1R.[74] Second, enhanced striatal diacylglycerol lipase activity was also reported.[75] And, enhancing 2-AG signaling by compound JZL184 (which inhibits the degradation of 2-AG by MAGL) normalized some of the behavioral abnormalities in *Fmr1*-KO mice, including hyperactivity and anxiety.[76] Altered ECS function in *Fmr1*-KO mice has also been revealed by acute pharmacological challenges with either the CB1R antagonist rimonabant, or agonist WIN 55,212-2. While the attenuation of fasting-induced food intake induced by acute rimonabant administration (3 mg/kg, intraperitoneal injection, *i.p.*) is preserved in *Fmr1*-KO—an effect dependent on peripheral CB1 activation,[77] their behavioral response was markedly reduced in the tetrad test (hypolocomotion, hypothermia, analgesia and catalepsy) in response to an acute injection of WIN 55,212-2 (3 mg/kg).[78]

When the impact of endocannabinoids receptors manipulations in *Fmr1*-KO mice was examined in detail, the effects appeared critically dependent on the behavioral alterations. Manipulations designed to suppress CB1R activation—either by pharmacological blocked or additional CB1R gene deletion (i.e., double-knockout mouse for genes CN1R and FMR1) are effective in normalizing the phenotypes of cognitive impairments, nociceptive alterations and susceptibility to audiogenic seizures in *Fmr1*-KO mice to wild-type levels. On the other hand, normalization of the anxiety phenotype in *Fmr1*-KO mice may be achieved by the blockade of CB2R.[79] Some of these beneficial effects were accompanied by the normalization of overactivated mTOR signaling and were similarly observed following pharmacological inhibition of mTOR or mGluR5 in *Fmr1*-KO mice. These have led some to suggest that mTOR pathway is a key mediator that may explain the role of ECS in ASD.[79] The same authors also demonstrated the therapeutic potential of rimonabant for fragile-X syndrome and ASD based on the beneficial effects on the memory function in *Fmr1*-KO mice.[80] However, due to risk of severe adverse effects despite efforts to lower the doses rimonabant,[80] targeting CB1 may not be a viable strategy. Instead, AEA has attracted more interest as a means to normalize ECS function in the *Fmr1*-KO model. Inhibitors of FAAH designed to elevate AEA levels have yielded positive effects against cognitive[81] and social deficits in *Fmr1*-KO mice.[82]

The BTBR T^{+} *tf*/J mouse model also yielded support for a role of AEA dysregulation in ASD.[23,24] BTBR T^{+} *tf*/J mice are marked by low levels of sociability, which were raised by the FAAH inhibitor, compound URB597.[82] There is also evidence for enhanced CB1R activities in cortical[83] and hippocampal areas[84] in BTBR T^{+} *tf*/J mice. Some authors suggested that CB1R upregulation could be related to the hyper-responsiveness of the hypothalamic-pituitary-adrenal axis characteristic in these mice.[84] The role of oxytocin has been highlighted by others,[82] as this neuropeptide is able to drive anandamide mobilization in some brain areas, and notably in the nucleus accumbens, which is implicated in the regulation of motivation.[85] Interestingly, AEA mobilization in this brain area is potentiated by social contact and inhibited by social isolation in mice.[85] To our knowledge, only one study investigated the behavioral response of BTBR T^{+} *tf*/J to CB1 agonists, which yielded results (e.g., less sensitive to sedative effects of high doses of THC)[86] that are in keeping with the impressions obtained in *Fmr1*-KO mice summarized earlier.

While evidence derived from the *Fmr1*-KO and BTBR T^{+} *tf*/J mouse model tend to converge with comparable ECS alterations, the "candidate gene" models based on mutations of the cell-adhesion molecule (e.g., *NLGN3*(R451C) substitution) seem to disagree. CB1-dependent modulation of GABAergic transmission in the hippocampus[87] and cortex[88] was weaker in these mutants. A partial rescue of LTD deficit was achieved by exogenous activation of cannabinoid CB1 receptors, suggesting a reduced cannabinoid drive in these mutants.[89] In line with these findings, a significant reduction of aggressive behavior in *NLGN3*(R451C) mice was achieved following non-sedating doses of CB1R agonists.[90]

Finally, evidence for the involvement of ECS in ASD was also supported by the valproic acid model—a model based on an environmental risk.[28,29] Relevant data are however restricted to rats. Prenatal valproic acid exposure led to altered expression of phosphorylated CB1R in the amygdala, hippocampus and dorsal striatum.[91,92] Although baseline concentrations of AEA and 2-AG were not affected, behavioral improvement could be obtained by enhancing AEA levels by inhibition of AEA degradation.[91,93] This is in keeping with the results obtained from some mouse models of ASD described earlier.

In conclusion, most mouse disease models lend support for the presence of ECS alterations involving all its functional constituents with varying degrees. The contributions of the ECS and AEA emerge as the most consistent across multiple models (including ASD patients), and therefore are candidates of common etiopathological pathways in ASD. Lower plasma AEA levels in autistic patients[66,67] suggested that circulating endocannabinoid levels may be explored as biological indices for demarcation of clinically significant subgroups along the autistic spectrum. Such objectively quantifiable indices could have substantial implications on diseases management. Not only can they provide crucial objectives diagnostic and prognostic information, but also shed lights on the intrinsic heterogeneity of the diagnostic criteria and separation of biologically meaningful homogeneous subgroups. At the clinical levels, they may also facilitate early diagnosis and early behavioral intervention, especially for at-risk individuals. There is already evidence that AEA levels are inversely correlated with the age in ASD patients.[66,67] Finally, blood-based markers like circulating endocannabinoids are convenient for monitoring of treatment responsiveness and optimizing drugs development including treatments aimed at specific core symptoms that may be titrate for individual disease profile.

Given the central roles of endocannabinoids in ASD etiopathology presented here, it is not surprising that the therapeutic potential of some phytocannabinoids, such as cannabidiol, has been subjected to investigation.[94] Cannabidiol has a low affinity for CB1 receptors, so its therapeutic effects are primarily based on its ability to inhibit FAAH.[55] It is expected to compensate for the low levels of AEA found in ASD patients.[67] Preliminary data showed that oral cannabidiol administration improved symptoms of hyperactivity, self-injurious behaviors, anxiety and sleep disturbance in ASD children.[95] Recent clinical studies have highlighted not only the efficacy of cannabidiol for ASD, but also cognitive symptoms in fragile-X syndrome patients, and importantly no adverse effects were observed.[96,97]

Indeed, their low affinity for CB1R enables cannabidiol and its derived compounds, e.g., cannabidivarin, to avoid the noxious psychotropic side effects typical of THC or rimonabant. In addition, cannabidivarin has powerful anti-inflammatory properties,[98] probably due to its action at transient receptor potential channels.[99] Several animal models of ASD have been used to preclinical screening of cannabidivarin. The ability of chronic cannabidivarin to rescue social deficits and reverse brain alterations has been demonstrated in *Mecp2*$^{308/y}$ mice,[100] a validated model of Rett syndrome.

Rett syndrome is a monogenic disorder caused by deficiency of methyl-CpG-binding protein-2 (Mecp2) and it is characterized by ASD deficits. Mice carrying null mutation for Mecp2,[101] as well as various polymorphisms found in Rett patients,[102] display ASD-like phenotypes, including altered emission of USVs at infancy,[103] and deficits in social and cognitive function in adulthood.[104] Cannabidivarin was able to reverse the cognitive deficits of *Mecp2*$^{308/y}$ mice, and reverse their upregulation of brain CB1R and CB2R.[105] It was similarly effective against cognitive, social and perseverative deficits in the valproic acid rat model, where the drug also normalized hippocampal endocannabinoid signaling and countered the neuroinflammation caused by prenatal valproic acid exposure.[106] These findings are consistent with the hypothesized roles of ECS across a wide spectrum of ASD conditions.[105] Our unpublished data obtained in *Fmr1*-KO mice also supported efficacy of chronic cannabidivarin, but also indicated that early intervention may be crucial. Initiation of cannabidivarin treatment at weaning yielded superior effects in *Fmr1*-KO mice in comparison with treatment that began at adult age. Hence, at least some ECS alterations may be irreversible by cannabidivarin beyond certain age of development. It remains to be tested if the therapeutic window we highlighted, including the adolescent age (4–7 weeks of age in rodents), may be generalized to other modalities targeting the ECS. Adolescence is an early life phase that is indeed characterized by high levels of neurobehavioral plasticity in rodents as well as in humans, and is easier to target with pharmacological interventions compared to earlier developmental phases (where the mother-pup interaction has a major confounding impact).

4. ASD-like phenotypes induced by manipulations of the ECS: Relevance for designing animal models of ASD

The recent excitement generated by research focusing on genetic mouse models of ASD has somewhat overshadowed the complementary pharmacological approach. Characterization of ASD-related behavior following pharmacological intervention at the ECS in wild-type animals is instrumental in dissecting the complex regulation of this extensive neuromodulatory system against the background of normal development.

Interests in the effects of cannabinoids on social behaviors can be traced back centuries to the recreational use marijuana and hashish as psychotropic and euphoria-inducing agents. The discovery of CB1R has sparked intense interests into the molecular mechanism underlying the neurophysiological and behavioral effects of these cannabinoids as potent CB1R agonists. Cannabis intoxication in humans and laboratory animals do share some resemblance with ASD symptoms, including cognitive and somatosensory impairments, social withdrawal and repetitive behavior.[107] Yet, little is still known regarding the recreational use of cannabinoids at non-intoxicating doses, specifically in terms of the control of social behavior. Imaging studies in healthy volunteers suggest that THC may modulate the perception of social threats.[108] At recreational doses, THC can significantly suppress amygdala reaction to threatening social cues without discernible impact on the activity in the primary visual and motor cortices.[108]

Studies in mice with acute (non-intoxicating) administration of crude marijuana extracts, pure THC or synthetic cannabinoids suspensions (e.g., WIN 55,212-2[109]) have yielded contradictory data. Depending on the precise dose and the exact test of social behavior used, such agents can induce either pro-social or antisocial effects, and at times even aggression.[110–114] The differential expression or selective activation of CB1R in specific neuronal populations may be partly responsible for the observed inconsistency across studies. An elegant series of experiments have demonstrated that the biphasic effects of THC on feeding and anxiety are linked to opposing CBR1-mediated effects at different doses on separate populations of neurons (glutamatergic vs GABAergic).[115,116]

A similar situation may apply to the modulation of sociability by THC. Conditional deletions of CB1R in distinct neuronal types are associated with abnormalities in different aspects of social behavior.[117–119] By contrast, interpretation of non-selective pharmacological activation of the ECS, for instance by elevation of brain AEA levels, seems to be more straightforward compared with selective CB1R stimulation. Inhibition of AEA hydrolysis by compound URB597—either systemically or administered directly into the amygdala or nucleus accumbens—increased social activity in rats[62] and reduced anxiety induced by social defeat in mice.[120]

More consistent outcomes are usually seen after blockade of CB1R, especially by comparison with simulation of the same receptor by agonist drugs. Animal data almost universally agree that CB1R antagonists induce with antisocial behavior. CB1R blockade therefore has been widely used to induce ASD-like states in animals, including non-human primates. Systemic blockade of CB1R by compound AM251 in adult male capuchin monkeys suppressed social behavior and concomitantly increased self-directed behavior. The treatment also induced resistance to extinction of aversive memory trace, but left locomotor activity or vigilance behavior unchanged.[121] In laboratory rodents, AM251 similarly impairs diverse social functioning: the drug has been reported to reduce social interest also in rats,[122] and disrupt social preference as well as social memory in mice.[85,123]

We have also investigated the ability of the CB1R antagonist, rimonabant, to induce ASD-like behavior in wild-type C57BL/6 mice—the background strain used to generate *Fmr1*-KO mice. A dose-dependent study of the effects of acute systemic rimonabant on male-female interaction was conducted. The test is a highly sensitive to *Fmr1*-KO mutation. As illustrated in Fig. 2, rimonabant suppressed anogenital sniffing and the number of USVs, and the suppression of these affiliative behaviors was accompanied by the intensification of repetitive self-grooming behavior. The overall ASD-like behavioral profile resembles the phenotypes of *Fmr1*-KOs, and all effects were obtained in the absence of any confounding changes in locomotor activity.

The robust *antisocial* effects of CB1R blockade in wild-type animals are also seen in CB1R knockout mice: Impaired sociability[126] and social exploration[117,118] were observed, although the expression of these

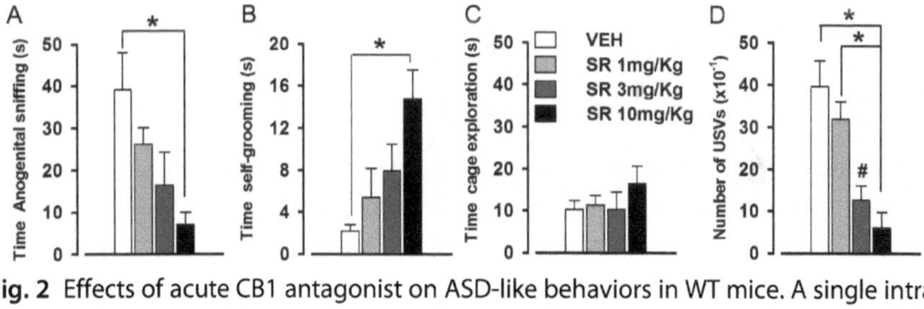

Fig. 2 Effects of acute CB1 antagonist on ASD-like behaviors in WT mice. A single intraperitoneal injection of the CB1R antagonist, rimonabant (a.k.a SR141716A), prepared as previously described,[51] was administered to C57BL/6 adult males (3–4 months-old), at 1, 3, or 10 mg/kg 30 min before testing. The test involved exposing the male mouse to an untreated adult female of NMRI strain in the non-oestrus phase for 3 min. The procedures and the quantification of direct social interaction behavior and ultrasonic vocalizations (USVs) have been fully described previously.[109,124,125] The results demonstrated in the male mice a clear dose-dependent reduction in social investigation (A) and a dose-dependent increase in self-grooming (B) without affect concurrent cage exploration (C). At doses above 3 mg/kg, rimonabant also significantly suppressed USVs emitted by the male (D). A significant treatment effect emerged from the ANOVA of the time spent in anogenital sniffing [$F(3,18)=5.44$, $P<0.01$], time spent in self-grooming [$F(3,18)=5.66$, $P<0.01$], and the number of USVs [$F(3,18)=12.64$, $P<0.01$]. Overall, the data suggested that the highest dose here (10 mg/kg, *i.p.*) would be the most effective in modeling ASD-like behavioral abnormalities. Keys: SR = rimonabant (a.k.a. SR141716A); *$P<0.05$ from Tuckey-Kramer *post-hoc* tests; # indicates comparison versus vehicle control (VEH) and SR 1 mg/kg.

phenotypes was sensitive to test context[83] and choice of social stimulus[119] employed. In addition, social memory was impaired.[123] We were able to observe alterations in USVs in CB1R-KO mice during the first post-natal week as well as later in adulthood.[127] These findings strengthen the widespread use of CB1R-KO mice to model ASD. Suffice it to say, CB1R-KO mice may not consistently generate phenotypes related to repetitive behaviors, such as enhanced self-grooming.[123,128] Another notable difference in comparison with pharmacological blockade is the impression that the ASD-like phenotypes observed in the CB1R-KO mice have smaller effect sizes than the corresponding pharmacological models (as demonstrated in wild-type mice). This may reflect a degree of developmental compensation, but this should not limit the application of the model albeit with some reduction in statistical power. Indeed, the CB1R-KO mice have been very instrumental in consolidating the contribution of ECS dysregulation to ASD pathogenesis. The available set of cell-specific CB1R deletion models has enabled researchers to dissect the relative roles of CB1R in distinct pools of neurons in the control of social behaviors and other ASD-relevant symptoms through the modulation of the classical neurotransmitter systems (glutamatergic, GABAergic, serotonergic or dopaminergic).[117–119,123,128]

5. Conclusions and perspectives

Clinical and preclinical data strongly support the involvement of the ECS in the etiopathology of ASD and its relevance to drug discovery. Evidence from ASD-related mouse lines as well as pharmacological manipulations targeting the ECS in wild-type animals suggests that an imbalance in ECS signaling represents one potential common etiopathological path of this complex disorder within the theoretical framework of "many genes, common pathways" proposed by Daniel Geschwind.[35] This is in keeping with the significant modulatory influence over neuronal functions and brain development by the ECS. Some authors have emphasized, in particular, the need to elucidate how ECS dysfunction lead to derailment in brain development and as it remains a relatively understudied area.[129] To better differentiate the relevance of ECS in abnormal brain development from mature synaptic functionality,[129] the utilization of neurons derived from induced pluripotent stem cells is expected to yield new insights, as it has already toward our understanding of the pathogenesis of several neuropsychiatric disorders.[130]

To a large extent, the complexity of ASD etiopathology is reflected in the complex and heterogeneous network of the ECS itself. Revelation that endocannabinoids may be relevant to ASD pathogenesis has given new impetus to develop safe pharmacological manipulation of the ECS to treat ASD. We have also summarized available evidence here supportive of the therapeutic potential of phytocannabinoids, such as cannabidiol and cannabidivarin. These compounds are highly promising because they can effectively raise endocannabinoid levels without the adverse effects associated with direct CB1R stimulation, and they possess additional anti-inflammatory properties. As our knowledge of the complexity of the ECS grows, so do the number of strategies for successful clinical interventions. Discovery that endocannabinoids can activate different novel receptors as well as modulate other mediators through their shared metabolic pathways has drastically enlarged the ECS system. An expanded signaling system, known as the endocannabinoidome, has been recognized.[54] This expanded system includes several key mediators that are biochemically related to the endocannabinoids. Together with their receptors and metabolic enzymes, they are expected to yield multiple new targets to drive ASD drug discovery in the near future.[54]

In this Chapter, we aim to convincingly argue that the role of ECS in ASD is almost an inescapable conclusion given by the extensive scope of evidence summarized here. However, we do not intend to leave an impression that the ECS system alone may provide a unitary explanation of ASD etiopathology. Instead, we are convinced, as other commentators in the field, that any full account of ASD must incorporate other parallel pathological factors. The pathogenesis of this complex developmental disorder is ultimately the result of intricate interactions between parallel systems overseeing brain development.[7,131] The ECS may offer a key to unlock the identities of the other key players and complexity involved. In keeping with this emphasis is search for new molecular switches for multi-target pharmacology in drug design.[132,133] It complements the ongoing effort to design safer and more refined pharmacological manipulations targeting the ECS.

References

1. Kanner L. Autistic disturbances of affective contact. *Nerv Child.* 1943;2:217–250.
2. Rutter M. Concepts of autism: a review of research. *J Child Psychol Psychiatry.* 1968;9(1):1–25.
3. American psychiatric association. *Diagnostic and Statistical Manual of Mental Disorders (DSM-III).* Washington, DC: American psychiatric association; 1980.

4. American psychiatric association. *Diagnostic and Statistical Manual of Mental Disorders (DSM-III-R)*. Washington, DC: American psychiatric association; 1987.
5. American psychiatric association. *Diagnostic and Statistical Manual of Mental Disorders (DSM-V)*. Washington, DC; 2000.
6. Rapin I. Autistic children: diagnosis and clinical features. *Pediatrics*. 1991;87(5 Pt 2): 751–760.
7. Happe F, Ronald A, Plomin R. Time to give up on a single explanation for autism. *Nat Neurosci*. 2006;9(10):1218–1220.
8. Nadler JJ, Moy SS, Dold G, et al. Automated apparatus for quantitation of social approach behaviors in mice. *Genes Brain Behav*. 2004;3(5):303–314.
9. Wohr M. Ultrasonic vocalizations in shank mouse models for autism spectrum disorders: detailed spectrographic analyses and developmental profiles. *Neurosci Biobehav Rev*. 2014;43:199–212.
10. Wohr M, Scattoni ML. Behavioural methods used in rodent models of autism spectrum disorders: current standards and new developments. *Behav Brain Res*. 2013;251:5–17.
11. Branchi I, Santucci D, Alleva E. Ultrasonic vocalisation emitted by infant rodents: a tool for assessment of neurobehavioural development. *Behav Brain Res*. 2001; 125(1–2):49–56.
12. D'Amato FR, Scalera E, Sarli C, Moles A. Pups call, mothers rush: does maternal responsiveness affect the amount of ultrasonic vocalizations in mouse pups? *Behav Genet*. 2005;35(1):103–112.
13. Holy TE, Guo Z. Ultrasonic songs of male mice. *PLoS Biol*. 2005;3(12):e386.
14. Moles A, Costantini F, Garbugino L, Zanettini C, D'Amato FR. Ultrasonic vocalizations emitted during dyadic interactions in female mice: a possible index of sociability? *Behav Brain Res*. 2007;182(2):223–230.
15. Moy SS, Nadler JJ, Poe MD, et al. Development of a mouse test for repetitive, restricted behaviors: relevance to autism. *Behav Brain Res*. 2008;188(1):178–194.
16. Crawley JN. Mouse behavioral assays relevant to the symptoms of autism. *Brain Pathol*. 2007;17(4):448–459.
17. Crawley JN. Designing mouse behavioral tasks relevant to autistic-like behaviors. *Ment Retard Dev Disabil Res Rev*. 2004;10(4):248–258.
18. Moy SS, Nadler JJ, Magnuson TR, Crawley JN. Mouse models of autism spectrum disorders: the challenge for behavioral genetics. *Am J Med Genet C Semin Med Genet*. 2006;142C(1):40–51.
19. Radde R, Duma C, Goedert M, Jucker M. The value of incomplete mouse models of Alzheimer's disease. *Eur J Nucl Med Mol Imaging*. 2008;35(Suppl 1):S70–S74.
20. Banerjee S, Riordan M, Bhat MA. Genetic aspects of autism spectrum disorders: insights from animal models. *Front Cell Neurosci*. 2014;8:58.
21. Bey AL, Jiang YH. Overview of mouse models of autism spectrum disorders. *Curr Protoc Pharmacol*. 2014;66, 5.66.1-26.
22. Ellegood J, Crawley JN. Behavioral and neuroanatomical phenotypes in mouse models of autism. *Neurotherapeutics*. 2015;12(3):521–533.
23. McFarlane HG, Kusek GK, Yang M, Phoenix JL, Bolivar VJ, Crawley JN. Autism-like behavioral phenotypes in BTBR T+tf/J mice. *Genes Brain Behav*. 2008;7(2):152–163.
24. Meyza KZ, Defensor EB, Jensen AL, et al. The BTBR T+ tf/J mouse model for autism spectrum disorders-in search of biomarkers. *Behav Brain Res*. 2013;251:25–34.
25. Oddi D, Crusio WE, D'Amato FR, Pietropaolo S. Monogenic mouse models of social dysfunction: implications for autism. *Behav Brain Res*. 2013;251:75–84.
26. Dufour-Rainfray D, Vourc'h P, Tourlet S, Guilloteau D, Chalon S, Andres CR. Fetal exposure to teratogens: evidence of genes involved in autism. *Neurosci Biobehav Rev*. 2011;35(5):1254–1265.

27. Patterson PH. Maternal infection and immune involvement in autism. *Trends Mol Med.* 2011;17(7):389–394.
28. Ingram JL, Peckham SM, Tisdale B, Rodier PM. Prenatal exposure of rats to valproic acid reproduces the cerebellar anomalies associated with autism. *Neurotoxicol Teratol.* 2000;22(3):319–324.
29. Schneider T, Przewlocki R. Behavioral alterations in rats prenatally exposed to valproic acid: animal model of autism. *Neuropsychopharmacology.* 2005;30(1):80–89.
30. Chapman JB, Cutler MG. Sodium valproate: effects on social behaviour and physical development in the mouse. *Psychopharmacology (Berl).* 1984;83(4):390–396.
31. Roux S, Bailly Y, Bossu JL. Regional and sex-dependent alterations in Purkinje cell density in the valproate mouse model of autism. *Neuroreport.* 2019;30(2):82–88.
32. Abrahams BS, Geschwind DH. Advances in autism genetics: on the threshold of a new neurobiology. *Nat Rev Genet.* 2008;9(5):341–355.
33. Persico AM, Bourgeron T. Searching for ways out of the autism maze: genetic, epigenetic and environmental clues. *Trends Neurosci.* 2006;29(7):349–358.
34. Buxbaum JD, Betancur C, Bozdagi O, Dorr NP, Elder GA, Hof PR. Optimizing the phenotyping of rodent ASD models: enrichment analysis of mouse and human neurobiological phenotypes associated with high-risk autism genes identifies morphological, electrophysiological, neurological, and behavioral features. *Mol Autism.* 2012;3(1):1.
35. Geschwind DH. Autism: many genes, common pathways? *Cell.* 2008;135(3):391–395.
36. Ey E, Leblond CS, Bourgeron T. Behavioral profiles of mouse models for autism spectrum disorders. *Autism Res.* 2011;4(1):5–16.
37. The Dutch-Belgian Fragile X Consortium. Fmr1 knockout mice: a model to study fragile X mental retardation. *Cell.* 1994;78(1):23–33.
38. Pieretti M, Zhang FP, Fu YH, et al. Absence of expression of the FMR-1 gene in fragile X syndrome. *Cell.* 1991;66(4):817–822.
39. Pietropaolo S, Subashi E. Mouse models of fragile X syndrome. In: Pietropaolo S, Sluyter F, Crusio WE, eds. *Behavioral Genetics of the Mouse.* Cambridge: Cambridge University Press; 2014:146–163. vol. 2.
40. Budimirovic DB, Kaufmann WE. What can we learn about autism from studying fragile X syndrome? *Dev Neurosci.* 2011;33(5):379–394.
41. Pardo CA, Eberhart CG. The neurobiology of autism. *Brain Pathol.* 2007; 17(4):434–447.
42. McOmish CE, Burrows EL, Hannan AJ. Identifying novel interventional strategies for psychiatric disorders: integrating genomics, 'enviromics' and gene-environment interactions in valid preclinical models. *Br J Pharmacol.* 2014;171(20):4719–4728.
43. Nithianantharajah J, Hannan AJ. Enriched environments, experience-dependent plasticity and disorders of the nervous system. *Nat Rev Neurosci.* 2006;7(9):697–709.
44. Zoghbi HY, Bear MF. Synaptic dysfunction in neurodevelopmental disorders associated with autism and intellectual disabilities. *Cold Spring Harb Perspect Biol.* 2012;4(3), pii: a009886.
45. Delorme R, Ey E, Toro R, Leboyer M, Gillberg C, Bourgeron T. Progress toward treatments for synaptic defects in autism. *Nat Med.* 2013;19(6):685–694.
46. Kleijer KTE, Schmeisser M, Krueger DD, et al. Neurobiology of autism gene products: towards pathogenesis and drug targets. *Psychopharmacology (Berl).* 2014;231:1037–1062.
47. Mackie K. Distribution of cannabinoid receptors in the central and peripheral nervous system. *Handb Exp Pharmacol.* 2005;168:299–325.
48. Berghuis P, Rajnicek AM, Morozov YM, et al. Hardwiring the brain: endocannabinoids shape neuronal connectivity. *Science.* 2007;316(5828):1212–1216.
49. Freund TF, Katona I, Piomelli D. Role of endogenous cannabinoids in synaptic signaling. *Physiol Rev.* 2003;83(3):1017–1066.

50. Monory K, Lutz B. Genetic models of the endocannabinoid system. *Curr Top Behav Neurosci*. 2009;1:111–139.
51. Bellocchio L, Soria-Gomez E, Quarta C, et al. Activation of the sympathetic nervous system mediates hypophagic and anxiety-like effects of CB(1) receptor blockade. *Proc Natl Acad Sci U S A*. 2013;110(12):4786–4791.
52. Monory K, Massa F, Egertova M, et al. The endocannabinoid system controls key epileptogenic circuits in the hippocampus. *Neuron*. 2006;51(4):455–466.
53. Busquets-Garcia A, Bains J, Marsicano G. CB1 receptor signaling in the brain: extracting specificity from ubiquity. *Neuropsychopharmacology*. 2018;43(1):4–20.
54. Cristino L, Bisogno T, Di Marzo V. Cannabinoids and the expanded endocannabinoid system in neurological disorders. *Nat Rev Neurol*. 2020;16(1):9–29.
55. Di Marzo V, Piscitelli F. The endocannabinoid system and its modulation by phytocannabinoids. *Neurotherapeutics*. 2015;12(4):692–698.
56. Jordan CJ, Xi ZX. Progress in brain cannabinoid CB2 receptor research: from genes to behavior. *Neurosci Biobehav Rev*. 2019;98:208–220.
57. Turcotte C, Blanchet MR, Laviolette M, Flamand N. The CB2 receptor and its role as a regulator of inflammation. *Cell Mol Life Sci*. 2016;73(23):4449–4470.
58. Pistis M, Melis M. From surface to nuclear receptors: the endocannabinoid family extends its assets. *Curr Med Chem*. 2010;17(14):1450–1467.
59. Rubino T, Zamberletti E, Parolaro D. Endocannabinoids and mental disorders. *Handb Exp Pharmacol*. 2015;231:261–283.
60. Iannotti FA, Di Marzo V, Petrosino S. Endocannabinoids and endocannabinoid-related mediators: targets, metabolism and role in neurological disorders. *Prog Lipid Res*. 2016;62:107–128.
61. Marco EM, Laviola G. The endocannabinoid system in the regulation of emotions throughout lifespan: a discussion on therapeutic perspectives. *J Psychopharmacol*. 2012; 26(1):150–163.
62. Trezza V, Damsteegt R, Manduca A, et al. Endocannabinoids in amygdala and nucleus accumbens mediate social play reward in adolescent rats. *J Neurosci*. 2012;32(43): 14899–14908.
63. Vaughn LK, Denning G, Stuhr KL, de Wit H, Hill MN, Hillard CJ. Endocannabinoid signalling: has it got rhythm? *Br J Pharmacol*. 2010;160(3):530–543.
64. Chakrabarti B, Kent L, Suckling J, Bullmore E, Baron-Cohen S. Variations in the human cannabinoid receptor (CNR1) gene modulate striatal responses to happy faces. *Eur J Neurosci*. 2006;23(7):1944–1948.
65. Purcell AE, Jeon OH, Zimmerman AW, Blue ME, Pevsner J. Postmortem brain abnormalities of the glutamate neurotransmitter system in autism. *Neurology*. 2001; 57(9):1618–1628.
66. Karhson DS, Krasinska KM, Dallaire JA, et al. Plasma anandamide concentrations are lower in children with autism spectrum disorder. *Mol Autism*. 2018;9:18.
67. Aran A, Eylon M, Harel M, et al. Lower circulating endocannabinoid levels in children with autism spectrum disorder. *Mol Autism*. 2019;10:2.
68. Smith DR, Stanley CM, Foss T, Boles RG, McKernan K. Rare genetic variants in the endocannabinoid system genes CNR1 and DAGLA are associated with neurological phenotypes in humans. *PLoS One*. 2017;12(11):e0187926.
69. Siniscalco D, Sapone A, Giordano C, et al. Cannabinoid receptor type 2, but not type 1, is up-regulated in peripheral blood mononuclear cells of children affected by autistic disorders. *J Autism Dev Disord*. 2013;43(11):2686–2695.
70. Basu S, Dittel BN. Unraveling the complexities of cannabinoid receptor 2 (CB2) immune regulation in health and disease. *Immunol Res*. 2011;51(1):26–38.

71. Ormstad H, Bryn V, Saugstad OD, Skjeldal O, Maes M. Role of the immune system in autism Spectrum disorders (ASD). *CNS Neurol Disord Drug Targets*. 2018; 17(7):489–495.
72. Siniscalco D, Schultz S, Brigida AL, Antonucci N. Inflammation and neuro-immune dysregulations in autism spectrum disorders. *Pharmaceuticals (Basel)*. 2018; 11(2), pii: E56.
73. Zamberletti E, Gabaglio M, Parolaro D. The endocannabinoid system and autism spectrum disorders: insights from animal models. *Int J Mol Sci*. 2017;18(9), pii: E1916.
74. Zhang L, Alger BE. Enhanced endocannabinoid signaling elevates neuronal excitability in fragile X syndrome. *J Neurosci*. 2010;30(16):5724–5729.
75. Maccarrone M, Rossi S, Bari M, et al. Abnormal mGlu 5 receptor/endocannabinoid coupling in mice lacking FMRP and BC1 RNA. *Neuropsychopharmacology*. 2010;35(7):1500–1509.
76. Jung KM, Sepers M, Henstridge CM, et al. Uncoupling of the endocannabinoid signalling complex in a mouse model of fragile X syndrome. *Nat Commun*. 2012;3:1080.
77. Gomez R, Navarro M, Ferrer B, et al. A peripheral mechanism for CB1 cannabinoid receptor-dependent modulation of feeding. *J Neurosci*. 2002;22(21):9612–9617.
78. Pietropaolo S, Lemaire-Mayo V, Bellocchio L, Subashi E, Crusio WE, Marsicano G. The endocannabinoid system critically modulates the expression of social behaviors in adult male mice. In: *Society for Neuroscience (SFN) Annual Meeting*; 2012, New Orleans, USA.
79. Busquets-Garcia A, Maldonado R, Ozaita A. New insights into the molecular pathophysiology of fragile X syndrome and therapeutic perspectives from the animal model. *Int J Biochem Cell Biol*. 2014;53:121–126.
80. Gomis-Gonzalez M, Busquets-Garcia A, Matute C, Maldonado R, Mato S, Ozaita A. Possible therapeutic doses of cannabinoid type 1 receptor antagonist reverses key alterations in fragile X syndrome mouse model. *Genes (Basel)*. 2016;7(9), pii: E56.
81. Qin M, Zeidler Z, Moulton K, Krych L, Xia Z, Smith CB. Endocannabinoid-mediated improvement on a test of aversive memory in a mouse model of fragile X syndrome. *Behav Brain Res*. 2015;291:164–171.
82. Wei D, Dinh D, Lee D, et al. Enhancement of anandamide-mediated endocannabinoid signaling corrects autism-related social impairment. *Cannabis Cannabinoid Res*. 2016; 1(1):81–89.
83. Gould GG, Seillier A, Weiss G, et al. Acetaminophen differentially enhances social behavior and cortical cannabinoid levels in inbred mice. *Prog Neuropsychopharmacol Biol Psychiatry*. 2012;38(2):260–269.
84. Gould GG, Burke TF, Osorio MD, et al. Enhanced novelty-induced corticosterone spike and upregulated serotonin 5-HT1A and cannabinoid CB1 receptors in adolescent BTBR mice. *Psychoneuroendocrinology*. 2014;39:158–169.
85. Wei D, Lee D, Cox CD, et al. Endocannabinoid signaling mediates oxytocin-driven social reward. *Proc Natl Acad Sci U S A*. 2015;112(45):14084–14089.
86. Onaivi ES, Benno R, Halpern T, et al. Consequences of cannabinoid and monoaminergic system disruption in a mouse model of autism spectrum disorders. *Curr Neuropharmacol*. 2011;9(1):209–214.
87. Foldy C, Malenka RC, Sudhof TC. Autism-associated neuroligin-3 mutations commonly disrupt tonic endocannabinoid signaling. *Neuron*. 2013;78(3):498–509.
88. Speed HE, Masiulis I, Gibson JR, Powell CM. Increased cortical inhibition in autism-linked neuroligin-3R451C mice is due in part to loss of endocannabinoid signaling. *PLoS One*. 2015;10(10):e0140638.

89. Martella G, Meringolo M, Trobiani L, De Jaco A, Pisani A, Bonsi P. The neurobiological bases of autism spectrum disorders: the R451C-neuroligin 3 mutation hampers the expression of long-term synaptic depression in the dorsal striatum. *Eur J Neurosci.* 2018;47(6):701–708.
90. Hosie S, Malone DT, Liu S, et al. Altered amygdala excitation and CB1 receptor modulation of aggressive behavior in the neuroligin-3(R451C) mouse model of autism. *Front Cell Neurosci.* 2018;12:234.
91. Servadio M, Melancia F, Manduca A, et al. Targeting anandamide metabolism rescues core and associated autistic-like symptoms in rats prenatally exposed to valproic acid. *Transl Psychiatry.* 2016;6(9):e902.
92. Kerr DM, Downey L, Conboy M, Finn DP, Roche M. Alterations in the endocannabinoid system in the rat valproic acid model of autism. *Behav Brain Res.* 2013;249:124–132.
93. Kerr DM, Gilmartin A, Roche M. Pharmacological inhibition of fatty acid amide hydrolase attenuates social behavioural deficits in male rats prenatally exposed to valproic acid. *Pharmacol Res.* 2016;113(Pt A):228–235.
94. Poleg S, Golubchik P, Offen D, Weizman A. Cannabidiol as a suggested candidate for treatment of autism spectrum disorder. *Prog Neuropsychopharmacol Biol Psychiatry.* 2019;89:90–96.
95. Barchel D, Stolar O, De-Haan T, et al. Oral Cannabidiol use in children with autism spectrum disorder to treat related symptoms and co-morbidities. *Front Pharmacol.* 2018;9:1521.
96. Heussler H, Cohen J, Silove N, et al. A phase 1/2, open-label assessment of the safety, tolerability, and efficacy of transdermal cannabidiol (ZYN002) for the treatment of pediatric fragile X syndrome. *J Neurodev Disord.* 2019;11(1):16.
97. Tartaglia N, Bonn-Miller M, Hagerman R. Treatment of fragile X syndrome with cannabidiol: a case series study and brief review of the literature. *Cannabis Cannabinoid Res.* 2019;4(1):3–9.
98. Burstein S. Cannabidiol (CBD) and its analogs: a review of their effects on inflammation. *Bioorg Med Chem.* 2015;23(7):1377–1385.
99. De Petrocellis L, Ligresti A, Moriello AS, et al. Effects of cannabinoids and cannabinoid-enriched Cannabis extracts on TRP channels and endocannabinoid metabolic enzymes. *Br J Pharmacol.* 2011;163(7):1479–1494.
100. Vigli D, Cosentino L, Raggi C, Laviola G, Woolley-Roberts M, De Filippis B. Chronic treatment with the phytocannabinoid Cannabidivarin (CBDV) rescues behavioural alterations and brain atrophy in a mouse model of Rett syndrome. *Neuropharmacology.* 2018;140:121–129.
101. Chen RZ, Akbarian S, Tudor M, Jaenisch R. Deficiency of methyl-CpG binding protein-2 in CNS neurons results in a Rett-like phenotype in mice. *Nat Genet.* 2001;27(3):327–331.
102. Shahbazian M, Young J, Yuva-Paylor L, et al. Mice with truncated MeCP2 recapitulate many Rett syndrome features and display hyperacetylation of histone H3. *Neuron.* 2002;35(2):243–254.
103. Picker JD, Yang R, Ricceri L, Berger-Sweeney J. An altered neonatal behavioral phenotype in Mecp2 mutant mice. *Neuroreport.* 2006;17(5):541–544.
104. Moretti P, Bouwknecht JA, Teague R, Paylor R, Zoghbi HY. Abnormalities of social interactions and home-cage behavior in a mouse model of Rett syndrome. *Hum Mol Genet.* 2005;14(2):205–220.
105. Zamberletti E, Gabaglio M, Piscitelli F, et al. Cannabidivarin completely rescues cognitive deficits and delays neurological and motor defects in male Mecp2 mutant mice. *J Psychopharmacol.* 2019;33(7):894–907.

106. Zamberletti E, Gabaglio M, Woolley-Roberts M, Bingham S, Rubino T, Parolaro D. Cannabidivarin treatment ameliorates autism-like behaviors and restores hippocampal endocannabinoid system and glia alterations induced by prenatal Valproic acid exposure in rats. *Front Cell Neurosci*. 2019;13:367.
107. Zehra A, Burns J, Liu CK, et al. Cannabis addiction and the brain: a review. *J Neuroimmune Pharmacol*. 2018;13(4):438–452.
108. Phan KL, Angstadt M, Golden J, Onyewuenyi I, Popovska A, de Wit H. Cannabinoid modulation of amygdala reactivity to social signals of threat in humans. *J Neurosci*. 2008;28(10):2313–2319.
109. Pietropaolo S, Bellocchio L, Ruiz-Calvo A, et al. Chronic cannabinoid receptor stimulation selectively prevents motor impairments in a mouse model of Huntington's disease. *Neuropharmacology*. 2015;89:368–374.
110. Cutler MG, Mackintosh JH. Cannabis and delta-9-tetrahydrocannabinol. Effects on elements of social behaviour in mice. *Neuropharmacology*. 1984;23(9):1091–1097.
111. Dorr M, Steinberg H. Effects of delta9-tetrahydrocannabinol on social behaviour in mice: comparison between two vehicles. *Psychopharmacology (Berl)*. 1976;47(1):87–91.
112. Ely DL, Henry JP, Jarosz CJ. Effects of marihuana (delta9-THC) on behavior patterns and social roles in colonies of CBA mice. *Behav Biol*. 1975;13(3):263–276.
113. Long LE, Chesworth R, Arnold JC, Karl T. A follow-up study: acute behavioural effects of Delta(9)-THC in female heterozygous neuregulin 1 transmembrane domain mutant mice. *Psychopharmacology (Berl)*. 2010;211(3):277–289.
114. Sieber B, Frischknecht HR, Waser PG. Behavioral effects of hashish in mice. IV. Social dominance, food dominance, and sexual behavior within a group of males. *Psychopharmacology (Berl)*. 1981;73(2):142–146.
115. Bellocchio L, Lafenetre P, Cannich A, et al. Bimodal control of stimulated food intake by the endocannabinoid system. *Nat Neurosci*. 2010;13(3):281–283.
116. Ruehle S, Rey AA, Remmers F, Lutz B. The endocannabinoid system in anxiety, fear memory and habituation. *J Psychopharmacol*. 2012;26(1):23–39.
117. Haring M, Enk V, Aparisi Rey A, et al. Cannabinoid type-1 receptor signaling in central serotonergic neurons regulates anxiety-like behavior and sociability. *Front Behav Neurosci*. 2015;9:235.
118. Haring M, Kaiser N, Monory K, Lutz B. Circuit specific functions of cannabinoid CB1 receptor in the balance of investigatory drive and exploration. *PLoS One*. 2011; 6(11):e26617.
119. Terzian AL, Micale V, Wotjak CT. Cannabinoid receptor type 1 receptors on GABAergic vs. glutamatergic neurons differentially gate sex-dependent social interest in mice. *Eur J Neurosci*. 2014;40(1):2293–2298.
120. Rossi S, De Chiara V, Musella A, et al. Preservation of striatal cannabinoid CB1 receptor function correlates with the antianxiety effects of fatty acid amide hydrolase inhibition. *Mol Pharmacol*. 2010;78(2):260–268.
121. Gonczarowska N, Tomaz C, Caixeta FV, et al. CB1 receptor antagonism in capuchin monkeys alters social interaction and aversive memory extinction. *Psychopharmacology (Berl)*. 2019;236(12):3413–3419.
122. Seillier A, Martinez AA, Giuffrida A. Phencyclidine-induced social withdrawal results from deficient stimulation of cannabinoid CB(1) receptors: implications for schizophrenia. *Neuropsychopharmacology*. 2013;38(9):1816–1824.
123. Litvin Y, Phan A, Hill MN, Pfaff DW, McEwen BS. CB1 receptor signaling regulates social anxiety and memory. *Genes Brain Behav*. 2013;12(5):479–489.
124. Gaudissard J, Ginger M, Premoli M, Memo M, Frick A, Pietropaolo S. Behavioral abnormalities in the Fmr1-KO2 mouse model of fragile X syndrome: The relevance of early life phases. *Autism Res*. 2017;10(10):1584–1596.

125. Oddi D, Subashi E, Middei S, et al. Early social enrichment rescues adult behavioral and brain abnormalities in a mouse model of fragile X syndrome. *Neuropsychopharmacology*. 2015;40(5):1113–1122.
126. Haller J, Varga B, Ledent C, Freund TF. CB1 cannabinoid receptors mediate anxiolytic effects: convergent genetic and pharmacological evidence with CB1-specific agents. *Behav Pharmacol*. 2004;15(4):299–304.
127. Fyke W, Premoli M, Lemaire-Mayo V, et al. Ultrasonic communication in the CB1 knockout mouse line during development and at adulthood. In: *European Brain and Behaviour Society Annual Meeting*; 2019, Prague, Czech Republic.
128. Terzian AL, Drago F, Wotjak CT, Micale V. The dopamine and cannabinoid interaction in the modulation of emotions and cognition: assessing the role of cannabinoid CB1 receptor in neurons expressing dopamine D1 receptors. *Front Behav Neurosci*. 2011;5:49.
129. Yeh ML, Levine ES. Perspectives on the role of endocannabinoids in autism spectrum disorders. *OBM Neurobiol*. 2017;1(2), pii: 005.
130. Wen Z, Nguyen HN, Guo Z, et al. Synaptic dysregulation in a human iPS cell model of mental disorders. *Nature*. 2014;515(7527):414–418.
131. Happe F. Cognition in autism: one deficit or many? *Novartis Found Symp*. 2003; 251:198–207, discussion 207–112, 281–197.
132. Pietropaolo S, Crusio WE, D'Amato FR. Treatment approaches in rodent models for autism spectrum disorder. *Curr Top Behav Neurosci*. 2017;30:325–340.
133. Gross C, Berry-Kravis EM, Bassell GJ. Therapeutic strategies in fragile X syndrome: dysregulated mGluR signaling and beyond. *Neuropsychopharmacology*. 2012; 37(1):178–195.

CHAPTER EIGHT

The effects of oxytocin administration on individuals with ASD: Neuroimaging and behavioral evidence

Elizabeth Baker, Katherine K.M. Stavropoulos*
University of California, Graduate School of Education, Riverside, CA, United States
*Corresponding author: e-mail address: katherine.stavropoulos@ucr.edu

Contents

Abstract

Autism spectrum disorder (ASD) is a neurodevelopmental disorder characterized by difficulties in social communication and the presence of restricted interests and repetitive behaviors. Although behavioral interventions are numerous, there are no Federal Drug Administration approved pharmacological treatments for the core symptoms of ASD. The neuropeptide oxytocin has been studied in animals for decades, and is involved in pair bonding and social affiliation. Given oxytocin's involvement in social communication in animals, researchers have begun exploring whether oxytocin administration in humans affects social behaviors and attachment. Particular attention has been paid to whether oxytocin has therapeutic benefits for improving social behaviors in individuals with ASD. Research on oxytocin administration in ASD has utilized both behavioral and brain-based outcomes. This chapter reviews the effects of oxytocin administration in ASD, with a focus on functional outcomes from neuroimaging investigations. Evidence of potential therapeutic benefits are reviewed, as well as limitations of extant research. A proposed model for future research into the therapeutic benefits of

Progress in Molecular Biology and Translational Science, Volume 173
ISSN 1877-1173
https://doi.org/10.1016/bs.pmbts.2020.04.009

oxytocin includes combining pharmacological (e.g. oxytocin) and behavioral (e.g. evidence-based behavioral interventions) techniques to improve social communication skills in ASD.

1. Introduction

It is estimated that 1 in 59 children in the United States has a diagnosis of Autism Spectrum Disorder (ASD),[1] which is characterized by life-long social communication deficits and the presence of restricted interests and repetitive behaviors.[2] In the first few years of life, commonly observed social communication deficits include poor imitation skills, reduced social orienting, and limited joint attention.[3,4] As children get older, challenges exist in maintaining social interactions and sustained conversations with their peers.[5,6] Restricted and repetitive interests, including resistance to change, repetitive motor movements, and strong interests in unusual topics (e.g. traffic lights, drain pipes), may further complicate and hinder social behaviors.[7,8] Despite shared core symptoms, ASD covers a wide spectrum and behavioral manifestations are varied between individuals to produce a heterogenous phenotype.

Given the heterogenous behavioral manifestations of ASD, it is unsurprising that multiple behavioral interventions have been developed to improve social communication and decrease challenging behaviors.[9–11] In terms of "treatments" for ASD, behavioral intervention is currently one of the most effective methods for improving social-communication skills. In contrast to other conditions in which both behavioral and medical treatments are supported by empirical evidence, no medication is currently approved to improve core symptoms of ASD. Although two pharmaceutical medications, risperidone and aripiprazole, have been approved by the Federal Drug Administration for use in ASD, they do not address core symptoms. Rather, they have been approved to treat ancillary symptoms of ASD, such as aggression and irritability.[12]

Many challenges exist in both the production and efficacy-testing of novel pharmaceutical medications, particularly in children.[13] For example, methodologies necessary to test treatment efficacy within clinical trials may require invasive techniques, such as blood samples, which may not be well-tolerated by children. Alternatives to blood samples, such as urine testing, tend to be less reproduceable, thus weakening the potential to identify significant outcomes. Additionally, unintended psychological harm may exist during active clinical trials or long after the trial has be completed that may be difficult to detect in children.[14,15] Risk monitoring should be implemented in any clinical trial involving pediatric populations in order to continually assess well-being of the sample. This requires specialist-level training and experience with children. These specialists are often underutilized in clinical trials with children, making the detection and assessment of well-being and psychological state more difficult.[15]

Due to the lack of pharmacological interventions to address the core symptoms of ASD, there has been a focus on compounds that may affect social communication directly. Though many have been discussed, one that has received considerable attention in recent decades is oxytocin, a hormone associated with prosocial behaviors in humans and animals.[16,17] Given oxytocin's role in prosocial behaviors, exogenous administration of the neuropeptide has potential noteworthy therapeutic benefits in ASD.

2. Oxytocin

Oxytocin, a neuropeptide produced in the hypothalamus,[18] has been extensively studied and utilized for its effects on social cognition and prosocial behaviors.[19–21] During childbirth, oxytocin is critical for uterine contractions and is necessary for the production of milk.[22,23] Oxytocin is developmentally regulated and its receptors are malleable, particularly in response to parent-child interactions.[24,25] Colloquially referred to as the "love hormone," oxytocin has been implicated in a variety of species, including rats and prairie voles, to be at increased levels during mating and pair-bonding.[26,27] In rats, development appears to play a significant role in both the number and location of oxytocin receptors. Two periods of development have been recognized as important for oxytocin expression: the week which precedes weaning, and puberty.[28] There is evidence that some brain areas which contain high densities of oxytocin binding sites in early life (e.g. cingulate and retrosplenial cortex, and regions of both the basal ganglia and limbic system), have low numbers of binding sites in the adult rat. The contrary has been found as well: some brain areas which do not contain high densities of oxytocin binding sites in early life begin to contain high densities of binding sites around puberty. Though studies of oxytocin binding sites in rats cannot be generalized to humans, this work sheds light on both the complexity of the oxytocin system, and the role that development plays in the maturation of the oxytocin receptors.[29]

In humans, oxytocin can be safely administered exogenously either intranasally or intravenously. Traces of oxytocin that are administered peripherally (e.g. intravenously) are found in cerebrospinal fluid, though uncertainty remains regarding the extent to which oxytocin is able to cross the blood-brain barrier.[30,31] A study in rhesus macaques provided evidence that intranasal oxytocin was detectable in cerebrospinal fluid, however this investigation included only six nonhuman primates who were given a higher dosage of oxytocin than is usually administered in humans.[32] Administration of oxytocin is able to pass the blood-brain barrier, though measurable effects may diminish over time.[33] In healthy individuals, increases in pro-social behaviors are seen after administration of oxytocin, including improved emotion-identification of human faces, increased attention to the eye-region of faces, and enhanced theory of mind abilities.[19,34–37] In the section below, we briefly review studies using neuroscience methodology that investigate both brain and behavioral effects of oxytocin in neurotypical participants.

2.1 Brain-based studies of oxytocin administration in neurotypical individuals

A seminal neuroimaging study examined neural mechanisms of prosocial behaviors that have widely been observed in behavioral investigations of oxytocin administration.[37] Results showed reduced amygdala activation in response to threatening stimuli after oxytocin administration compared to placebo in healthy adults.[37] This response modulation provided the first evidence that increased prosocial behaviors observed after oxytocin administration may be a result of reduced amygdala response. Neuroimaging methods provide evidence that oxytocin administration consistently dampens amygdala activation in response to emotional stimuli. For example, Domes and colleagues[19] found that intranasal oxytocin administration in adult males attenuated amygdala activity to emotional faces, including angry, fearful, and happy faces. Thus, amygdala activation was globally reduced despite the faces' emotional valence.[36]

Using electrophysiological methods, a recent study in neurotypical adults provided evidence that oxytocin administration also affects brain activity related to imitation and social action perception. The authors observed enhanced mu rhythm suppression in response to both performing and viewing tasks of social biological motion compared to nonsocial motion.[38] As such, the mirror neuron system underlying imitation and action perception, which are thought to be impaired in social-communicative disorders such as ASD, appear to be more readily engaged after oxytocin administration.

Prosocial effects of oxytocin administration can both elucidate neural mechanisms underlying social behavior and have implications for treatment of psychiatric disorders that are characterized by social deficits, such as ASD.[39] The subsequent section will review previous work related to oxytocin in ASD.

3. Oxytocin and ASD

Early theories about the relation between oxytocin and ASD began in the late 20th century.[40,41] Modahl and colleagues were the first to provide evidence of reduced levels of oxytocin in children with ASD. Specifically, the authors found that compared to their neurotypical peers, oxytocin does not increase prior to the onset of puberty in individuals with ASD. This suggests that oxytocin is less available to individuals with ASD during development.[42] Our understanding of oxytocin in individuals with ASD has since been shaped by research that suggests deficits in oxytocin may reveal the pathogenesis of ASD. No known mechanistic pathway exists to form a substantial link between the neuropeptide and development of ASD, but studies support that lower concentrations of oxytocin are strongly related to social impairments.[43] As such, oxytocin administration may enhance social communication in ASD.

Measurement of oxytocin levels in individuals with ASD continues to be explored, as binding sites, receptor function, and the production of oxytocin are challenging to study in humans, given limitations of in-vivo neurometabolic investigation. Oxytocin-related genes have been implicated in ASD, to the extent that polymorphisms in the oxytocin receptor genes are related to symptom severity.[44–46] Early studies of the endocrine system have found that oxytocin release occurs differently in children with ASD compared to their neurotypical peers, suggesting that oxytocin may be disrupted early in development.[47] Oxytocin blood plasma levels increase throughout neurotypical development, while children with ASD exhibit lower levels of plasma oxytocin[48] that are stable over time.[42] Therefore, disruptions of the oxytocin system may possibly occur early in life in individuals with ASD, resulting in cascading consequences.

An investigation of early life effects of oxytocin found that when oxytocin was administered to human mothers to induce labor, no negative effects were identified in offspring 20 years later.[49] Specifically, oxytocin exposure early in life did not influence behavioral problems or symptoms of ASD. However, levels of plasma oxytocin are positively correlated with social behaviors in individuals with ASD, their neurotypical siblings, and neurotypical controls.[44] Parker and colleagues found that oxytocin levels were largely heritable between individuals with ASD and their siblings, providing further evidence of oxytocin's role in the biology of social behavior and its heritability.[44] That is, low oxytocin levels are observed in siblings of individuals with ASD compared to neurotypical children. This is particularly important, as approximately 20% of siblings of children with ASD will meet criteria for ASD by age three[50] and these siblings are at a significantly higher risk of developmental delay and general social deficits, even if they do not meet criteria for ASD.[51] Thus, first-degree relatives of siblings with ASD appear to also have reduced oxytocin levels compared to individuals who have siblings without ASD.

Sex differences and comorbidities should also be considered when measuring oxytocin levels in ASD, as oxytocin serum concentrations in females with both ASD and Intellectual Disabilities (ID) have been shown to be positively correlated with increased nonverbal skills and repetitive behaviors.[52] Conversely, in males with ASD and ID, oxytocin concentration has been shown to be negatively correlated with measures of repetitive behaviors.[52] Previous work suggests that males with ASD and ID exhibit greater restricted and repetitive interests than females with ASD and ID,[53] though this may be modulated by oxytocin concentration between sexes. Endogenous oxytocin levels may serve as a potential biomarker for later behavioral manifestations of ASD in females who also have ID and may act as a protective factor in males. That is, additional levels of oxytocin may not always contribute to decreased symptoms of ASD when individual characteristics, such as sex and cognitive ability, are considered. A recent investigation showed that increased variants in the oxytocin receptor gene were associated with greater connectivity between the brain regions related to reward and social cognition (including the nucleus accumbens and frontal pole, superior frontal gyrus, frontal medial and orbital cortex, paracingulate and cingulate cortex, caudate, and putamen) within females with ASD.[54] This investigation highlights the relationship between genetic risk specific to the oxytocin receptor and brain connectivity in females. Interesting, the authors also found that the reward network connectivity with frontal brain regions in females with ASD is similar to that of neurotypical males.[54] This underscores the complexity of oxytocin within individuals with ASD.

It is important to note that findings from studies of oxytocin in ASD tend to be inconsistent. Recent studies with larger samples report no differences in levels of plasma oxytocin between individuals with ASD and neurotypical controls.[44,55,56] Unique findings across studies may relate to how plasma oxytocin is measured and whether examinations of the peripheral or central nervous system influence measurement.[57,58] Using a pharmacologically optimized approach to measure oxytocin receptor binding with postmortem human brains, individuals with ASD were found to have more oxytocin binding receptor sites in the nucleus basalis of Meynert (NBM), but reduced binding receptor sites in the ventral pallidum compared to neurotypical controls.[59] The NBM is located within the basal forebrain and is involved in the regulation of visual attention, gaze, and memory. As individuals with ASD have more binding sites in the NBM than expected, this may lead to atypical attentional allocation to social stimuli.[60] The ventral pallidum is located within the basal ganglia and is associated with the reward system in humans. A dearth of binding sites (e.g. lower density of binding sites) in brain areas associated with the reward system may account for why individuals with ASD have disruptions in social communication, which has implications for how social stimuli are processed.

According to the social motivation hypothesis, individuals with ASD may experience social interactions as less rewarding than their neurotypical peers, which leads to reduced social initiation.[61] Accordingly, it has been hypothesized that the reward system in individuals with ASD is less responsive to social stimuli compared to individuals without ASD (though it is important to note that research findings related to the neural reward system in ASD are mixed). Given oxytocin's role in social affiliation and bonding, it is not entirely surprising that this neuropeptide has been hypothesized to be relevant to potential disruptions in the reward system in individuals with ASD.[62,63] As refined methodologies are developed to locate and identify altered oxytocin receptors in ASD, the field moves closer to understanding the etiology of ASD and how behavior represents the presentation of biological aberrances. Particular attention should be paid to the distribution and density of oxytocin receptors in the reward system in both individuals with and without ASD, as this may clarify the role of oxytocin in the reward system more broadly. Given the hypothesized relation between oxytocin levels and symptoms of ASD, investigators have studied whether oxytocin administration improves social behaviors in this population.

4. Oxytocin administration in individuals with ASD

4.1 Behavioral findings of oxytocin administration in individuals with ASD

Though research on oxytocin in ASD is relatively new, literature on the topic has grown significantly in recent years. The earliest empirical studies of oxytocin administration in individuals with ASD occurred in the early

21st century, when Hollander and colleagues were the first to show that repetitive behaviors were reduced after intravenous oxytocin administration.[64] Subsequent studies noted that oxytocin is generally well tolerated amongst individuals with ASD with few side-effects (see Refs. 65–67), though some negative side-effects have been observed.[68] For example, three participants in an oxytocin trial experienced hyperactivity and aggression, though side-effects ceased once administration was terminated.[68] Outcomes of oxytocin administration in individuals with ASD include increased sustained eye gaze and inferring emotion, social cooperation, improved emotion recognition, and a decrease in symptom severity.[48,69–71] These findings are critical, as the aforementioned behaviors (e.g. eye gaze, ability to infer emotions of others) set the stage for high-order social interactions. These findings are promising, though improvements in social behaviors in isolation may not impact functional social communication skills that make up core deficits of ASD, including the ability to integrate verbal and nonverbal communication skills to effectively socialize with others. In addition to findings related to social behaviors, reductions in repetitive behaviors have been observed after long-term intranasal oxytocin administration (4–6 weeks, once to twice a day) in adults with ASD compared to placebo.[67,72] Interestingly, some of the aforementioned studies failed to observe improvements in social communication after oxytocin administration compared to placebo.[61,66] In both of those studies, oxytocin administration decreased repetitive behaviors but did not have the hypothesized effects on social communicative behaviors.

It is important to note that some studies have failed to observe any significant benefits of oxytocin administration compared to placebo for participants with ASD. For example, Dadds and colleagues[66] conducted a double-blind randomized control trial with 38 male youths with ASD (7–16 years old) during which participated were given either oxytocin or placebo one per day for four consecutive days during parent—child interaction training sessions. Compared to placebo, oxytocin did not improve emotion recognition or social interaction abilities. Similarly, Guastella and colleagues[65] conducted a double-blind placebo-controlled trial in which participants with ASD (12–18 years old) received either oxytocin or placebo twice daily for 8 weeks. Participants were assessed before, during, and after treatment, as well as at a 3-month follow-up. No benefits of oxytocin were observed following treatment. Notably, parents who believed their children received oxytocin reported greater improvements on parent-rating scales compared to those who believed their child received placebo. This has important implications for how open-label trials should be interpreted and provides strong evidence that studies of oxytocin versus placebo should utilize double-blind designs whenever possible.

Several factors must be considered when evaluating the clinical impact of oxytocin administration, including the heterogeneity observed in ASD, comorbid disorders, and participant age.[73,74] Additionally, it is challenging to ascertain the behavioral effects of oxytocin in ASD given the variety of outcome measures that are utilized. Many measures exist to evaluate symptoms of ASD, including questionnaires, surveys, and interviews, though few studies utilize the same measures both before and after oxytocin administration. The use of objective and sensitive measures both before and after oxytocin administration can further our ability to measure the effects and utility of oxytocin in individuals with ASD. Investigations can be supplemented by including neuroimaging findings to guide both future clinical trials and neurobiological studies, respectively.[55,75,76]

4.2 Brain-based findings of oxytocin administration in individuals with ASD

This section will summarize existing studies that have examined the effects of oxytocin on brain function in individuals with ASD (see Table 1). The use of neuroscience measures may not only elucidate the neural basis of neuropeptidergic functioning, but can also serve as a sensitive measure of the effects of oxytocin administration, as brain-based measures are objective and less susceptible to bias than self-report or caregiver-report.

The first investigation of the neural effects of oxytocin administration in ASD was conducted by Domes and colleagues.[77] Participants saw pictures of both neutral faces and houses, and were asked to determine whether the images were the same or different. Though no behavioral changes were observed in response to oxytocin administration, fMRI findings indicated increased activity in the right posterior amygdala in response to faces. The authors posit that amygdala activation may reflect stimulation of a social saliency network in ASD in response to facial images.[77] Thus, amygdala activation is not solely based on the emotional valence of the stimuli, but in individuals with ASD, increased response in this region may reflect increased attentional allocation to social stimuli more broadly. Alternatively, amygdala activation may represent oversensitivity to faces in general, such that neutral faces may be perceived with heightened emotion in individuals with ASD after oxytocin administration. Interestingly, this is in contrast with previous findings on the effects of oxytocin in neurotypical individuals.[19,37,78] That is, neuroscience studies with neurotypical individuals demonstrate that oxytocin administration causes a decrease in amygdala activity in response to

Table 1 Summary of published neuroimaging studies, main findings on the effects of oxytocin administration on individuals with autism spectrum disorder (ASD).

References	Participants	Oxytocin administration	Stimuli/paradigm	Time delay and task length	Major findings
[77]	14 males with Asperger syndrome • M_{age} = 24.0 years, SD_{age} = 6.9 years 14 male neurotypical controls • M_{age} = 24.3 years, SD_{age} = 5.4 years	Intranasally 3 puffs per nostril (Syntocinon), each puff contained 4 IU of oxytocin, for a total of 24 IU. Placebo was administered in the same dosage and contained same ingredients except for the peptide	fMRI 36 black and white still images of neutral faces with direct or averted gaze. 36 black and white images of houses, presented frontally or averted at 45 degrees. Images were paired within each condition. Participants indicated if the pictures of faces or houses were the same or different	45 min between oxytocin administration and fMRI procedure; 45 min in the scanner; Task was roughly 8.20 min long	Behavior: No effect of oxytocin on face matching in ASD Brain: Increased activity in the right posterior amygdala in ASD in response to faces
[84]	21 children and adolescents (3 females; 18 males) with ASD • M_{age} = 13.2 years, SD_{age} = 2.7 years	Intranasally Older participants (16–19 years) received a dose of 24 IU (4 puffs per nostril); 15 year olds received 18 IU (3 puffs per nostril); Younger participants (7–11 years) received 12 IUs, (1 puff per nostril); or placebo. Testing was repeated on consecutive study visits	fMRI Reading the Mind in the Eyes Test (RMET). Participants were instructed to label the mental state of each facial picture, or label the category of automobile images	45 min between oxytocin administration and fMRI procedure; fMRI task was 5.1 min in length	Social condition: Enhanced activity in the dorsal and ventral striatum, premotor cortex, posterior cingulate, inferior parietal lobule, and posterior STS in response to oxytocin

[80]	14 males with Asperger syndrome • $M_{age}=24.0$ years • $SD_{age}=6.9$ years 14 male neurotypical controls • $M_{age}=23.6$ years, $SD_{age}=5.4$ years	Intranasally 3 puffs per nostril (Syntocinon), each puff contained 4 IU of oxytocin, for a total of 24 IU. Placebo was administered in the same dosage and contained same ingredients except for the peptide	fMRI Black and white still images of eyes or mouths of males and females displaying six facial emotions (anger, fear, disgust, happiness, sadness, and surprise). A correct or incorrect emotion label appeared after the images. Participants were instructed to indicate label accuracy	45 min between oxytocin administration and fMRI procedure; 45 min in the scanner	Behavior: Improved emotional labeling after oxytocin administration Brain: Oxytocin administration increased left amygdala activation
[94]	*Experiment 1*: 17 males with ASD • $M_{age}=29.6$ years • $SD_{age}=8.0$ years 17 male neurotypical controls • $M_{age}=30.4$ years • $SD_{age}=5.6$ years *Experiment 2*: 20 males with ASD • $M_{age}=30.8$ years, $SD_{age}=6.0$ years	*Experiment 1*: No oxytocin was administered *Experiment 2:* Intranasal administration: 24 IU of oxytocin (Syntocinon) or placebo. Participants were seen again 1 week after the first study visit. Placebo or oxytocin was randomized at the first visit and participants received the opposite nasal spray at the second visit	*Experiment 1*: Performed Sally-Ann task[93] to infer social emotions and beliefs of others during fMRI *Experiment 2*: Same fMRI task, but performed after oxytocin or placebo administration on two study visits	*Experiment 1*: n/a *Experiment 2*: 45 min between oxytocin administration and fMRI procedure; Sally-Ann task length was 16 min	*Experiment 1*: Lower accuracy at inferring social emotions of others. Lower activation in right anterior insula and superior temporal sulcus during social emotional conditions. Decreased activity was seen in the dorsomedial prefrontal cortex when inferring beliefs of others *Experiment 2*: Improved ability to infer social emotions of others without the presence of emotional cues and enhanced activity in the right anterior insula in individuals with ASD after oxytocin administration

Continued

Table 1 Summary of published neuroimaging studies, main findings on the effects of oxytocin administration on individuals with autism spectrum disorder (ASD).—cont'd

References	Participants	Oxytocin administration	Stimuli/paradigm	Time delay and task length	Major findings
[85]	40 males with ASD • $M_{age}=28.5$ years, $SD_{age}=5.9$ years	Intranasally 24 IU of oxytocin (Syntocinon) or placebo containing the same inactive ingredients	fMRI Friend or Foe judgment task. Congruent condition: Positive facial/positive vocal expression or negative facial/negative vocal expression. Incongruent condition: Facial and vocal expressions of opposite the opposite valence were paired	40 min between oxytocin administration and fMRI procedure	Behavior: Increased response time to both conditions after oxytocin administration
					Brain: Oxytocin enhanced activation in the anterior cingulate cortex (ACC) and dorsomedial prefrontal cortex (dmPFC). Coordinated activity between the ACC and dmPFC mediated the effects of oxytocin on behavioral deficits
[87]	31 males with ASD • $M_{age}=28.8$ years, $SD_{age}=6.0$ years	Intranasally 24 IU of oxytocin (Syntocinon) or placebo containing the same inactive ingredients	fMRI task and ^{1}H-magnetic resonance spectroscopy (^{1}H-MRS; proton nuclear) to measure *N*-acetylaspartate Task: Friend or Foe judgment task, see Ref. [85]	40 min between oxytocin administration and fMRI procedure. The ^{1}H-MRS scan followed the fMRI; time elapsed between oxytocin administration and ^{1}H-MRS scan was 87.5 min on average	*N*-acetylaspartate level, which represents neuronal energy demands, reliably predicted activation of the ventromedial prefrontal cortex (vmPFC) and ACC after oxytocin administration. This finding was independent of fMRI task demands

[82]	32 males with ASD • $M_{age}=22.7$ years • $SD_{age}=4.8$ years 30 male neurotypical controls • $M_{age}=22.6$ years • $SD_{age}=3.2$ years	Intranasally 3 puffs per nostril (Syntocinon), each puff contained 4 IU of oxytocin, for a total of 24 IU. Nostrils were alternated between puffs. Placebo was a saline solution	Event-related potential (ERP) and concurrent electrocardiogram (ECG) Participants viewed images from the International Affective Picture System[83]	60 min between oxytocin administration and start of ERP task. IAPS task was 45 min in length	No oxytocin treatment effects were found between ASD and TD groups, such that ERP responses to images did not differ between groups
[96]	20 participants with ASD (3 females) • $M_{age}=13.2$ years • $SD_{age}=2.8$ years	Intranasally Older participants (16–19 years.) received a dose of 24 IU (4 puffs per nostril); 15 year olds received 18 IU (3 puffs per nostril); Younger participants (7–11 years) received 12 IUs, (1 puff per nostril); or placebo. Testing was repeated on consecutive study visits. Placebo or oxytocin was randomized at the first visit and participants received the opposite nasal spray at the second visit	fMRI. Participants passively viewed a biological motion paradigm (human motion and scrambled motion) and listened to a vocal affect perception paradigm (angry voices and happy voices)	45 min between oxytocin administration and start of fMRI scan	Biological motion condition: enhanced response in the right posterior superior temporal sulcus (pSTS) after oxytocin administration Negative vocal affect condition: Enhanced activation in right brainstem and right amygdala after oxytocin administration

Continued

Table 1 Summary of published neuroimaging studies, main findings on the effects of oxytocin administration on individuals with autism spectrum disorder (ASD).—cont'd

References	Participants	Oxytocin administration	Stimuli/paradigm	Time delay and task length	Major findings
[79]	20 participants with ASD (1 female, 19 males) • $M_{age}=26.4$ years, $SD_{age}=8.5$ years	Intranasally. 3 puffs per nostril (Syntocinon), each puff contained 4 IU of oxytocin, for a total of 24 IU, or placebo	fMRI. 1. Cooperative social ball-tossing task (with fair and unfair fictious players) 2. Face-matching task (some of the faces were observed by participants previously and some were novel. Geometric shapes were also presented)	40–45 min between oxytocin administration and start of fMRI scan. Social-ball task was 13 min in length; face-matching task was 8 min in length	Cooperative game: Increased activation in the anterior OFC. Reduced activation in the amygdala and hippocampus for participants with ASD who received oxytocin. Recognition of social unjustness led to activation of the right insula
					Face matching: Increased activation in the inferior occipital gyrus and fusiform gyrus after oxytocin in the ASD group

[90]	28 children and adolescents with ASD (2 females, 26 males) • $M_{age}=13.4$ years, $SD_{age}=2.4$ years	Intranasally 3 puffs per nostril (Syntocinon), each puff contained 4 IU of oxytocin, for a total of 24 IU, or placebo containing the same inactive ingredients. Nostrils were alternated between puffs over the course of several minutes	fMRI Incentive delay task: nonsocial (money) or social rewards (smiling face)	Time between oxytocin administration and fMRI scan was not reported. Incentive delay task was 32 min in length	Anticipation of nonsocial reward: Increased activity in right nucleus accumbens (NAcc), right frontal pole, left ACC, left superior frontal cortex, bilateral orbital frontal cortex (OFC) after oxytocin administration. Increased functional connectivity during nonsocial reward anticipation (between the right NAcc and the right FP)
					Nonsocial reward outcome: Decreased frontostriatal functional connectivity between left ACC, bilateral postcentral gyrus, left inferior front gyrus, left precentral gyrus, and left medial frontal gyrus

Continued

Table 1 Summary of published neuroimaging studies, main findings on the effects of oxytocin administration on individuals with autism spectrum disorder (ASD).—cont'd

References	Participants	Oxytocin administration	Stimuli/paradigm	Time delay and task length	Major findings
[62]	15 males with ASD • $M_{age}=21.8$ years, $SD_{age}=2.6$ years 24 male neurotypical controls • $M_{age}=22.1$ years, $SD_{age}=1.9$ years	Intranasally 10 puffs for a total of 20 IU (Syntocinon) or placebo	fMRI Probabilistic social reinforcement learning task. Participants were instructed to categorize learning targets. Social and nonsocial feedback was provided	Average of 48 min between oxytocin administration and start of fMRI scan. Task length was 45 min	Behavior: Improvement of social learning in the participants with ASD during social feedback after oxytocin administration Brain: Positive correlation between reward prediction error (RPE) signal and NAcc activation during social feedback was found
[88]	38 males with ASD • $M_{age}=24.4$ years, $SD_{age}=5.2$ years	Intranasally 3 puffs per nostril (Syntocinon), each puff contained 4 IU of oxytocin, for a total of 24 IU. Or placebo (saline natrium-chloride solution)	fMRI resting state to measure amygdala-hippocampal connectivity fMRI was performed before nasal spray (oxytocin or placebo) and after nasal spray administration	30 min between oxytocin administration and start of fMRI scan. Resting state lasted 7 min	Negative correlation between endogenous oxytocin and functional connectivity between amygdala and hippocampal regions; administration of oxytocin reduced amygdala-hippocampal connectivity even further in individuals with ASD

IU, International Units.

emotional stimuli, whereas results from Domes and colleagues found evidence that oxytocin administration enhanced amygdala activation to faces in individuals with ASD. Though these findings appear at odds with one another, it is possible that activation patterns relate to differences between tasks and/or stimuli.

A later study measured brain activity in both a cooperative social game (ball tossing with fair and unfair fictitious computer players) and a face or shape matching task.[79] In the face matching task, participants with ASD who had received oxytocin compared to placebo demonstrated enhanced activity in the inferior occipital gyrus and fusiform gyrus. During the social cooperation game, increased activity in the anterior orbito-frontal cortex (OFC), and reduced activity in the amygdala and hippocampus were observed in participants with ASD after receiving oxytocin.[79] The dampening of these two brain structures may reflect a stress-reducing effect of oxytocin when individuals with ASD are faced with social stimuli.[79] Thus, reduced activation found in brain areas associated with stress maintenance (e.g., the amygdala and the hippocampus) may facilitate improved social communication within certain social contexts. However, as noted above, other findings show increased activation of both right posterior and left amygdala to social stimuli after oxytocin treatment, indicating heightened emotional arousal.[77,80] As previously mentioned, these mixed findings may relate to task variation between studies (e.g. viewing neutral faces versus memory for whether a face is novel or familiar), or to sample characteristics.[81] Additionally, it is important to note that oxytocin administration does not always result in measurable brain-based differences. An electrophysiological investigation of oxytocin's effects in ASD did not differ from placebo in response to neutral, pleasant, and unpleasant images that displayed social and nonsocial information.[82,83]

During the Reading the Mind in the Eyes Test (RMET)—a task which involves "reading" other people's emotions based solely on their eyes—participants with ASD who had been given oxytocin displayed enhanced activity in areas of the brain associated with reward processing (dorsal and ventral striatum) and areas of social attention and perception (premotor cortex, posterior cingulate, inferior parietal lobule, posterior superior temporal sulcus) compared to placebo controls.[84] Though no differences were observed in behavioral accuracy on the RMET between individuals who received oxytocin and those in the placebo group, individuals who received oxytocin displayed increased activation within the medial prefrontal cortex, which is closely associated with the ability to detect and decode information about mental states.[84] This finding underscores the utility of measuring both brain activity and behavior in oxytocin studies, as neural measures may be more sensitive to subtle changes that are undetected with behavioral measures alone.

An fMRI task of social judgment—known as Friend or Foe—requires participants to judge emotions of a human actor who displayed verbal (positive or negative word) and nonverbal (positive or negative facial expression) information in congruent and incongruent conditions.[85] Enhanced activation in the anterior cingulate cortex (ACC) and dorsomedial prefrontal cortex (dmPFC) was positively correlated with improvements in behavioral performance after oxytocin administration compared to placebo.[85] In a previous investigation using a similar task but without oxytocin administration, the ACC and dmPFC were not activated in individuals with ASD.[86] Therefore, both behavioral performance and brain activation can be enhanced after oxytocin administration during a task requiring sophisticated social communication skills.

Using ^{1}H-magnetic resonance spectroscopy (H-MRS), investigators provided evidence of a possible mechanistic pathway underlying ACC and ventromedial prefrontal cortex (vmPFC) activation after oxytocin administration.[87] Independent of the fMRI task, neuronal energy, measured by *N*-acetylaspartate (NAA) level, was positively correlated with both ACC and vmPFC activity.[87] A path analysis further supported the authors' hypothesis that increased NAA levels likely lead to observed increases in brain activity in the ACC and vmPFC during fMRI tasks. The path analysis also suggested that increases in brain activity observed during fMRI relate to observable changes in behavior. Overall, it appears that increased NAA levels lead to increased activation in the ACC and vmPFC during social-cognitive tasks, which then results in observed behavioral improvements on such tasks.[87] This finding is critical, as it is the first to shed light on plausible neural mechanisms underlying changes in both brain activity and behavior in ASD after oxytocin administration.

Explorations into functional connectivity have also been utilized to further understand neurobiological effects of oxytocin administration. Functional connectivity methods can assess the strength of neural connections between brain regions and increase our understanding of how brain regions communicate with one another. Both increases in endogenous and exogenous levels of oxytocin have been shown to be associated with reduction in amygdala-hippocampal connectivity during resting state fMRI.[88] This is consistent with the fMRI findings from Andari and colleagues described above,[70] as they observed decreased activation in both the amygdala and hippocampal regions in individuals with ASD after oxytocin administration. Taken together, both the fMRI and functional connectivity findings suggest that hyperactive connectivity in the amygdala-hippocampal pathway contributes to attenuated social responses seen in individuals with ASD (see Ref. 89). It is important to note that decreased activation in these areas was evidenced by two different fMRI circumstances: resting-state and an task-dependent, which may suggest that brain-based findings are reflective of the testing environment.

Previous research suggests that oxytocin also affects brain regions associated with both learning and reward sensitivity. In a probabilistic reinforcement learning task with adult males with ASD, oxytocin administration increased activity in the nucleus accumbens (NAcc) during social feedback compared to nonsocial feedback.[62] Effects of oxytocin are not limited to social stimuli, but also extend to nonsocial stimuli. During an incentive delay task, where participants received social (smiling face) or nonsocial (money) rewards, both increased and decreased connectivity were observed in nonsocial conditions.[90] Decreased functional connectivity in individuals with ASD was observed in response to nonsocial feedback (between left ACC, bilateral postcentral gyrus, left inferior front gyrus, left precentral gyrus, and left medial frontal gyrus) and increased functional connectivity during anticipation of nonsocial feedback (between the right NAcc and the right FP).[90] This highlights the role of the mesocorticolimbic system as a possible mediator between social stimuli, nonsocial stimuli, and neurobiological function within socially impaired populations.

The anterior insula, a brain area known for its role in social cognition,[91] is less activated in individuals with ASD in response to social stimuli compared to neurotypical controls (see Ref. 92) Aoki and colleagues found that oxytocin administration enhanced activity in the right anterior insula and the ability to make social-emotional inferences during a Sally-Ann theory of mind task[93] in individuals with ASD.[94] The anterior insula appears sensitive to oxytocin across participant populations, as performance in this task is also enhanced in neurotypical controls after oxytocin administration.[35,78,95] This provides notable evidence that social deficits in ASD can be augmented both behaviorally and through brain activation with exogenous oxytocin treatment.

Brain activity in ASD can even resemble observed brain activity in neurotypical controls after oxytocin administration. For instance, Gordon and colleagues[96] measured brain function in children and teens with ASD in response to biological motion. After oxytocin administration, the participants with ASD displayed more activation in the right posterior temporal sulcus when viewing human motion compared to scrambled motion.[96] Such activation has also been observed in neurotypical populations.[97] Participants with ASD also displayed increased connectivity between the NAcc and primary auditory cortex when listening to happy voices; similar connectivity and activation levels have been observed in neurotypical adults when listening to preferred sounds.[98] Thus, augmentation of the mesolimbic pathway has been observed in individuals with ASD after oxytocin administration, and these changes resemble brain activity observed in neurotypical participants.

The mesolimbic pathway forms the basis the dopaminergic reward pathway. It appears that oxytocin affects this pathway in individuals with ASD.[99] Although mechanistic adaptations of neural structures, brain function, and brain connectivity remain unknown, there is evidence that oxytocin can facilitate reward responses in ASD. This is not to say that brain function is "normalized" after oxytocin administration, but rather that the brain may be more readily able to process social information. Overall, neuroimaging studies have accelerated our understanding of how oxytocin can bolster social behaviors in ASD and have provided key insights into how brain areas function in response to this neuropeptide.

5. Discussion

Much remains unknown about how to effectively measure the effects of oxytocin in the human brain. In order to better understand the effects of oxytocin administration, it is crucial to both understand how endogenous levels of oxytocin relate to social behavior, and to obtain more accurate information about dosing and timing of oxytocin administration. In ASD, individual variability exists in oxytocin levels, processing, and receptor sites. Some of this variability may stem from genetic alterations or may be the result of disrupted development. Future studies should attempt to clarify the origins of such variability.

It may be the case that oxytocin administration in ASD is effective once the dose exceeds endogenous levels (e.g. once the dose is greater than the amount of oxytocin available naturally). The bioavailability of oxytocin is quite short and is likely influenced by the dosage of oxytocin. Given the individual variability of oxytocin levels in those with ASD, future research should attempt to create more personalized dosing "schedules" to maximize effectiveness. As more is uncovered about both the development and processing of oxytocin in the brain, we can begin to identify specific neural mechanisms that may relate to the ASD endophenotype. Additionally, once neural mechanisms are better understood, more precise trials of exogenous oxytocin administration can be implemented with specific neural targets. The development of personalized treatments based on genetics, neurobiology, behavior, and the environment can then be developed.[100–102]

Importantly, other factors must be taken into consideration when evaluating current research into the benefits of oxytocin in ASD. It has become common practice to allot 45 min between oxytocin administration and experimental tasks.[77,79,80,84] However, this time delay is largely based on literature of neurotypical individuals rather than those with ASD (see Ref. 103). It will be important for future studies to measure levels of oxytocin in participants with ASD before administration, after administration, and after a time delay in order to clarify both the level of oxytocin that is available after administration and the rate at which oxytocin concentration declines over time. This is important, as the time between oxytocin administration and experimental tasks, length of experimental task, and the amount of oxytocin that successfully crosses the blood-brain barrier may influence outcome measures. A current barrier to generalizing findings related to oxytocin and brain activity in ASD is between-study differences in task length. For example, if Lab A gives each participant 24 IU of oxytocin, waits 45 min, and then has participants complete a 10 min fMRI task, and Lab B uses the same dosage and time delay but has participants complete a 45 min fMRI task, comparing results between studies is problematic. It is likely that the concentration of oxytocin in a given participant at the end of Lab A's fMRI task is significantly higher than the concentration in a participant of oxytocin at the conclusion of Lab B's paradigm.

Furthermore, differences in age, ASD symptom severity, and individual oxytocin-related pathophysiology should be considered in future studies. Future studies should implement a data-driven approach to determine the appropriate dosage and time between administration and behavioral or neuroscientific measurement. Factors such as sex, BMI, medication-use, and comorbid disorders must be taken into account when considering treatment effectiveness and the potential for adverse-events.

5.1 Future directions

Much promise exists in the potential therapeutic use of oxytocin in ASD. Future research should move beyond attempting to improve specific social behaviors (e.g. eye contact) after a one-time oxytocin administration, and instead combine oxytocin administration with behavioral intervention. Initial investigations have begun to take this approach, by pairing oxytocin administration with behavioral intervention,[66] though it is not yet clear whether combining oxytocin and behavioral interventions leads to better outcomes compared to behavioral intervention or oxytocin alone. Dadds and colleagues paired oxytocin administration with Parent-Child Interaction Training over the course of 4 days in children with ASD, such that oxytocin was given before each training session.[66] No differences between the oxytocin and placebo groups were observed on measures of emotion recognition or social skills. However, the timing between oxytocin administration and behavioral training differed across treatment days and nearly half the sample also had attention-deficit/hyperactivity disorder.[66]

We posit that there is a potential to augment long-term social communication outcomes beyond that observed with either single-dose oxytocin administration or behavioral intervention in isolation. We hypothesize that outcomes will be enhanced when combining pharmacological (e.g. oxytocin administration) and behavioral (e.g. evidence-based behavioral interventions) techniques. Although novel in ASD research, the idea of combining medication and therapy to improve outcomes is not novel in research on psychological disorders more broadly. For example, individuals with anxiety or depression often experience the greatest clinical gains when medication is prescribed in conjunction with evidence-based therapy.[104] The studies reviewed above suggest that oxytocin bolsters social behavior by altering the function of specific brain structures, including the amygdala and NAcc. If oxytocin allows individuals with ASD to find social stimuli more rewarding and increase attention to the social environment, behavioral intervention can be implemented after oxytocin administration to leverage these temporary neural changes. Timing behavioral interventions to coincide with neural enhancements of social reward in individuals with ASD might produce the greatest benefits. Fig. 1 depicts a model of how such research may be conducted.

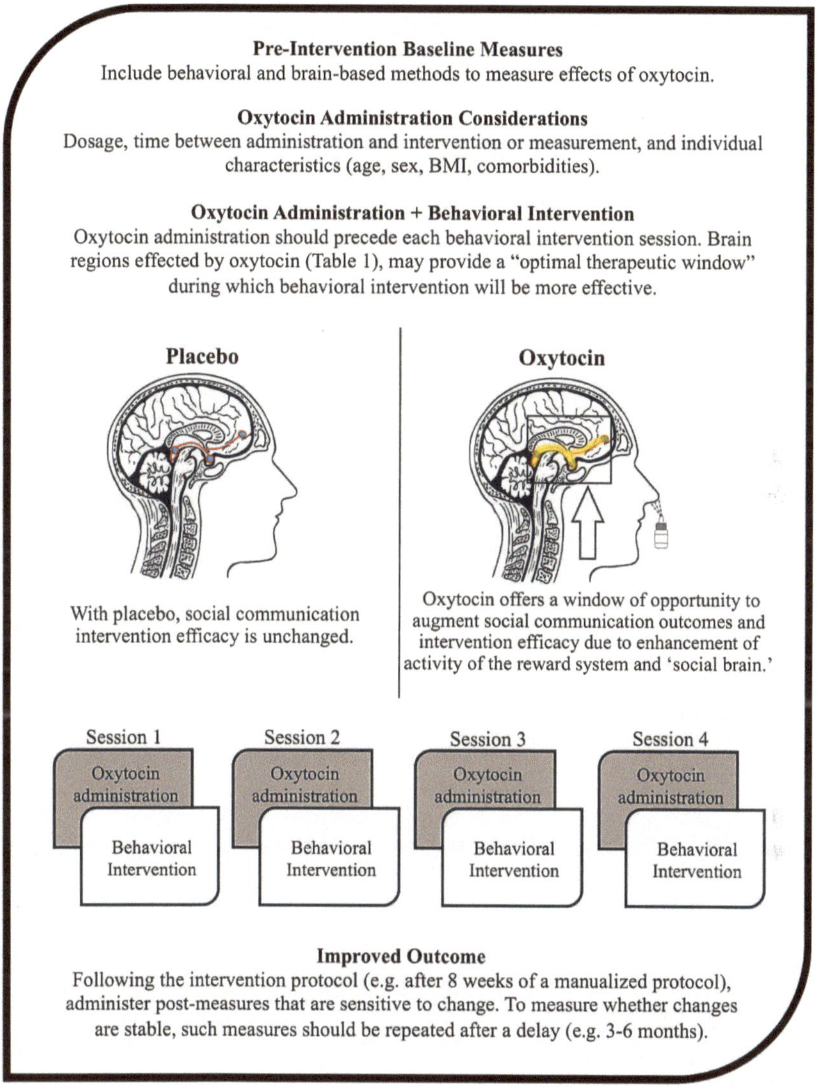

Fig. 1 Suggested model for study design that includes simultaneous behavioral intervention and oxytocin administration. Hypothesized role of the neural reward system on intervention outcomes are shown.

6. Conclusions

In sum, both behavioral and brain-based studies of oxytocin administration in individuals with ASD provide evidence for biological changes related to the enhanced processing of social information. Integrating genetic, neuroscientific, and behavioral investigations will further contribute to the characterization of the ASD endophenotype. With more nuanced information about how oxytocin affects neural circuitry related to social cognition and the brain's reward systems, treatment outcomes in ASD may be improved by leveraging a window of opportunity during which the brain is primed to more efficiently process social information.

References

1. Christensen DL, Braun KVN, Baio J, et al. Prevalence and characteristics of autism spectrum disorder among children aged 8 years—autism and developmental disabilities monitoring network, 11 sites, United States, 2012. *MMWR Surveill Summ.* 2018;65(13):1–23. https://doi.org/10.15585/mmwr.ss6513a1.
2. American Psychiatric Association, ed. *Diagnostic and Statistical Manual of Mental Disorders*. 5th ed. Washington, DC: American Psychiatric Publishing; 2013.
3. Dawson G, Toth K, Abbott R, et al. Early social attention impairments in autism: social orienting, joint attention, and attention to distress. *Dev Psychol.* 2004;40(2):271–283. https://doi.org/10.1037/0012-1649.40.2.271.
4. Zwaigenbaum L, Bryson S, Rogers T, Roberts W, Brian J, Szatmari P. Behavioral manifestations of autism in the first year of life. *Int J Dev Neurosci.* 2005;23(2–3):143–152. https://doi.org/10.1016/j.ijdevneu.2004.05.001.
5. Whitehouse AJO, Durkin K, Jaquet E, Ziatas K. Friendship, loneliness and depression in adolescents with Asperger's syndrome. *J Adolesc.* 2009;32(2):309–322. https://doi.org/10.1016/j.adolescence.2008.03.004.
6. Bauminger N, Kasari C. Loneliness and friendship in high-functioning children with autism. *Child Dev.* 2000;71(2):447–456. https://doi.org/10.1111/1467-8624.00156.
7. Honey E, Leekam S, Turner M, McConachie H. Repetitive behaviour and play in typically developing children and children with autism spectrum disorders. *J Autism Dev Disord.* 2007;37(6):1107–1115. https://doi.org/10.1007/s10803-006-0253-4.
8. Loftin RL, Odom SL, Lantz JF. Social interaction and repetitive motor behaviors. *J Autism Dev Disord.* 2008;38(6):1124–1135. https://doi.org/10.1007/s10803-007-0499-5.
9. Eldevik S, Hastings RP, Hughes JC, Jahr E, Eikeseth S, Cross S. Meta-analysis of early intensive behavioral intervention for children with autism. *J Clin Child Adolesc Psychol.* 2009;38(3):439–450. https://doi.org/10.1080/15374410902851739.
10. Kasari C, Lawton K. New directions in behavioral treatment of autism spectrum disorders. *Curr Opin Neurol.* 2010;23(2):137–143. https://doi.org/10.1097/WCO.0b013e32833775cd.
11. Dawson G, Rogers S, Munson J, et al. Randomized, controlled trial of an intervention for toddlers with autism: The early start Denver model. *Pediatrics.* 2010;125(1):e17–e23. https://doi.org/10.1542/peds.2009-0958.
12. LeClerc S, Easley D. Pharmacological therapies for autism spectrum disorder: a review. *P T.* 2015;40(6):389–397.
13. Pica N, Bourgeois F. Discontinuation and nonpublication of randomized clinical trials conducted in children. *Pediatrics.* 2016;138(3):e20160223. https://doi.org/10.1542/peds.2016-0223.
14. Sammons H. Ethical issues of clinical trials in children: a European perspective. *Arch Dis Child.* 2009;94(6):474–477. https://doi.org/10.1136/adc.2008.149898.
15. Sammons HM, Starkey ES. Ethical issues of clinical trials in children. *Paediatr Child Health.* 2012;22(2):47–50. https://doi.org/10.1016/j.paed.2011.04.011.

16. Green JJ, Hollander E. Autism and oxytocin: new developments in translational approaches to therapeutics. *Neurotherapeutics*. 2010;7(3):250–257. https://doi.org/10.1016/j.nurt.2010.05.006.
17. Guastella AJ, Hickie IB. Oxytocin treatment, circuitry, and autism: a critical review of the literature placing oxytocin into the autism context. *Biol Psychiatry*. 2016;79(3):234–242. https://doi.org/10.1016/j.biopsych.2015.06.028.
18. Gainer H, Altstein M, Whitnall MH. The cell biology and development of vasopressinergic and oxytocinergic neurons. *Prog Brain Res*. 1987;72:153–161. https://doi.org/10.1016/s0079-6123(08)60204-6.
19. Domes G, Heinrichs M, Michel A, Berger C, Herpertz SC. Oxytocin improves "mind-reading" in humans. *Biol Psychiatry*. 2007;61(6):731–733. https://doi.org/10.1016/j.biopsych.2006.07.015.
20. Hurlemann R, Patin A, Onur OA, et al. Oxytocin enhances amygdala-dependent, socially reinforced learning and emotional empathy in humans. *J Neurosci*. 2010;30(14):4999–5007. https://doi.org/10.1523/JNEUROSCI.5538-09.2010.
21. Kosfeld M, Heinrichs M, Zak PJ, Fischbacher U, Fehr E. Oxytocin increases trust in humans. *Nature*. 2005;435(7042):673–676. https://doi.org/10.1038/nature03701.
22. Zingg HH, Laporte SA. The oxytocin receptor. *Trends Endocrinol Metab*. 2003;14(5):222–227. https://doi.org/10.1016/S1043-2760(03)00080-8.
23. Caldeyro-Barcia R, Poseiro JJ. Oxytocin and contractility of the pregnant human uterus. *Ann N Y Acad Sci*. 1959;75(2):813–830. https://doi.org/10.1111/j.1749-6632.1959.tb44593.x.
24. Shapiro LE, Insel TR. Ontogeny of oxytocin receptors in rat forebrain: a quantitative study. *Synapse*. 1989;4(3):259–266. https://doi.org/10.1002/syn.890040312.
25. Krol KM, Moulder RG, Lillard TS, Grossmann T, Connelly JJ. Epigenetic dynamics in infancy and the impact of maternal engagement. *Sci Adv*. 2019;5(10). :eaay0680https://doi.org/10.1126/sciadv.aay0680.
26. Magon N, Kalra S. The orgasmic history of oxytocin: love, lust, and labor. *Indian J Endocrinol Metab*. 2011;15(suppl 3):S156–S161. https://doi.org/10.4103/2230-8210.84851.
27. Carter CS, Porges SW. The biochemistry of love: an oxytocin hypothesis. *EMBO Rep*. 2013;14(1):12–16. https://doi.org/10.1038/embor.2012.191.
28. Gimpl G, Fahrenholz F. The oxytocin receptor system: structure, function, and regulation. *Physiol Rev*. 2001;81(2):629–683. https://doi.org/10.1152/physrev.2001.81.2.629.
29. Tribollet E, Dubois-Dauphin M, Dreifuss JJ, Barberis C, Jard S. Oxytocin receptors in the central nervous system. *Ann N Y Acad Sci*. 1992;652(1):29–38. https://doi.org/10.1111/j.1749-6632.1992.tb34343.x.
30. Mens WB, Witter A, van Wimersma Greidanus TB. Penetration of neurohypophyseal hormones from plasma into cerebrospinal fluid (CSF): half-times of disappearance of these neuropeptides from CSF. *Brain Res*. 1983;262(1):143–149. https://doi.org/10.1016/0006-8993(83)90478-x.
31. Kang YS, Park JH. Brain uptake and the analgesic effect of oxytocin—its usefulness as an analgesic agent. *Arch Pharm Res*. 2000;23(4):391–395. https://doi.org/10.1007/bf02975453.
32. Lee MR, Scheidweiler KB, Diao XX, et al. Oxytocin by intranasal and intravenous routes reaches the cerebrospinal fluid in rhesus macaques: determination using a novel oxytocin assay. *Mol Psychiatry*. 2018;23(1):115–122. https://doi.org/10.1038/mp.2017.27.
33. Landgraf R, Neumann ID. Vasopressin and oxytocin release within the brain: a dynamic concept of multiple and variable modes of neuropeptide communication. *Front Neuroendocrinol*. 2004;25(3):150–176. https://doi.org/10.1016/j.yfrne.2004.05.001.

34. Guastella AJ, Mitchell PB, Dadds MR. Oxytocin increases gaze to the eye region of human faces. *Biol Psychiatry*. 2008;63(1):3–5. https://doi.org/10.1016/j.biopsych.2007.06.026.
35. Baumgartner T, Heinrichs M, Vonlanthen A, Fischbacher U, Fehr E. Oxytocin shapes the neural circuitry of trust and trust adaptation in humans. *Neuron*. 2008;58(4):639–650. https://doi.org/10.1016/j.neuron.2008.04.009.
36. Petrovic P, Kalisch R, Singer T, Dolan RJ. Oxytocin attenuates affective evaluations of conditioned faces and amygdala activity. *J Neurosci*. 2008;28(26):6607–6615. https://doi.org/10.1523/JNEUROSCI.4572-07.2008.
37. Kirsch P, Esslinger C, Chen Q, et al. Oxytocin modulates neural circuitry for social cognition and fear in humans. *J Neurosci*. 2005;25(49):11489–11493. https://doi.org/10.1523/JNEUROSCI.3984-05.2005.
38. Festante F, Ferrari PF, Thorpe SG, Buchanan RW, Fox NA. Intranasal oxytocin enhances EEG mu rhythm desynchronization during execution and observation of social action: an exploratory study. *Psychoneuroendocrinology*. 2020;111:104467. https://doi.org/10.1016/j.psyneuen.2019.104467.
39. Liu Y, Li S, Lin W, et al. Oxytocin modulates social value representations in the amygdala. *Nat Neurosci*. 2019;22(4):633–641. https://doi.org/10.1038/s41593-019-0351-1.
40. Modahl C, Fein D, Waterhouse L, Newton N. Does oxytocin deficiency mediate social deficits in autism? *J Autism Dev Disord*. 1992;22(3):449–451. https://doi.org/10.1007/BF01048246.
41. Waterhouse L, Fein D, Modahl C. Neurofunctional mechanisms in autism. *Psychol Rev*. 1996;103(3):457–489. https://doi.org/10.1037/0033-295X.103.3.457.
42. Modahl C, Green L, Fein D, et al. Plasma oxytocin levels in autistic children. *Biol Psychiatry*. 1998;43(4):270–277. https://doi.org/10.1016/s0006-3223(97)00439-3.
43. Parker KJ, Oztan O, Libove RA, et al. Intranasal oxytocin treatment for social deficits and biomarkers of response in children with autism. *Proc Natl Acad Sci USA*. 2017;114(30):8119–8124. https://doi.org/10.1073/pnas.1705521114.
44. Parker KJ, Garner JP, Libove RA, et al. Plasma oxytocin concentrations and OXTR polymorphisms predict social impairments in children with and without autism spectrum disorder. *Proc Natl Acad Sci USA*. 2014;111(33):12258–12263. https://doi.org/10.1073/pnas.1402236111.
45. Wu S, Jia M, Ruan Y, et al. Positive association of the oxytocin receptor gene (OXTR) with autism in the Chinese Han population. *Biol Psychiatry*. 2005;58(1):74–77. https://doi.org/10.1016/j.biopsych.2005.03.013.
46. Campbell DB, Datta D, Jones ST, et al. Association of oxytocin receptor (OXTR) gene variants with multiple phenotype domains of autism spectrum disorder. *J Neurodev Disord*. 2011;3(2):101–112. https://doi.org/10.1007/s11689-010-9071-2.
47. Green L, Fein D, Modahl C, Feinstein C, Waterhouse L, Morris M. Oxytocin and autistic disorder: alterations in peptide forms. *Biol Psychiatry*. 2001;50(8):609–613. https://doi.org/10.1016/S0006-3223(01)01139-8.
48. Andari E, Duhamel J-R, Zalla T, Herbrecht E, Leboyer M, Sirigu A. Promoting social behavior with oxytocin in high-functioning autism spectrum disorders. *Proc Natl Acad Sci USA*. 2010;107(9):4389–4394. https://doi.org/10.1073/pnas.0910249107.
49. Guastella AJ, Cooper MN, White CRH, White MK, Pennell CE, Whitehouse AJO. Does perinatal exposure to exogenous oxytocin influence child behavioural problems and autistic-like behaviours to 20 years of age? *J Child Psychol Psychiatry*. 2018;59(12):1323–1332. https://doi.org/10.1111/jcpp.12924.
50. Ozonoff S, Young GS, Carter A, et al. Recurrence risk for autism spectrum disorders: a baby siblings research consortium study. *Pediatrics*. 2011;128(3):e488–e495. https://doi.org/10.1542/peds.2010-2825.

51. Charman T, Young GS, Brian J, et al. Non-ASD outcomes at 36 months in siblings at familial risk for autism spectrum disorder (ASD): a baby siblings research consortium (BSRC) study: non-ASD outcomes at 36 months in high-risk siblings. *Autism Res.* 2017;10(1):169–178. https://doi.org/10.1002/aur.1669.
52. Aita C, Mizoguchi Y, Yamamoto M, et al. Oxytocin levels and sex differences in autism spectrum disorder with severe intellectual disabilities. *Psychiatry Res.* 2019;273:67–74. https://doi.org/10.1016/j.psychres.2018.12.139.
53. Hattier MA, Matson JL, Tureck K, Horovitz M. The effects of gender and age on repetitive and/or restricted behaviors and interests in adults with autism spectrum disorders and intellectual disability. *Res Dev Disabil.* 2011;32(6):2346–2351. https://doi.org/10.1016/j.ridd.2011.07.028.
54. Hernandez LM, Lawrence KE, Padgaonkar NT, et al. Imaging-genetics of sex differences in ASD: distinct effects of OXTR variants on brain connectivity. *Transl Psychiatry.* 2020;10(1):1–12. https://doi.org/10.1038/s41398-020-0750-9.
55. Miller M, Bales KL, Taylor SL, et al. Oxytocin and vasopressin in children and adolescents with autism spectrum disorders: sex differences and associations with symptoms. *Autism Res.* 2013;6(2):91–102. https://doi.org/10.1002/aur.1270.
56. Taurines R, Schwenck C, Lyttwin B, et al. Oxytocin plasma concentrations in children and adolescents with autism spectrum disorder: correlation with autistic symptomatology. *Atten Defic Hyperact Disord.* 2014;6(3):231–239. https://doi.org/10.1007/s12402-014-0145-y.
57. Leng G, Ludwig M. Intranasal oxytocin: myths and delusions. *Biol Psychiatry.* 2016;79(3):243–250. https://doi.org/10.1016/j.biopsych.2015.05.003.
58. Leng G, Sabatier N. Measuring oxytocin and vasopressin: bioassays, immunoassays and random numbers. *J Neuroendocrinol.* 2016;28(10):1–13. https://doi.org/10.1111/jne.12413.
59. Freeman SM, Palumbo MC, Lawrence RH, Smith AL, Goodman MM, Bales KL. Effect of age and autism spectrum disorder on oxytocin receptor density in the human basal forebrain and midbrain. *Transl Psychiatry.* 2018;8(1):1–11. https://doi.org/10.1038/s41398-018-0315-3.
60. Jones W, Carr K, Klin A. Absence of preferential looking to the eyes of approaching adults predicts level of social disability in 2-year-old toddlers with autism spectrum disorder. *Arch Gen Psychiatry.* 2008;65(8):946–954. https://doi.org/10.1001/archpsyc.65.8.946.
61. Chevallier C, Kohls G, Troiani V, Brodkin ES, Schultz RT. The social motivation theory of autism. *Trends Cogn Sci.* 2012;16(4):231–239. https://doi.org/10.1016/j.tics.2012.02.007.
62. Kruppa JA, Gossen A, Weiß EO, et al. Neural modulation of social reinforcement learning by intranasal oxytocin in male adults with high-functioning autism spectrum disorder: a randomized trial. *Neuropsychopharmacology.* 2019;44(4):749–756. https://doi.org/10.1038/s41386-018-0258-7.
63. Stavropoulos KKM, Carver LJ. Research review: social motivation and oxytocin in autism—implications for joint attention development and intervention. *J Child Psychol Psychiatry.* 2013;54(6):603–618. https://doi.org/10.1111/jcpp.12061.
64. Hollander E, Novotny S, Hanratty M, et al. Oxytocin infusion reduces repetitive behaviors in adults with autistic and asperger's disorders. *Neuropsychopharmacology.* 2003;28(1):193–198. https://doi.org/10.1038/sj.npp.1300021.
65. Guastella AJ, Gray KM, Rinehart NJ, et al. The effects of a course of intranasal oxytocin on social behaviors in youth diagnosed with autism spectrum disorders: a randomized controlled trial. *J Child Psychol Psychiatry.* 2015;56(4):444–452. https://doi.org/10.1111/jcpp.12305.

66. Dadds MR, MacDonald E, Cauchi A, Williams K, Levy F, Brennan J. Nasal oxytocin for social deficits in childhood autism: a randomized controlled trial. *J Autism Dev Disord*. 2014;44(3):521–531. https://doi.org/10.1007/s10803-013-1899-3.
67. Yamasue H, Okada T, Munesue T, et al. Effect of intranasal oxytocin on the core social symptoms of autism spectrum disorder: a randomized clinical trial. *Mol Psychiatry*. 2018;1–10. https://doi.org/10.1038/s41380-018-0097-2.
68. Yatawara CJ, Einfeld SL, Hickie IB, Davenport TA, Guastella AJ. The effect of oxytocin nasal spray on social interaction deficits observed in young children with autism: a randomized clinical crossover trial. *Mol Psychiatry*. 2016;21(9):1225–1231. https://doi.org/10.1038/mp.2015.162.
69. Hollander E, Bartz J, Chaplin W, et al. Oxytocin increases retention of social cognition in autism. *Biol Psychiatry*. 2007;61(4):498–503. https://doi.org/10.1016/j.biopsych.2006.05.030.
70. Guastella AJ, Einfeld SL, Gray KM, et al. Intranasal oxytocin improves emotion recognition for youth with autism spectrum disorders. *Biol Psychiatry*. 2010;67(7):692–694. https://doi.org/10.1016/j.biopsych.2009.09.020.
71. Anagnostou E, Soorya L, Chaplin W, et al. Intranasal oxytocin versus placebo in the treatment of adults with autism spectrum disorders: a randomized controlled trial. *Mol Autism*. 2012;3(1):16. https://doi.org/10.1186/2040-2392-3-16.
72. Bernaerts S, Boets B, Bosmans G, Steyaert J, Alaerts K. Behavioral effects of multiple-dose oxytocin treatment in autism: a randomized, placebo-controlled trial with long-term follow-up. *Mol Autism*. 2020;11(1):6. https://doi.org/10.1186/s13229-020-0313-1.
73. Wang Y, Wang M-J, Rong Y, He H-Z, Yang C-J. Oxytocin therapy for core symptoms in autism spectrum disorder: an updated meta-analysis of randomized controlled trials. *Res Autism Spectr Disord*. 2019;64:63–75. https://doi.org/10.1016/j.rasd.2019.03.007.
74. Higashida H, Munesue T, Kosaka H, Yamasue H, Yokoyama S, Kikuchi M. Social interaction improved by oxytocin in the subclass of autism with comorbid intellectual disabilities. *Diseases*. 2019;7(1):24. https://doi.org/10.3390/diseases7010024.
75. Insel TR. The challenge of translation in social neuroscience: a review of oxytocin, vasopressin, and affiliative behavior. *Neuron*. 2010;65(6):768–779. https://doi.org/10.1016/j.neuron.2010.03.005.
76. Yamasue H, Domes G. Oxytocin and autism spectrum disorders. In: Hurlemann R, Grinevich V, eds. *Behavioral Pharmacology of Neuropeptides: Oxytocin*. Cham: Springer International Publishing; 2018:449–465. Current Topics in Behavioral Neurosciences. https://doi.org/10.1007/7854_2017_24.
77. Domes G, Heinrichs M, Kumbier E, Grossmann A, Hauenstein K, Herpertz SC. Effects of intranasal oxytocin on the neural basis of face processing in autism spectrum disorder. *Biol Psychiatry*. 2013;74(3):164–171. https://doi.org/10.1016/j.biopsych.2013.02.007.
78. Striepens N, Scheele D, Kendrick KM, et al. Oxytocin facilitates protective responses to aversive social stimuli in males. *Proc Natl Acad Sci USA*. 2012;109(44):18144–18149. https://doi.org/10.1073/pnas.1208852109.
79. Andari E, Richard N, Leboyer M, Sirigu A. Adaptive coding of the value of social cues with oxytocin, an fMRI study in autism spectrum disorder. *Cortex*. 2016;76:79–88. https://doi.org/10.1016/j.cortex.2015.12.010.
80. Domes G, Kumbier E, Heinrichs M, Herpertz SC. Oxytocin promotes facial emotion recognition and amygdala reactivity in adults with asperger syndrome. *Neuropsychopharmacology*. 2014;39(3):698–706. https://doi.org/10.1038/npp.2013.254.
81. Bartz JA, Zaki J, Bolger N, Ochsner KN. Social effects of oxytocin in humans: context and person matter. *Trends Cogn Sci*. 2011;15:301–309. S1364661311000830. https://doi.org/10.1016/j.tics.2011.05.002.

82. Althaus M, Groen Y, Wijers AA, Noltes H, Tucha O, Hoekstra PJ. Oxytocin enhances orienting to social information in a selective group of high-functioning male adults with autism spectrum disorder. *Neuropsychologia*. 2015;79:53–69. https://doi.org/10.1016/j.neuropsychologia.2015.10.025.
83. Lang PJ, Bradley MM, Cuthbert BN. International affective picture system (IAPS): technical manual and affective ratings. *Tech Rep*. 2008;1:39–58.
84. Gordon I, Wyk BCV, Bennett RH, et al. Oxytocin enhances brain function in children with autism. *Proc Natl Acad Sci USA*. 2013;110(52):20953–20958. https://doi.org/10.1073/pnas.1312857110.
85. Watanabe T, Abe O, Kuwabara H, et al. Mitigation of sociocommunicational deficits of autism through oxytocin-induced recovery of medial prefrontal activity: a randomized trial. *JAMA Psychiat*. 2014;71(2):166–175. https://doi.org/10.1001/jamapsychiatry.2013.3181.
86. Watanabe T, Yahata N, Abe O, et al. Diminished medial prefrontal activity behind autistic social judgments of incongruent information. *PLoS One*. 2012;7(6). : e39561https://doi.org/10.1371/journal.pone.0039561.
87. Aoki Y, Watanabe T, Abe O, et al. Oxytocin's neurochemical effects in the medial prefrontal cortex underlie recovery of task-specific brain activity in autism: a randomized controlled trial. *Mol Psychiatry*. 2015;20(4):447–453. https://doi.org/10.1038/mp.2014.74.
88. Alaerts K, Bernaerts S, Vanaudenaerde B, Daniels N, Wenderoth N. Amygdala–hippocampal connectivity is associated with endogenous levels of oxytocin and can be altered by exogenously administered oxytocin in adults with autism. *Biol Psychiatry Cogn Neurosci Neuroimaging*. 2019;4(7):655–663. https://doi.org/10.1016/j.bpsc.2019.01.008.
89. Sweeten TL, Posey DJ, Shekhar A, McDougle CJ. The amygdala and related structures in the pathophysiology of autism. *Pharmacol Biochem Behav*. 2002;71(3):449–455. https://doi.org/10.1016/S0091-3057(01)00697-9.
90. Greene RK, Spanos M, Alderman C, et al. The effects of intranasal oxytocin on reward circuitry responses in children with autism spectrum disorder. *J Neurodev Disord*. 2018;10(1):12. https://doi.org/10.1186/s11689-018-9228-y.
91. Carr L, Iacoboni M, Dubeau M-C, Mazziotta JC, Lenzi GL. Neural mechanisms of empathy in humans: a relay from neural systems for imitation to limbic areas. *Proc Natl Acad Sci USA*. 2003;100(9):5497–5502. https://doi.org/10.1073/pnas.0935845100.
92. Di Martino A, Ross K, Uddin LQ, Sklar AB, Castellanos FX, Milham MP. Functional brain correlates of social and nonsocial processes in autism spectrum disorders: an activation likelihood estimation meta-analysis. *Biol Psychiatry*. 2009;65(1):63–74. https://doi.org/10.1016/j.biopsych.2008.09.022.
93. Baron-Cohen S, Leslie AM, Frith U. Does the autistic child have a "theory of mind"? *Cognition*. 1985;21(1):37–46. https://doi.org/10.1016/0010-0277(85)90022-8.
94. Aoki Y, Yahata N, Watanabe T, et al. Oxytocin improves behavioural and neural deficits in inferring others' social emotions in autism. *Brain*. 2014;137(11):3073–3086. https://doi.org/10.1093/brain/awu231.
95. Wittfoth-Schardt D, Gründing J, Wittfoth M, et al. Oxytocin modulates neural reactivity to children's faces as a function of social salience. *Neuropsychopharmacology*. 2012;37(8):1799–1807. https://doi.org/10.1038/npp.2012.47.
96. Gordon I, Jack A, Pretzsch CM, et al. Intranasal oxytocin enhances connectivity in the neural circuitry supporting social motivation and social perception in children with autism. *Sci Rep*. 2016;6(1):35054. https://doi.org/10.1038/srep35054.
97. Pelphrey KA, Morris JP. Brain mechanisms for interpreting the actions of others from biological-motion cues. *Curr Dir Psychol Sci*. 2006;15(3):136–140. https://doi.org/10.1111/j.0963-7214.2006.00423.x.

98. Salimpoor VN, van den Bosch I, Kovacevic N, McIntosh AR, Dagher A, Zatorre RJ. Interactions between the nucleus accumbens and auditory cortices predict music reward value. *Science*. 2013;340(6129):216–219. https://doi.org/10.1126/science.1231059.
99. Love TM. Oxytocin, motivation and the role of dopamine. *Pharmacol Biochem Behav*. 2014;119:49–60. https://doi.org/10.1016/j.pbb.2013.06.011.
100. Geschwind DH, State MW. Gene hunting in autism spectrum disorder: on the path to precision medicine. *Lancet Neurol*. 2015;14(11):1109–1120. https://doi.org/10.1016/S1474-4422(15)00044-7.
101. Loth E, Murphy DG, Spooren W. Defining precision medicine approaches to autism spectrum disorders: concepts and challenges. *Front Psych*. 2016;7. https://doi.org/10.3389/fpsyt.2016.00188.
102. Sahin M, Sur M. Genes, circuits, and precision therapies for autism and related neurodevelopmental disorders. *Science*. 2015;350(6263):aab3897-1–aab3897-8. https://doi.org/10.1126/science.aab3897.
103. Bethlehem RAI, van Honk J, Auyeung B, Baron-Cohen S. Oxytocin, brain physiology, and functional connectivity: a review of intranasal oxytocin fMRI studies. *Psychoneuroendocrinology*. 2013;38(7):962–974. https://doi.org/10.1016/j.psyneuen.2012.10.011.
104. Cuijpers P, Sijbrandij M, Koole SL, Andersson G, Beekman AT, Reynolds CF. Adding psychotherapy to antidepressant medication in depression and anxiety disorders: a meta-analysis. *Focus*. 2014;12(3):347–358. https://doi.org/10.1176/appi.focus.12.3.347.

CHAPTER NINE

Microglia in animal models of autism spectrum disorders

Megumi Andoh, Yuji Ikegaya, Ryuta Koyama*
Laboratory of Chemical Pharmacology, Graduate School of Pharmaceutical Sciences, The University of Tokyo, Bunkyo-ku, Tokyo, Japan
*Corresponding author: e-mail address: rkoyama@mol.f.u-tokyo.ac.jp

Contents

Abstract

Various genetic and environmental factors have been suggested to cause autism spectrum disorders (ASDs). A variety of animal models of ASDs have been developed and used to investigate the mechanisms underlying the pathogenesis of ASDs. These animal models have contributed to clarifying that abnormalities in neuronal morphology and neurotransmission are responsible for the onset of ASDs. In recent years, researchers have started to focus not only on neurons but also on glial cells, particularly microglia. This is because microglial malfunction is strongly associated with structural and functional abnormalities of neurons, as well as the inflammation that is commonly observed both in the brains of patients with ASDs and in animal models of ASDs. In this chapter, we first introduce a list of commonly available animal models of ASDs and describe the validity of each model from the viewpoint of behaviors and neuroanatomy. We next detail the malfunction of microglia that has been reported in animal models of ASDs and discuss the roles of microglia in ASD pathogenesis. We will further propose possible therapeutic strategies to tackle ASDs by controlling microglial functions.

Progress in Molecular Biology and Translational Science, Volume 173
ISSN 1877-1173
https://doi.org/10.1016/bs.pmbts.2020.04.012

1. Introduction

A variety of genetic and environmental factors have been suggested to cause autism spectrum disorders (ASDs).[1] However, studies using human autopsy brains from individuals with a history of ASD, when available, are not adequate to test whether genetic and environmental factors contribute to the development of ASDs. In order to elucidate the mechanisms underlying the pathogenesis of ASDs, it is essential to make use of appropriate animal models.

To date, a number of animal models of ASDs, which have been proposed to mimic the genetic and environmental factors of ASDs, have been developed and used for studying the mechanisms underlying ASD pathogenesis.[2] The surface validity of these ASD models has been evaluated primarily using behavioral tests that assess the main symptoms of ASDs. Tests such as the number of contacts with other mice in a three-chamber test (social interaction), the properties of ultrasonic vocalization (communication) and the length of grooming behavior or persistence in the same environment (repetitive or persistent behavior) are often used as an index.[2] In addition, whether model animals exhibit complications of ASDs, such as epilepsy, anxiety and neuroanatomical changes (decreased cerebellar Purkinje cell number, atrophy of the prefrontal cortex and hippocampus, increased spine density) is also taken into account.[2]

Common features found in the brains of ASD patients and in ASD animal models can be roughly classified into two phenotypes. The first phenotype includes morphological and functional abnormalities of synapses, which affects neural circuits, possibly triggering ASD-related behaviors. The second phenotype is brain inflammation, which is largely observed in the environmental factor-induced ASD models that mimic maternal immune activation (MIA) and infection. The expression levels of various inflammatory cytokines (tumor necrosis factor α [TNFα], interleukin [IL]-6, IL-18, etc.) tend to increase in the environmental factor-induced ASD models. Because of these pathological changes that are characterized by inflammation, the involvement of microglia, the brain-resident immune cells, in the pathogenesis of ASDs has been highlighted and enthusiastically investigated. As immune cells, microglia produce and release various inflammatory mediators such as IL-1β, TNFα, IL-10, TGF-β; additionally, they have phagocytic capabilities.

In the recent field of neuroscience, microglia are particularly recognized for their phagocytic ability. An increasing number of research studies have focused on activated microglia, which phagocytose dead cells and pathogens under inflammatory conditions caused by infection or trauma.[3] Morphologically activated microglia are often found in the autopsied brain tissue of patients who had ASD, and thus inflammation in ASD animal models has been targeted as a way to study the properties of activated microglia.[4] Furthermore, in recent years, it has been suggested that the phagocytic activity of microglia actively contributes not only to inflammation but also to normal brain development.[5] For example, it was shown that microglia participate in the maturation of functional neural circuits by pruning synapses that are less active in order to maintain more active synapses during development.[6] This discovery has led us to propose the hypothesis that a failure in microglial synaptic pruning causes ASDs.[7]

In this chapter, we first introduce the major animal models of ASDs. Second, we introduce the findings of studies that examined the involvement of microglia in ASDs and discuss the role of microglia in the pathogenesis of ASDs. Finally, we discuss the potential of microglial manipulation as a possible therapeutic strategy for ASDs.

2. Animal models of ASDs

2.1 Genetic animal models (Table 1)

Patients with mental development disorders such as Fragile X syndrome and Rett syndrome often manifest autistic behaviors (symptomatic autism). These diseases are attributed to mutations in a single gene; mice in which each gene is mutated have been used as ASD models. In addition, new ASD models that target other genes, such as *Neuroligin* and *Shank*, have been developed.

2.1.1 Fmr1 (fragile X mental retardation 1)

Fmr1, which is found on the X chromosome, is the gene responsible for Fragile X syndrome. Fmr1 protein (FMRP) regulates mRNA trafficking, dendritic maturation, and synaptic plasticity. When FMRP is not synthesized because of an increase in the number of codon repeats in *Fmr1*, normal brain development is affected. *Fmr1* knockout mice exhibit decreased sociality and persistent behavior.[8,9] In addition, neuroanatomical changes such as

Table 1 Genetically-induced animal models of ASDs.

Gene	Related disease; protein function	Rodent model of gene mutation	Behavioral abnormalities	Neuroanatomical abnormalities	Microglial involvement	Reference
Fmr1	Fragile X syndrome; mRNA trafficking, dendritic maturation, synaptic plasticity	*Fmr1* I304N knock in C57BL/6J mice	Persistence, seizure, hypoesthesia, increased anxiety	Enhanced LTD in hippocampal CA1	–	8
		Fmr1 knockout Sprague-Dawley rats	Decreased sociability, persistence	–	–	9
		Fmr1 knockout C57BL/6 mice	Decreased sociability, persistence	Enhanced LTD in hippocampus, increased spine density in hippocampal CA1	–	10
		Fmr1 knockout C57BL/6 mice	–	Immature spine and increased spine density in hippocampal CA1	Decreased PSD95 phagocytosis	11
Mecp2	Rett syndrome; transcription factor	$Mecp2^{308}$ mutation C57BL/6J mice	Decreased sociability, persistence, seizure, increased anxiety, motor dysfunction	–	–	12
		Viaat-$Mecp2^{-/y}$ C57BL/6J mice	Abnormal sociability, persistence, increased anxiety, motor dysfunction, enhanced prepulse inhibition	Inhibitory interneuron dysfunction, impaired LTP	–	13

		Mecp2$^{-/y}$ C57BL/6J mice	Increased anxiety, motor dysfunction	–	Decreased phagocytosis, improved behaviors by microglia replacement	14
		Mecp2$^{-/y}$ C57BL/6J mice	Motor dysfunction	–	No change in behaviors by microglia replacement	15
		Mecp2$^{-/y}$ *Cx3cr1*^{CreER}; *Mecp2*$^{fl/y}$ *Cx3cr1*^{CreER}; *Mecp2*$^{LSL/y}$ C57BL/6J mice	Motor dysfunction	–	Increased phagocytosis of the presynapse in the lateral geniculate nucleus in *Mecp2* full knockout	16
Tsc1/2	Tuberous sclerosis; inhibition of mTOR signaling	*Tsc1* knockout in cerebellum C57Bl/6J, BALB/c mice	Decreased sociability, persistence, motor dysfunction	Decreased density, abnormal morphology and decreased excitability of cerebellum Purkinje cell	–	17
		Tsc2 missense mutation C57BL/6J mice	Decreased sociability, persistence, impaired communication	–	–	18
Shank3	ASDs; scaffold protein at the postsynapse	*Shank3* knockout C57 mice	Decreased sociability, persistence	Abnormal morphology of striatum neuron, striatum hypertrophy, impaired synaptic transmission between cortex and striatum	–	19

Continued

Table 1 Genetically-induced animal models of ASDs.—cont'd

Gene	Related disease; protein function	Rodent model of gene mutation	Behavioral abnormalities	Neuroanatomical abnormalities	Microglial involvement	Reference
Pten	ASDs, cancer; inhibition of mTOR activation	*Pten* knockout in cortical and hippocampal neurons C57/BL6 mice	Decreased sociability, decreased prepulse inhibition, increased anxiety, seizure	Cortical and hippocampal hypertrophy, abnormal morphology of dentate granule cells	–	20
Cntnap2	CDFE syndrome; synaptic adhesion protein at the presynapse	*Cntnap2* knockout C57BL/6J mice	Decreased sociability, impaired communication, persistence, seizure	Impaired migration of cortical neurons, decreased inhibitory interneuron density	–	21
Scn1a	Dravet syndrome; codes for Nav1.1 (α subunit of sodium ion channel)	*Scn1a* hetero knockout C57BL/6J mice	Decreased sociability, persistence	Decreased number and dysfunction of inhibitory interneuron	–	22
15q11–13 (chromosome region)	ASDs, Prader-Willi syndrome, Angelman syndrome; multiple functions	patDp hetero knockout C57BL/6 mice	Decreased sociability, impaired communication, enhanced anxiety	Increased serotonin activity	–	23
		patDp hetero knockout C57BL/6 mice	Impaired communication, enhanced anxiety	–	Decreased Iba1 intensity in basolateral amygdala	24
		Ube3a duplication C57BL/6 mice	Decreased sociability, impaired communication, persistence	Impaired excitatory synaptic transmission	–	25

Nlgn3/4	ASDs, Asperger syndrome; anchor protein for Neurexin and PSD95	*NL3R451C* knock in 129S2 • SvPasCrl mice	Decreased sociability	–	–	26
		NL4 knockout C57BL/6 mice	Decreased sociability, impaired communication	Decreased volume of cerebellum and brain stem	–	27
Nrx1	ASDs, SCZ, epilepsy, Pitt-Hopkins syndrome; synaptic adhesion protein at the presynapse	*Nrx1a* knockout C57BL/6J mice	Decreased sociability	–	–	28
Shank	ASDs; scaffold protein at the postsynapse	*Shank1* knockout C57BL/6J mice	Decreased sociability, persistence	–	–	29
		Shank2 knockout C57BL/6J mice	Decreased sociability, impaired communication, enhanced anxiety	NMDA receptor dysfunction	–	30
Parvalbumin	ASDs, SCZ, bipolar disorder; calcium-binding protein of inhibitory interneurond	*Parvalbumin* knockout C57BL/6J mice	Decreased sociability, impaired communication, persistence, seizure, hypoesthesia	Dysfunction of inhibitory interneuron		31

Continued

Table 1 Genetically-induced animal models of ASDs.—cont'd

Gene	Related disease; protein function	Rodent model of gene mutation	Behavioral abnormalities	Neuroanatomical abnormalities	Microglial involvement	Reference
Cx3cr1	Psychiatric diseases; fractalkine receptor	*Cx3cr1* knockout C57BL/6 mice	Decreased sociability, persistence	Increased mEPSC frequency of hippocampal CA1 pyramidal cells, decreased functional connectivity between mPFC and hippocampus	–	32
BTBR-strain	–	BTBR mice	Decreased sociability, persistence	–		33
			–	Increased levels of IgG antibody, IL-1β, IL-33 and IL-18	Increased MHC II positive microglia	34

Abbreviations: ASD, autism spectrum disorders; CDFE, cortical dysplasia-focal epilepsy; IgG, immunoglobulin G; LTD, long-term depression; mEPSC, miniature excitatory post-synaptic currents; MHC, major histocompatibility; mPFC, medial prefrontal cortex; NMDA, *N*-methyl-D-aspartate; SCZ, schizophrenia.

enhanced long-term depression (LTD) and enhanced phosphorylation of transcriptional regulators in the hippocampus have been reported.[10] However, it has also been also reported that anxiety, a typical complication of ASDs, was reduced in this animal model.

2.1.2 Mecp2 (methyl-CpG binding protein 2)

Mecp2 is the gene responsible for Rett syndrome. Because *Mecp2* is found on the X chromosome and is a gene, coding transcription factor that binds to methylated DNA cytosine, mutations in *Mecp2* can cause abnormal expression of other genes. Heterozygous *Mecp2* deficient female mice (heterozygous male and homozygous are embryonically lethal) exhibit seizures along with decreased social behavior and persistent behavior.[12] Mice in which *Mecp2* is specifically knocked out in inhibitory interneurons exhibited abnormal social behavior and persistent behavior.[13] This study also showed that the activity of inhibitory interneurons was decreased, which resulted in a change in the balance between excitatory (E) and inhibitory (I) synapses, i.e., the synaptic E/I balance was tipped toward excitation in the hippocampus. These results suggest that ASD is caused by a disruption in the synaptic E/I balance. However, it has been reported that anxiety was reduced in the *Mecp2* mutated mice.

2.1.3 Tsc1/2 (tuberous sclerosis proteins1/2)

Tsc1/2 are the causative genes of tuberous sclerosis. Mutations in *Tsc1/2* activate mammalian target of rapamycin complex 1 (mTOR1) and promote cell growth and proliferation. The mutations also cause intellectual disability, seizures, and ASDs-like symptoms. Mice in which *Tsc1* is specifically knocked out in the cerebellum, and mice who have a missense mutation in *Tsc2* exhibit all the main symptoms of ASDs.[17,18] Based on the structural abnormalities of the cerebellum seen in patients with ASDs, the former report strongly suggests that the cerebellum is a brain region involved in the pathogenesis of ASDs.

2.1.4 Shank3 (SH3 and multiple ankyrin repeat domains 3)

Shank3 is located on chromosome 22 in the 22q13 region and is considered to be a risk gene for the development of ASDs. Shank3 is a post-synaptic scaffold protein that regulates synaptic functions and neurotransmissions. *Shank3b* knockout mice showed reduced social behavior and persistent behavior.[19] In addition, the striatum was enlarged and the excitatory synaptic transmission between the cortex and the striatum was impaired.

2.1.5 Pten (phosphatase and tensin homolog deleted from chromosome 10)

Pten is a tumor suppressor gene that is considered to be a risk gene for development of ASDs. PTEN both negatively regulates the PIP3/Akt signaling pathway by dephosphorylating PIP3 and suppresses the activation of the downstream molecule, mTOR. *Pten*-mutated mice exhibit complication behaviors representative of those observed in patients with ASDs, including decreased social behavior, increased anxiety and epileptic seizures.[20] In addition, hypertrophy of the hippocampus and cortex, and abnormal morphologies of dentate granule cells, such as ectopic dendrites and increased spine density, were reported in these mice.

2.1.6 Cntnap2 (contactin-associated protein-like 2)

Cntnap2 is a member of the neurexin superfamily and is considered to be a risk gene for the development of ASDs. Cntnap2 is involved in synaptic junctions and is abundant in the cortical areas important for language development. *Cntnap2*-mutated mice exhibited decreased social behavior, persistent behavior and epileptic seizures.[21] These abnormalities have been suggested to be caused by both the impaired migration of cortical neurons and the decrease of inhibitory interneurons. Findings in these mice also suggest that there is a relationship between the breakdown of the E/I balance and the onset of ASDs.

2.1.7 Scn1a (sodium voltage-gated channel alpha subunit 1)

Scn1a is the causative gene of Dravet syndrome, an epilepsy syndrome in infants. *Scn1a* encodes Nav1.1, an α subunit of the sodium ion channel. *Scn1a* haploinsufficient mice exhibited impaired social behavior and persistent behavior.[22] This study further showed that administration of clonazepam, an antiepileptic drug, enhanced gamma-amino butyric acid (GABA) transmission and improved the social behavior of *Scn1a* haplosufficient mice. These results also indicate that the disruption of the E/I balance is involved in ASD pathogenesis.

2.1.8 15q11–13 duplicate

15q11–13 is the causative gene of two intellectual disorders, Prader-Willi syndrome and Angelman syndrome.[35] The duplication of this chromosomal region is most frequently observed in patients with ASDs.[35] 15q11–13 is the region where genomic imprinting occurs, and the phenotype differs depending on whether the genetic information is derived from the father or the mother.[35] Therefore, the 15q11–13 duplication is studied in mice using either paternally- or maternally-derived chromosome-duplicated mice. In the paternally-derived chromosome-duplicated mice, decreased social behavior, abnormal communication and increased anxiety were observed, whereas in the maternally-derived chromosome-duplicated mice, no abnormal behavior was reported.[23] 15q11–13 contains *ubiquitin-protein ligase E3A* (*Ube3a*), which is a risk gene for the development of ASDs that is transcribed from a maternal allele. In mice with the triploid *Ube3a*, impaired excitatory synaptic transmission, decreased social behavior and persistent behavior were observed.[25]

2.2 Genetic animal models targeting genes which are not related to symptomatic autism (Table 1)

Genetic analysis of patients with ASDs identified more than 100 risk genes for ASDs. Many of them are related to brain development and function; synapse-related genes are particularly attracting attention.[36] Mutations in synapse-related genes cause abnormalities in the structure of the synaptic junction, resulting in impaired synaptic transmission. Epilepsy, a typical complication of ASDs, is caused by a synaptic E/I imbalance toward higher excitability, suggesting that a disruption of the synaptic E/I balance underlies the mechanism of ASD development.[37] In the following section, we describe abnormalities in synaptic transmissions and behaviors reported in the animal models of ASDs in which genes differing from those that are causative of symptomatic autism are mutated.

2.2.1 Nlgn3/4 (Neuroligin3/4)

Neuroligin is a transmembrane protein present at post-synapses that acts as an anchor between pre-synaptic neurexin and post-synaptic proteins, thus regulating the structure and function of synaptic junctions. Among the five members of the neurogligin family, *Nlgn3* and *Nlgn4* have been identified as risk genes for ASDs. *Nlgn* (R451C) knock-in mice, which were created by mimicking the mutations found in ASD patients, have been reported to exhibit impaired social behavior, while spatial learning was improved and anxiety was reduced.[26] In *Nlgn4* knockout mice, decreased social behavior and impaired communication were reported.[27]

2.2.2 Nrx1 (neurexin1)

Neurexin is a protein that binds to post-synaptic neuroligin as described above. Among *Nrx1–3*, mutations of *Nrx1*, especially the α isoform, are associated with ASDs. *NRX1α* knockout mice showed decreased social behavior.[28]

2.2.3 Shank (SH3 and multiple ankyrin repeat domains protein)

Shank is an excitatory post-synaptic scaffold protein. It binds to the *N*-methyl-D-aspartate (NMDA) receptor (NMDAR) and the metabotropic glutamate receptor (mGluR), and is involved in dendritic spine maturation. In both *Shank1* and *Shank2* knockout mice decreased social behavior and impaired communication were reported.[29,30] In *Shank2* knockout mice, administration of an mGluR5 positive allosteric modulator, which increases the function of NMDAR via mGluR5 activation, improved social behavior.[30] These results indicate that Shank2 knockout possibly decreased NMDAR accumulation to spines, which may contribute to the development of ASDs.

2.2.4 Parvalbumin

Functional deficits of inhibitory interneurons can cause the disruption of the synaptic E/I balance by inducing hyperexcitability of neural circuits. Parvalbumin is a calcium-binding protein which is expressed by some inhibitory interneurons. In mice with a heterozygous or homozygous parvalbumin deficiency, decreased sociality, persistent behavior and impaired communication have been reported.[31] It has also been suggested that these behavioral abnormalities were caused by functional impairment of inhibitory interneurons.

As described above, various genetic models have been established and used to elucidate the pathogenesis of ASDs. However, considering that only about 2% of patients with ASDs have mutations in these genes, they may affect the onset of ASDs by interacting with each other or with environmental factors.[38]

2.3 Animal models created by environmental manipulations (Table 2)

2.3.1 Valproic acid administration

Valproic acid (VPA) is a widely used antiepileptic drug that is also used to treat bipolar disorder, migraines and neuropathic pain. However, the inhibition of histone deacetylation by VPA may have teratogenic effects such as neural tube defects, cardiovascular malformations and delayed neurodevelopment. Prospective and retrospective studies have shown that exposure of pregnant mothers to VPA, especially during the first 3 months of pregnancy, may cause cognitive decline and increase risk of ASDs in their children. Based on these findings, attempts have been made to produce animal models of ASDs by exposing them to VPA. In rodents, offspring exposed to VPA during pregnancy exhibited decreased social behavior and persistent behavior.[39,40] The VPA model mimics not only the behavioral phenotypes but also the neuroanatomical changes that are observed in patients with ASDs, such as a reduced volume of the cerebellum, frontal cortex and hippocampus, as well as abnormal levels of serotonin in the brain.[41–43]

In general, VPA exposure is conducted at embryonic days 8–15 (E8–15), when the fetal tissue formation takes place; the risk of ASD development by VPA exposure is highest around E12. The dose of VPA used in animal models of ASDs is generally 600 mg/kg. However, it should be noted that clinical VPA doses range from approximately 3 to 55 mg/kg (plasma concentration 100 μg/mL), meaning the dose used in animal models is 10–20 times higher.[2] Few studies have mentioned the issue of different VPA doses between these animal models of ASDs and humans. The main argument to justify the differing doses would be that the pharmacokinetics of VPA is different between species: high bioavailability in humans and rapid clearance in rodents. Additionally, rodents receive a single dose whereas humans generally receive multiple doses. Some studies have exposed rodents to VPA postnatally; one such rat model showed impaired social behavior and motor ability after VPA exposure.[44]

2.3.2 Maternal immune activation

Epidemiological studies have shown that fever and severe infection during pregnancy increase the risk of ASDs in offspring. This is likely related to the immune response of mother; MIA has been shown to greatly affect central nervous system development in offspring.[59] Many MIA models have focused on the process of the inflammatory response rather than the effects of specific pathogens. In this article, we outline the lipopolysaccharide (LPS) model that mimics bacterial infection and the polyinosinic-polycytidylic acid (poly[I:C]) model that mimics viral infection.

2.3.2.1 LPS administration

LPS is a component of the outer membrane of the cell wall of Gram-negative bacteria. LPS elicits an immune response by activating toll-like receptor 4 (TLR4). It has been shown that in rodents, LPS exposure during pregnancy induces decreased social behavior and impaired communication in the offspring.[45,46]

2.3.2.2 Poly(I:C) administration

Poly(I:C) is a double-stranded RNA that elicits an immune response by activating toll-like receptor 3 (TLR3). When poly(I:C) is intraperitoneally administered to pregnant mice and rats, the offspring showed the main symptoms of ASDs including decreased social behavior, impaired communication and persistent behavior.[48,49] In addition, behavioral abnormalities related to ASDs such as increased anxiety, learning disability, paresthesia and seizures have been observed in the poly(I:C) model. The poly(I:C) model mimics not only the behaviors observed in ASDs but also the neuroanatomical changes such as disordered cerebellar Purkinje cell location and decreased expression of inhibitory interneuron-related proteins.[50,51]

The MIA model is also used to study other psychiatric diseases such as schizophrenia; it has been controversial as to whether it is appropriate to consider it an "ASD model." Indeed, infection during pregnancy may be associated with various central nervous system disorders. However, it would be pointless to argue whether the MIA model mimics a specific disease. Rather, we should not narrow the usefulness of the MIA model and remember that it is important to study multiple aspects of behavioral changes in each disease by properly choosing animal models.

3. Animal models of ASD exhibiting possible microglia involvement

In the following section, we introduce studies that used animal models to examine the involvement of microglia in ASDs, specifically focusing on abnormalities in inflammatory responses and synaptic morphology and function.

3.1 Genetic animal models (Table 1)

3.1.1 *Fmr1* knockout

Jawaid et al. conducted a study that focused on synaptic and microglial changes in *Fmr1* knockout mice.[11] At postnatal day 60 (P60), the spine density of CA1 pyramidal cells increased in *Fmr1* knockout mice compared to the control. They observed the spine morphology in these mice and found that mature spines decreased and immature spines increased, which suggests that spine maturation was impaired. The authors attributed the increased spine density in *Fmr1* knockout mice to a failure in microglial synaptic pruning. They measured postsynaptic density protein 95 (PSD95) phagocytosis by microglia in the hippocampus at P21, finding that microglial phagocytosis of PSD95 was significantly decreased in *Fmr1* knockout mice. Thus, it is possible that pyramidal cells in *Fmr1* knockout mice have abnormalities in synaptic plasticity and axonal terminal maturation, which may inhibit neuronal activity-dependent synaptic pruning by microglia.

3.1.2 *Cx3cr1* knockout

CX3CR1 is a fractalkine receptor which is expressed specifically by microglia in the brain parenchyma. CX3CR1 signaling, which is activated by its ligand CX3CL1, regulates adult neurogenesis in the hippocampus, a phenomenon related to several psychiatric disease. Zhan et al. found that *Cx3cr1* knockout mice exhibited decreased social behavior in the three-chamber test and persistent behavior (grooming).[32] They recorded miniature excitatory post-synaptic currents (mEPSCs) from CA1 pyramidal cells in hippocampal acute slices at P15 and P40. While the frequency of mEPSCs decreased from P15 to P40 in wild-type mice, this decrease was not observed in *Cx3cr1* knockout mice. These results suggest that synaptic pruning by microglia was impaired in *Cx3cr1* knockout mice. Using local field potential recordings and functional magnetic resonance imaging analysis from multiple brain regions, they also found that *Cx3cr1* knockout mice had reduced functional connectivity between the frontal cortex and the hippocampus, as well as in the connection between the left and right hippocampus. Furthermore, it was shown that there was a positive correlation between the strength of the frontal cortex–hippocampus functional connectivity and the duration of social interaction. These findings suggest that these brain regions are important for social behavior.

3.1.3 BTBR mice

BTBR is a mouse strain which exhibits ASD-like behaviors such as reduced social behavior in the three-chamber test and persistent behavior in the water maze task.[33] Heo et al. compared BTBR mice with other mouse strains to test whether the inflammatory response is abnormal in BTBR mice.[34] They found that the levels of serum immunoglobin G (IgG) and IgE and the levels of IgG in the brain were increased in BTBR mice compared to B6 mice. In addition, they found that BTBR mice showed an increase in the expression levels of IL-33, IL-18 and IL-1β as well as an increase in the number of major histocompatibility (MHC) II positive microglia, which indicates that both microglial activation and inflammatory response were promoted in BTBR mice.

3.1.4 patDp/+ mice

In male mice having paternal duplication (patDp/+) of the mouse chromosome region corresponding to human chromosome 15q11–13, abnormal communication (increased ultrasonic vocalization) and increased anxiety in the open field test have been observed.[24] In this study, microglial properties were examined in the amygdala, a brain region that regulates social behavior and anxiety. In the basolateral amygdala, the number of microglia did not change, but the expression levels of Iba1, a calcium-binding protein whose expression is increased in activated microglia, were decreased. Furthermore, administration of minocycline, which is widely used to inhibit microglial activation, from E17 to P21 did not alter the communication ability, but improved anxiety behavior in patDp/+ mice.

3.1.5 *Mecp2*

Derecki et al. tested the involvement of microglia in abnormal behaviors of male *Mecp2*$^{-/y}$ mice.[14] As mentioned in Section 2.1.2, increased social behavior, persistent behavior, increased anxiety and impaired motor ability have been confirmed in this mouse. Microglia were removed from the *Mecp2*$^{-/y}$ mice at P28 by γ-irradiation, followed by intravenous transplantation of bone marrow-derived wild-type microglia. The transplant improved dyskinesia and anxiety, and prolonged the lifetime of *Mecp2*$^{-/y}$ mice. The ex vivo capability for phagocytosis of dead cells by microglia isolated from the cortex of *Mecp2*$^{-/y}$ mice was reduced compared with wild-type. These results indicate that reduced phagocytosis by microglia in *Mecp2*$^{-/y}$ mice may lead to the development of Rett syndrome. On the contrary, another study reported that wild-type microglial transplantation neither prolonged the lifespan nor improved motor function in *Mecp2*$^{-/y}$ mice.[15]

Schafer et al. focused on neural circuit regulation by microglia in *Mecp2*$^{-/y}$ mice.[16] They found that in the lateral geniculate nucleus, phagocytosis of pre-synapses by microglia was significantly increased in *Mecp2*$^{-/y}$ mice at P56 when brain development was finished. However, when Mecp2 was gained or lost in a microglia-specific manner using *CX3CR1*^{CreER} mice, the abnormalities in synaptic phagocytosis and the decreased motor function in *Mecp2*$^{-/y}$ mice were not changed. These results suggest that the changes in microglial function caused by *Mecp2* knockout are not a direct cause of Rett syndrome.

3.2 Animal models using environmental manipulations to induce disease (Table 2)

Many studies have analyzed microglial morphology and function using animal models of ASDs that mimic MIA during pregnancy (for details and their references, please see Table 2). In the following, we introduce studies that verify ASD-like behaviors and microglial changes. Although not specifically

Table 2 Reagent-induced animal models of ASDs.

Reagent	Animal	Administration conditions (timing, dose, route)	Behavioral abnormalities	Neuroanatomical abnormalities	Microglial involvement	Reference
Valproic acid	C57BL/6Hsd (B6) mice	E13 600 mg/kg subcutaneous administration	Decreased sociability, impaired communication, persistence, decreased prepulse inhibition	Decreased power of gamma-wave	–	39
	Wistar rats	E12 600 mg/kg intraperitoneal administration	Decreased sociability, persistence, decreased prepulse inhibition, hypoesthesia, hyperesthesia	–	–	40
	Long Evans rats	E12 600 mg/kg intraperitoneal administration	–	Decreased cell number in cerebellum, decreased volume of cerebellum	–	41
	Sprague-Dawley rats	E12 500 mg/kg intraperitoneal administration	–	Decreased prefrontal cortex volume, hippocampal CA1, basolateral amygdala	–	42
	Wistar rats	E9 600 mg/kg subcutaneous administration	Decreased sociability	Decreased serotonin level in the hippocampus	–	43
	Long Evans rats	P6–12 150 mg/kg intraperitoneal administration	Decreased prepulse inhibition, hypoesthesia, motor dysfunction	–	–	44

Continued

Table 2 Reagent-induced animal models of ASDs.—cont'd

Reagent	Animal	Administration conditions (timing, dose, route)	Behavioral abnormalities	Neuroanatomical abnormalities	Microglial involvement	Reference
LPS	CD1 mice	E9 0.008 mg/kg intraperitoneal administration	Decreased sociability, impaired communication, persistence, enhanced anxiety	Increased volume of the cortex and subventricular zone, increased ROS level in the subventricular zone	Increased microglial number in the frontal cortex	45
	Sprague-Dawley rats	E15 0.25 mg/kg intraperitoneal administration	Decreased sociability	Increased levels of TNFα, IL-6, and IL-1β		46
	C57BL/6 mice	E15 100 μg/kg intraperitoneal administration	Decreased sociability, impaired communication, persistence	Increased spine density of hippocampal granule cells	Decreased level of CX3CR1 mRNA	47
Poly(I:C)	C57BL/6J mice	E10, E12, E14 5 mg/kg intraperitoneal administration	Decreased sociability, impaired communication, persistence	Abnormal cortical formation		48
	C57BL/6 mice	E12 20 mg/kg intraperitoneal administration	Decreased sociability, impaired communication, persistence	Increased levels of TNFα, IL-6, IL-1β, and IL-17		49

	C57BL/6J mice	E12 20 mg/kg intraperitoneal administration	–	Decreased Purkinje cell density in the cerebellum	–	50
	C57BL/6N mice	E9 1 mg/kg intravenous administration	–	Decreased PV interneuron number in the dentate gyrus, decreased expression level of reelin in hippocampal CA1/3	–	51
	C57BL/6 mice	E15 5 mg/kg intraperitoneal administration	Decreased sociability	Increased level of IL-6 in the hippocampus	Increased Iba1 intensity in hippocampus, decreased gene expression related phagocytosis, decreased phagocytic capacity, improved behaviors by minocycline	52
	C57BL/6J mice	E12, E17 3 mg/kg, 5 mg/kg intraperitoneal administration	Decreased sociability, persistence, enhanced anxiety	Increased synaptic density in hippocampal CA3	Decreased synaptic phagocytosis in hippocampal CA3	7
IL-6	C57BL/6 mice	E12–P0 5 μg intraperitoneal administration	Decreased sociability, enhanced anxiety	Impaired migration of GABA expressing interneurons	Abnormal morphology of cortical plate	53

Continued

Table 2 Reagent-induced animal models of ASDs.—cont'd

Reagent	Animal	Administration conditions (timing, dose, route)	Behavioral abnormalities	Neuroanatomical abnormalities	Microglial involvement	Reference
Diesel exhaust particle	C57BL/6 mice	E2, E5, E8, E12, E16 50 μg oropharyngeal aspiration	–	–	Abnormal morphology of hippocampal microglia, increased interaction between microglia and neuron in parietal cortex	54
Diesel exhaust particle + maternal stress	C57BL/6 mice	Diesel exhaust particle: E2-E17 (6 times in total) 50 μg oropharyngeal aspiration maternal stress: E12–E17	Impaired cognitive function and memory, enhanced anxiety	Increased level of IL-1β, decreased level of IL-10	–	55
			Decreased sociability, impaired communication, enhanced anxiety, impaired cognitive function and memory	Increased IL-1β/IL-10 ratio, increased expression of TLR4 and caspase1	Abnormal morphology of cortical and hippocampal microglia	56
Ovalubumin	C57BL/6J mice	Before mating 10 μg intraperitoneal administration E9, E12, E17 aerosol	–	–	Changed microglial gene expression pattern	57
			Decreased sociability, persistence	–	–	58

Abbreviations: E, embryonic day; GABA, gamma-amino butyric acid; IL, interleukin; LPS, lipopolysaccharide; P, postnatal day; poly(I:C), polyinosinic-polycytidylic acid; ROS, reactive oxygen species; TLR4, toll-like receptor 4.

mentioned here, it should be noted that researchers have adopted a variety of drugs, dosages, administration routes and administration timings for use in these models, and that there are studies conducted using MIA animal models that have reported no microglial changes.

3.2.1 IL-6 administration

Smith et al. exposed groups of pregnant mice to either restraint stress or IL-6 administration from E12 to E19 and examined the properties of microglia in embryonic and adult offspring.[53] Although the number of microglia in the cortical plate did not change, the number of multivacuolated microglia increased in the offspring of pregnant mice exposed to restraint stress compared to the offspring of untreated mice. This morphological abnormality was also observed in the offspring of pregnant mice exposed to IL-6 administration; the increase in multivacuolated microglia was suppressed when and IL-6 antibody was administered simultaneously with exposure to restraint stress. Further, in the adult offspring, microglia became highly ramified when exposure to restraint stress or IL-6 administration occurred during pregnancy. These results indicate that IL-6 affects microglial properties.

Offspring from pregnant mice subjected to restraint stress showed increased anxiety in the elevated plus maze and decreased social behavior in the three-chamber test; however, these behavioral changes were not improved by IL-6 antibody administration. They also examined the migration of GABAergic neurons to the cortical plate and showed that both restraint stress and IL-6 administration inhibited this migration, a phenomenon which was not improved by IL-6 antibody administration. These results suggest that the relationship between restraint stress-induced ASD-like behaviors and morphological changes in microglia remains questionable.

3.2.2 LPS administration

Le Belle et al. administered LPS intraperitoneally (0.008 mg/kg) to CD-1 mice at E9 to induce MIA.[45] Offspring showed impaired communication (decreased duration of ultrasonic vocalization), decreased social behavior in the three-chamber test, increased anxiety in the elevated plus maze and increased persistent behavior (grooming). At P0, increased thickness of the cerebral cortex, increased number of microglia in the forebrain and increased proliferation of progenitor cells in the subventricular zone (SVZ) were observed. The authors also showed that the production of reactive oxygen species in the SVZ was increased in the LPS group at P3.

Further, simultaneous administration of LPS and apocynin, a nitrogen oxide (NOX) inhibitor, suppressed these structural changes and improved persistent behavior. However, NOX is expressed not only in microglia but also in neurons and astrocytes. Thus, it is unclear whether the LPS-induced neurological changes are a microglia-dependent phenomenon.

Another researcher induced MIA by administering LPS intraperitoneally (100 μg/kg) to pregnant C57BL/6 mice at E15.[47] In this protocol, offspring exhibited impaired communication (reduced duration of ultrasonic vocalization), decreased social behavior in the three-chamber test and persistent behavior in the marble burying test. The authors hypothesized that MIA caused synaptic abnormalities and investigated the morphology of hippocampal granule and pyramidal cells using Golgi staining. They found that the spine density of the granule cells in male mice was significantly increased after MIA. Next, they examined the mRNA expression levels of molecules involved in synaptic pruning. They found that the *Cx3cr1* mRNA level was decreased in male offspring from the LPS treated group. These results indicate that environmental factors can affect microglia-dependent synaptic pruning.

3.2.3 Poly(I:C) administration

Mattei et al. administered poly(I:C) intraperitoneally (5 mg/kg) to pregnant C57BL/6 mice at E15 to induce MIA.[52] Offspring exhibited decreased social behavior in the three-chamber test at P120. Administration of minocycline to MIA offspring from P70–P80 for 5 weeks improved social behavior. In the hippocampus of the MIA offspring, the intensity of Iba1 staining was increased (fluorescent immunostaining) and the IL-6 concentration was elevated (ELISA), suggesting microglial activation in response to MIA. Additionally, transcriptome analysis of hippocampal microglia revealed that MIA decreased the expression levels of phagocytosis-related genes such as *Cx3cr1*, *Itgam* and *Mertk*, among others. These changes were reversed by minocycline administration. Furthermore, hippocampal microglia isolated from the offspring of the MIA group showed a reduced capacity to phagocytose latex beads compared to the control group; this activity was also rescued by minocycline administration.

In our study, pregnant C57BL/6J mice were administered poly(I:C) at E12.5 and E17.5 (3 mg/kg on E12.5 and 1.5 mg/kg on E17.5) to induce MIA.[7] MIA offspring showed decreased social behavior in the three-chamber test, persistent behavior (grooming) and increased anxiety in the novelty suppressed feeding test at P60. We found no changes in microglial morphology or density in the postnatal offspring. However, microglial phagocytosis of synapses in the hippocampal CA3 field was reduced at P18 in the offspring of the MIA group, resulting in surplus synaptic density. These results indicate that administration of poly(I:C) to pregnant mice suppresses the phagocytic activity of microglia and inhibits synaptic pruning in offspring, possibly inducing ASD-like behaviors.

3.2.4 Air pollution

To mimic exposure to air pollution, Bilbo and colleagues delivered diesel exhaust particles (DEP) to pregnant C57BL/6 mice by oropharyngeal aspiration at E2, E5, E8, E12 and E16.[54] In the hippocampal CA1 at E18, the number of immature microglia (round and stout morphology) was increased in offspring from pregnant mice exposed to DEP. This change was not observed in *Tlr4* KO mice, which suggests that DEP affects the brain structure via TLR4. At P30, the volume of microglia increased in the parietal cortex and the degree of contact between the cell bodies of microglia and neurons increased. These results suggest that microglia were activated by DEP, but it remains unclear how microglial functions were affected.

The authors have previously shown that DEP alone did not affect learning ability in the fear conditioning test or anxiety behavior assessed by elevated zero-maze in offspring at P60.[55] Next, they combined DEP with maternal stress (MS) by limiting bedding during pregnancy. These manipulations attenuated learning ability and increased anxiety in the offspring. In addition, abnormal communication (increased number of calls of ultrasonic vocalization) and decreased social behavior in the three-chamber social exploration test were confirmed in the offspring of the DEP + MS group.[56] The authors noted that the use of animal models with a single environmental stimulus could be too simplistic to assess the effects of the stimulus, suggesting that a single stimulus alone is not enough to elucidate the complex mechanisms underlying the development of disease, which is influenced by multiple environmental stimuli in the real world.

3.2.5 Ovalbumin administration

Vogel Ciernia et al. studied the mechanisms by which allergic asthma during pregnancy (maternal allergic asthma; MAA) increases the risk of ASD development in offspring.[57] First, ovalbumin (OVA) was intraperitoneally administered to C57BL/6J female mice before mating. Additional OVA was delivered by aerosol as secondary sensitization at E9, E12 and E17. In the male offspring of the MAA group, decreased social behavior in the three-chamber test and increased persistent behavior in the marble burying test were observed, though persistent behavior (grooming time) was reduced.[58] The authors isolated microglia from the whole brains of P35 offspring and carried out RNA sequencing.[57] They found that the overlap between the genes for which expression levels were reduced in the MAA offspring and those reduced in the brains of humans with ASDs was high. In addition, DNA methylation patterns of the MAA offspring and those of poly(I:C) offspring were similar. Although the gene expression pattern in MAA offspring differs from that of other MIA models (LPS and IL-6) and does not overlap with ASD risk genes, these results suggest that the MAA model is a helpful tool to elucidate the role of microglia in the ASD brain.

3.3 Possible animal models of ASD (Table 3)

In recent years, it has been found that mice with modified microglia-specific genes exhibit ASD-like behaviors and neurological changes related to ASD pathology.

3.3.1 Trem2 knockout

Triggering receptor expressed on myeloid cells 2 (Trem2) is an innate immune receptor that belongs to the immunoglobulin superfamily and is expressed specifically by microglia in the brain. *Trem2* gene mutations have been reported to be involved in the development of various neurodegenerative diseases such as Nasu-Hakola disease, Alzheimer's disease, Parkinson's disease and amyotrophic lateral sclerosis. *Trem2* knockout mice at P18–P20 showed increased mEPSC frequency in pyramidal cells and increased synaptic density in the hippocampal CA1.[60] The authors performed a synaptic phagocytosis assay in vivo using the neuron-microglia co-culture system, showing that *Trem2* knockout reduced synaptic phagocytosis by microglia. These findings are consistent with the mechanisms suggested as the possible cause of ASDs; in fact, P90 *Trem2* knockout mice showed ASD-like behaviors such as persistent behavior assessed by marble burying test, and increased grooming and decreased social behavior in the three-chamber test. Postmortem analysis in the brains of patients with ASDs has shown decreased expression of Trem2 protein.

3.3.2 Atg7 knockout

Kim et al. focused on autophagy using mice in which *Atg7*, an autophagy-related gene specifically expressed by bone marrow as well as microglia, is knocked out.[61] *Atg7* knockout mice exhibited decreased social behavior

Table 3 Possible animal models of ASDs.

Gene	Related disease; protein function	Rodent model of gene mutation	Behavioral abnormalities	Neuroanatomical abnormalities	Microglial involvement	Reference
Trem2	Alzheimer's disease, Parkinson's disease, amyotrophic lateral sclerosis, Nasu-Hakola disease; innate immune receptor belonging to immunoglobulin super family	*Trem2* knockout C57BL/6 mice	Decreased sociability, persistence	Increased mEPSC frequency and synaptic density of hippocampal CA1 pyramidal cells	Decreased synaptic phagocytosis	60
Atg7	Ubiquitin activation enzyme regulating autophagy	Bone marrow cell-specific *Atg7* knockout C57BL/6 mice	Decreased sociability, persistence	Increased spine density and synaptic protein expression in the sensory cortex	Increased PSD95 within microglia	61
Csf1	Hereditary diffuse leukoencephalopathy with spheroid; ligand of CSF1R; CSF1/CSF1R signaling is necessary for microglial survival and proliferation	Nestin expressing cell-specific *Csf1* knockout C57BL/6 mice	Decreased sociability, motor dysfunction	Decreased morphological complexity of Purkinje cells, increased synaptic density projecting to Purkinje cells, impaired migration of Purkinje cells	Decreased microglial density in the cerebellum	62

Abbreviation: mEPSC, miniature excitatory post-synaptic currents.

in the three-chamber test and persistent behavior in the marble burying test. Further, in *Atg7* knockout mice, spine density and expression levels of synapse-related proteins were increased in the sensory cortex, suggesting that synaptic phagocytosis by microglia was impaired; however, the level of PSD95 puncta contained in the microglia was increased. From these results, it was suggested that autophagy dysfunction impaired the digestion of synapses by microglia, leading to the inhibition of further synaptic phagocytosis.

3.3.3 CSF1-CSF1R signaling deficiency

Colony-stimulating factor 1 (CSF1) receptor (CSF1R) has two ligands, CSF1 and IL-34. It has been shown that signals mediated by CSF1R are involved in microglial differentiation and proliferation. Recently, an increasing number of studies have examined microglial functions by depleting microglia using CSF1R inhibitors. Kana et al. examined effects of nestin-positive cell-specific knockout of *Csf1*.[62] In these mice, IL-34 (CSF1R ligand) was expressed normally in the cortex and hippocampus, but was scarcely expressed in the cerebellum. Therefore, in nestin-positive cell-specific *Csf1* knockout mice, the microglial density was remarkably reduced in the cerebellum because CSF1 was knocked out in addition to the basically low level of IL-34. In addition, cerebellar Purkinje cells showed impaired migration as well as decreased dendritic length and complexity, and increased synaptic density near the cell body in these mice. These results suggest that microglia are required for normal development of Purkinje cells. The authors also revealed that nestin-positive cell-specific *Csf1* knockout mice exhibited decreased social behavior in the three-chamber test and impaired motor function.

3.4 Non-rodent animal models of ASDs (Table 4)

Many studies have used rodents to develop ASD models as they are widely available, relatively easy to handle, and can be genetically and pharmacologically manipulated. However, it is obviously difficult to study mental disorders using rodents that do not use language for communication. In addition, there are limitations to determining whether animal models successfully mimic human ASD phenotypes by merely using behavioral tests. For example, the three-chamber test used for the evaluation of sociability cannot completely eliminate the effects of memory; simultaneous performance of behavioral tests to verify memory loss is recommended. The most critical caveat in the use of rodent models is the anatomical differences in brain

Table 4 Non-rodent animal models of ASDs.

Animal	Reagent	Administration conditions (timing, dose, route)	Behavioral abnormalities	Neuroanatomical abnormalities	Microglial involvement	Reference
Marmoset	Valproic acid	GD60–66, 200 mg/kg, oral administration	Decreased sociability	–	–	63
			–	–	Abnormal morphology and density of cortical microglia	64
Pig	Reproductive and respiratory syndrome virus	GD76, 5×10^5 $TCID_{50}$, inoculation	Decreased sociability	–	No change (P27)	65
			–	Increased levels of TNFα, IL-1β, and GFAP in the hippocampus, decreased neuronal density in the dentate gyrus and subiculum	Increased level of MHC II in hippocampus (GD111)	66
			–	–	Increased level of MHC II (GD83, GD97), increased microglial density in amygdala (GD83)	67

Abbreviations: GD, gestational day; GFAP, glial fibrillary acidic protein; IL-1β, interleukin 1β; MHC, major histocompatibility complex; P, postnatal day; TCID, tissue culture infective dose; TNFα, tumor necrosis factor α.

structure between rodents and humans.[68–70] Furthermore, it has been recently reported that there are species-specific differences in microglial gene expression.[71] These issues make it difficult to apply the results of basic research to clinical therapeutics. This begs the question, what other species make suitable alternative options for the study of ASDs? In the next section, we will introduce studies in which the role of microglia was examined in ASD models using animals other than rodents.

3.4.1 Marmoset

Marmosets are New World monkeys that have attracted attention as primate model animals for the study of mental disorders. Their attractiveness comes from the fact that they have higher social cognitive functions and that they can be genetically manipulated. Yasue et al. orally administered VPA to pregnant marmosets 60–66 days after they conceived.[63] The authors used third-party reciprocal/non-reciprocal exchange to examine the social cognitive abilities of marmosets. In this behavioral test, marmosets watch two actors changing food and whether the marmosets receive food from the two actors was examined. In the control group, in non-reciprocal situations, the probability of receiving food from the actor who delivered the food increased, whereas in the VPA group, there was no difference in the preference of the actors, suggesting that social skills were impaired due to VPA exposure. The authors provided three interpretations of this result: the VPA group (1) might have impaired individual discrimination, (2) might not be able to distinguish between reciprocal and non-reciprocal behavior, and (3) might not have negative emotions for non-reciprocal behavior. Future studies are needed to verify that marmosets can distinguish between reciprocal and non-reciprocal situations.

Using a similar VPA exposure protocol, Sanagi et al. examined the morphology of microglia in young, adolescent and adult offspring.[64] At each stage of development, microglia in the VPA offspring group showed thin and fragmented processes. A decrease in the density of microglia and a decrease in the primary process number were also confirmed in the VPA offspring group. It is unclear whether these morphological changes represent changes in microglial function. Additional analysis of synaptic structures and neuronal functions in the VPA-treated marmoset brain may provide further insight.

3.4.2 Pig

Antonson et al. used a pig MIA model.[65–67] Reproductive and respiratory syndrome virus (PRRSV) was administered intranasally to pregnant pigs at gestational day (GD) 76. PRRSV is a single-stranded RNA virus that causes interstitial pneumonia by infecting alveolar macrophages. P27 piglets exhibited decreased social behavior in the three-chamber test.[65] Microglial analysis was also performed, but the inflammatory responses of microglia, including MHC II expression and the response to LPS (determined by the level of released cytokines), were not significantly different from control animals. A subsequent report from the same research group showed that TNFα, IL-β and glial fibrillary acidic protein levels were significantly increased in the hippocampus of fetuses at GD111, whereas MHC II levels were significantly reduced.[66] In addition, the density of neurons was significantly reduced in the dentate gyrus and hippocampal subiculum. In recent research by this group, microglia were analyzed at earlier time points, GD83 and GD97, and the percentage of microglia expressing MHC II was increased in the MIA offspring group.[67] In addition, immunostaining of tissue sections was performed, showing no change in microglial density in the hippocampus. In contrast, the microglial density was significantly increased at GD83 in the amygdala. This series of studies using a pig model of MIA suggests that transient changes in microglial properties caused by MIA could affect neuronal function and induce abnormal social behaviors in offspring.

4. Treatment of ASDs by controlling microglial functions

As mentioned above, the relationship between abnormalities in microglial functions and the pathogenesis of ASDs has been enthusiastically investigated. Besides being highlighted as a key player for elucidating the mechanism of ASD onset, microglia are beginning to attract attention as a new target for the treatment of ASDs. Microglia regulate neural circuit formation and plasticity, so strategies to modulate microglial functions can be used to repair neural circuits that have been degenerated during the development of ASDs. Possible methods to manipulate microglial functions for ASD treatments include drug-mediated activation and inactivation (or even spatiotemporal depletion) of microglia, microglial transplantation and non-invasive methods such as exercise and sensory stimulations. We have previously described the possible efficacy of drugs, cell transplantation and exercise on microglial manipulation in ASD treatments; interested readers can find further details of this by referring to our article.[72] In this section, we discuss whether non-invasive sensory stimulations are useful for the manipulation of microglial functions.

Tsai and colleagues have found that gamma-wave sensory stimulation modulates microglial functions. Iaccarino et al. gave a 40 Hz flickering to Alzheimer's disease model mice (5xFAD mice).[73] They found that microglia in the visual cortex exhibited cell body hypertrophy and processes shortening. Further, increased Aβ phagocytosis by microglia and decreased Aβ plaques in the brain were confirmed. In contrast, the visual stimulus did not affect the amount of synaptophysin, a possible target of synaptic phagocytosis by microglia, which suggests that whether 40 Hz stimulation enhances the phagocytic capacity of microglia is target-dependent. Next, the authors provided 40 Hz sound stimuli to the 5xFAD mice and were able to reproduce the microglial morphological changes and reduction of Aβ plaques not only in the auditory cortex but also in the hippocampus.[74] In addition, the sound stimuli improved cognitive and learning performance in 5xFAD mice. When visual and sound stimuli were simultaneously applied at 40 Hz to 5xFAD mice, microglial activation and accumulation around Aβ plaques were observed even in the medial prefrontal cortex. In their subsequent report, CK-p25 mice, an animal model of neurodegenerative disease, were treated with 40 Hz visual stimuli and microglia were collected from the primary visual cortex (V1) for RNA sequence analysis.[75] The 40 Hz visual stimuli reduced the expression levels of genes related to immune response and MHC I-dependent antigen presentation in CK-p25 mice. Furthermore, the increases in microglial density and hypertrophy in the V1, hippocampal CA1, sensory cortex and cingulate cortex were also suppressed. These results suggest that 40 Hz visual stimuli may normalize activated microglia. As a next step, it would be interesting to examine whether 40 Hz sensory stimuli would modulate the pathogenesis of various diseases, including ASDs. As an example of a possible effect on ASDs, we may expect that 40 Hz stimuli would suppress microglial activation while promoting the phagocytosis of excess synapses by microglia.

5. Conclusion

In this chapter, we first introduced various animal models of ASDs, specifically focusing on the changes in microglial properties, and discussed the role of microglia in the development of ASDs. The phenotypes commonly observed in the brains of patients with ASDs and in animal models of ASDs include morphological and functional abnormalities in synapses and accelerated inflammation. Thus, it is likely that microglia are responsible for neural circuit formation and inflammatory responses that contribute to the pathogenesis of ASDs. However, histological analysis of brain slices alone falls short of clearly revealing whether morphological and functional changes in microglia are the cause or result of ASDs. For this purpose, it is necessary to examine microglial properties at multiple time points, such as before and after ASD-like behaviors appear. Additionally, the role of other cell types, including neurons and astrocytes, must also be considered. It should be also noted that we should always pay attention to the differences between humans and animals, including monkeys. Indeed, when we use non-human primate models of ASDs, it is possible that we are confronted with higher order differences that make the interpretation of data more complex compared to rodent models. Thus, in addition to comparing the results from different animal models, the development and use of sophisticated and well-validated in vitro system to understand the cellular and molecular dynamics underlying the pathogenesis of ASDs should be pursued.

Finally, we mentioned the possibility of treating ASDs by controlling the function of microglia (also see Ref. 72). In particular, the use of non-invasive methods such as exercise and sensory stimuli would be a major breakthrough. Further clarifying how these stimuli affect microglial dynamics and ASD pathogenesis in animal models will forge a new path for the treatment of ASDs.

References

1. Chaste P, Leboyer M. Autism risk factors: genes, environment, and gene-environment interactions. *Dialogues Clin Neurosci*. 2012;14(3):281–292.
2. Ergaz Z, Weinstein-Fudim L, Ornoy A. Genetic and non-genetic animal models for autism spectrum disorders (ASD). *Reprod Toxicol*. 2016;64:116–140.
3. Yanuck SF. Microglial phagocytosis of neurons: diminishing neuronal loss in traumatic, infectious, inflammatory, and autoimmune CNS disorders. *Front Psych*. 2019;10:712.
4. Rodriguez JI, Kern JK. Evidence of microglial activation in autism and its possible role in brain underconnectivity. *Neuron Glia Biol*. 2011;7(2–4):205–213.
5. Lenz KM, Nelson LH. Microglia and beyond: innate immune cells as regulators of brain development and behavioral function. *Front Immunol*. 2018;9:698.
6. Schafer DP, Lehrman EK, Kautzman AG, et al. Microglia sculpt postnatal neural circuits in an activity and complement-dependent manner. *Neuron*. 2012;74(4):691–705.
7. Andoh M, Shibata K, Okamoto K, et al. Exercise reverses behavioral and synaptic abnormalities after maternal inflammation. *Cell Rep*. 2019;27(10):2817–25.e5.
8. Zang JB, Nosyreva ED, Spencer CM, et al. A mouse model of the human fragile X syndrome I304N mutation. *PLoS Genet*. 2009;5(12):e1000758.
9. Hamilton SM, Green JR, Veeraragavan S, et al. Fmr1 and Nlgn3 knockout rats: novel tools for investigating autism spectrum disorders. *Behav Neurosci*. 2014;128(2):103–109.
10. Bhattacharya A, Kaphzan H, Alvarez-Dieppa AC, Murphy JP, Pierre P, Klann E. Genetic removal of p70 S6 kinase 1 corrects molecular, synaptic, and behavioral phenotypes in fragile X syndrome mice. *Neuron*. 2012;76(2):325–337.

11. Jawaid S, Kidd GJ, Wang J, Swetlik C, Dutta R, Trapp BD. Alterations in CA1 hippocampal synapses in a mouse model of fragile X syndrome. *Glia.* 2018;66(4):789–800.
12. Shahbazian M, Young J, Yuva-Paylor L, et al. Mice with truncated MeCP2 recapitulate many Rett syndrome features and display hyperacetylation of histone H3. *Neuron.* 2002;35(2):243–254.
13. Chao HT, Chen H, Samaco RC, et al. Dysfunction in GABA signalling mediates autism-like stereotypies and Rett syndrome phenotypes. *Nature.* 2010;468(7321): 263–269.
14. Derecki NC, Cronk JC, Lu Z, et al. Wild-type microglia arrest pathology in a mouse model of Rett syndrome. *Nature.* 2012;484(7392):105–109.
15. Wang J, Wegener JE, Huang TW, et al. Wild-type microglia do not reverse pathology in mouse models of Rett syndrome. *Nature.* 2015;521(7552):E1–E4.
16. Schafer DP, Heller CT, Gunner G, et al. Microglia contribute to circuit defects in Mecp2 null mice independent of microglia-specific loss of Mecp2 expression. *Elife.* 2016;5:e15224.
17. Tsai PT, Hull C, Chu Y, et al. Autistic-like behaviour and cerebellar dysfunction in Purkinje cell Tsc1 mutant mice. *Nature.* 2012;488(7413):647–651.
18. Bader PL, Faizi M, Kim LH, et al. Mouse model of Timothy syndrome recapitulates triad of autistic traits. *Proc Natl Acad Sci U S A.* 2011;108(37):15432–15437.
19. Peça J, Feliciano C, Ting JT, et al. Shank3 mutant mice display autistic-like behaviours and striatal dysfunction. *Nature.* 2011;472(7344):437–442.
20. Kwon CH, Luikart BW, Powell CM, et al. Pten regulates neuronal arborization and social interaction in mice. *Neuron.* 2006;50(3):377–388.
21. Peñagarikano O, Abrahams BS, Herman EI, et al. Absence of CNTNAP2 leads to epilepsy, neuronal migration abnormalities, and core autism-related deficits. *Cell.* 2011;147(1):235–246.
22. Han S, Tai C, Westenbroek RE, et al. Autistic-like behaviour in Scn1a+/− mice and rescue by enhanced GABA-mediated neurotransmission. *Nature.* 2012;489(7416): 385–390.
23. Nakatani J, Tamada K, Hatanaka F, et al. Abnormal behavior in a chromosome-engineered mouse model for human 15q11-13 duplication seen in autism. *Cell.* 2009;137(7):1235–1246.
24. Shigemori T, Sakai A, Takumi T, Itoh Y, Suzuki H. Altered microglia in the amygdala are involved in anxiety-related behaviors of a copy number variation mouse model of autism. *J Nippon Med Sch.* 2015;82(2):92–99.
25. Smith SE, Zhou YD, Zhang G, Jin Z, Stoppel DC, Anderson MP. Increased gene dosage of Ube3a results in autism traits and decreased glutamate synaptic transmission in mice. *Sci Transl Med.* 2011;3(103):103ra97.
26. Jaramillo TC, Liu S, Pettersen A, Birnbaum SG, Powell CM. Autism-related neuroligin-3 mutation alters social behavior and spatial learning. *Autism Res.* 2014;7(2):264–272.
27. Jamain S, Radyushkin K, Hammerschmidt K, et al. Reduced social interaction and ultrasonic communication in a mouse model of monogenic heritable autism. *Proc Natl Acad Sci U S A.* 2008;105(5):1710–1715.
28. Grayton HM, Missler M, Collier DA, Fernandes C. Altered social behaviours in neurexin 1α knockout mice resemble core symptoms in neurodevelopmental disorders. *PLoS One.* 2013;8(6):e67114.
29. Sungur A, Vörckel KJ, Schwarting RK, Wöhr M. Repetitive behaviors in the Shank1 knockout mouse model for autism spectrum disorder: developmental aspects and effects of social context. *J Neurosci Methods.* 2014;234:92–100.
30. Won H, Lee HR, Gee HY, et al. Autistic-like social behaviour in Shank2-mutant mice improved by restoring NMDA receptor function. *Nature.* 2012;486(7402):261–265.

31. Wöhr M, Orduz D, Gregory P, et al. Lack of parvalbumin in mice leads to behavioral deficits relevant to all human autism core symptoms and related neural morphofunctional abnormalities. *Transl Psychiatry*. 2015;5:e525.
32. Zhan Y, Paolicelli RC, Sforazzini F, et al. Deficient neuron-microglia signaling results in impaired functional brain connectivity and social behavior. *Nat Neurosci*. 2014;17(3): 400–406.
33. Moy SS, Nadler JJ, Young NB, et al. Mouse behavioral tasks relevant to autism: phenotypes of 10 inbred strains. *Behav Brain Res*. 2007;176(1):4–20.
34. Heo Y, Zhang Y, Gao D, Miller VM, Lawrence DA. Aberrant immune responses in a mouse with behavioral disorders. *PLoS One*. 2011;6(7):e20912.
35. Nicholls RD, Knepper JL. Genome organization, function, and imprinting in Prader-Willi and Angelman syndromes. *Annu Rev Genomics Hum Genet*. 2001;2:153–175.
36. Guang S, Pang N, Deng X, et al. Synaptopathology involved in autism spectrum disorder. *Front Cell Neurosci*. 2018;12:470.
37. Canitano R. Epilepsy in autism spectrum disorders. *Eur Child Adolesc Psychiatry*. 2007;16(1):61–66.
38. Betancur C. Etiological heterogeneity in autism spectrum disorders: more than 100 genetic and genomic disorders and still counting. *Brain Res*. 2011;1380:42–77.
39. Gandal MJ, Edgar JC, Ehrlichman RS, Mehta M, Roberts TP, Siegel SJ. Validating γ oscillations and delayed auditory responses as translational biomarkers of autism. *Biol Psychiatry*. 2010;68(12):1100–1106.
40. Schneider T, Przewłocki R. Behavioral alterations in rats prenatally exposed to valproic acid: animal model of autism. *Neuropsychopharmacology*. 2005;30(1):80–89.
41. Ingram JL, Peckham SM, Tisdale B, Rodier PM. Prenatal exposure of rats to valproic acid reproduces the cerebellar anomalies associated with autism. *Neurotoxicol Teratol*. 2000;22(3):319–324.
42. Sosa-Díaz N, Bringas ME, Atzori M, Flores G. Prefrontal cortex, hippocampus, and basolateral amygdala plasticity in a rat model of autism spectrum. *Synapse*. 2014;68(10): 468–473.
43. Dufour-Rainfray D, Vourc'h P, Le Guisquet AM, et al. Behavior and serotonergic disorders in rats exposed prenatally to valproate: a model for autism. *Neurosci Lett*. 2010;470(1):55–59.
44. Reynolds S, Millette A, Devine DP. Sensory and motor characterization in the postnatal valproate rat model of autism. *Dev Neurosci*. 2012;34(2–3):258–267.
45. Le Belle JE, Sperry J, Ngo A, et al. Maternal inflammation contributes to brain overgrowth and autism-associated behaviors through altered redox signaling in stem and progenitor cells. *Stem Cell Reports*. 2014;3(5):725–734.
46. Oskvig DB, Elkahloun AG, Johnson KR, Phillips TM, Herkenham M. Maternal immune activation by LPS selectively alters specific gene expression profiles of interneuron migration and oxidative stress in the fetus without triggering a fetal immune response. *Brain Behav Immun*. 2012;26(4):623–634.
47. Fernández de Cossío L, Guzmán A, van der Veldt S, Luheshi GN. Prenatal infection leads to ASD-like behavior and altered synaptic pruning in the mouse offspring. *Brain Behav Immun*. 2017;63:88–98.
48. Malkova NV, Yu CZ, Hsiao EY, Moore MJ, Patterson PH. Maternal immune activation yields offspring displaying mouse versions of the three core symptoms of autism. *Brain Behav Immun*. 2012;26(4):607–616.
49. Choi GB, Yim YS, Wong H, et al. The maternal interleukin-17a pathway in mice promotes autism-like phenotypes in offspring. *Science*. 2016;351(6276):933–939.
50. Shi L, Smith SE, Malkova N, Tse D, Su Y, Patterson PH. Activation of the maternal immune system alters cerebellar development in the offspring. *Brain Behav Immun*. 2009;23(1):116–123.

51. Giovanoli S, Weber L, Meyer U. Single and combined effects of prenatal immune activation and peripubertal stress on parvalbumin and reelin expression in the hippocampal formation. *Brain Behav Immun*. 2014;40:48–54.
52. Mattei D, Ivanov A, Ferrai C, et al. Maternal immune activation results in complex microglial transcriptome signature in the adult offspring that is reversed by minocycline treatment. *Transl Psychiatry*. 2017;7(5):e1120.
53. Smith SE, Li J, Garbett K, Mirnics K, Patterson PH. Maternal immune activation alters fetal brain development through interleukin-6. *J Neurosci*. 2007;27(40):10695–10702.
54. Bolton JL, Marinero S, Hassanzadeh T, et al. Gestational exposure to air pollution alters cortical volume, microglial morphology, and microglia-neuron interactions in a sex-specific manner. *Front Synaptic Neurosci*. 2017;9:10.
55. Bolton JL, Huff NC, Smith SH, et al. Maternal stress and effects of prenatal air pollution on offspring mental health outcomes in mice. *Environ Health Perspect*. 2013;121(9): 1075–1082.
56. Bilbo SD, Block CL, Bolton JL, Hanamsagar R, Tran PK. Beyond infection—maternal immune activation by environmental factors, microglial development, and relevance for autism spectrum disorders. *Exp Neurol*. 2018;299(pt A):241–251.
57. Vogel Ciernia A, Careaga M, LaSalle JM, Ashwood P. Microglia from offspring of dams with allergic asthma exhibit epigenomic alterations in genes dysregulated in autism. *Glia*. 2018;66(3):505–521.
58. Schwartzer JJ, Careaga M, Chang C, Onore CE, Ashwood P. Allergic fetal priming leads to developmental, behavioral and neurobiological changes in mice. *Transl Psychiatry*. 2015;5:e543.
59. Knuesel I, Chicha L, Britschgi M, et al. Maternal immune activation and abnormal brain development across CNS disorders. *Nat Rev Neurol*. 2014;10(11):643–660.
60. Filipello F, Morini R, Corradini I, et al. The microglial innate immune receptor TREM2 is required for synapse elimination and normal brain connectivity. *Immunity*. 2018;48(5):979–91.e8.
61. Kim HJ, Cho MH, Shim WH, et al. Deficient autophagy in microglia impairs synaptic pruning and causes social behavioral defects. *Mol Psychiatry*. 2017;22(11):1576–1584.
62. Kana V, Desland FA, Casanova-Acebes M, et al. CSF-1 controls cerebellar microglia and is required for motor function and social interaction. *J Exp Med*. 2019;216(10): 2265–2281.
63. Yasue M, Nakagami A, Banno T, Nakagaki K, Ichinohe N, Kawai N. Indifference of marmosets with prenatal valproate exposure to third-party non-reciprocal interactions with otherwise avoided non-reciprocal individuals. *Behav Brain Res*. 2015;292:323–326.
64. Sanagi T, Sasaki T, Nakagaki K, Minamimoto T, Kohsaka S, Ichinohe N. Segmented Iba1-positive processes of microglia in autism model marmosets. *Front Cell Neurosci*. 2019;13:344.
65. Antonson AM, Radlowski EC, Lawson MA, Rytych JL, Johnson RW. Maternal viral infection during pregnancy elicits anti-social behavior in neonatal piglet offspring independent of postnatal microglial cell activation. *Brain Behav Immun*. 2017;59: 300–312.
66. Antonson AM, Balakrishnan B, Radlowski EC, Petr G, Johnson RW. Altered hippocampal gene expression and morphology in fetal piglets following maternal respiratory viral infection. *Dev Neurosci*. 2018;40(2):104–119.
67. Antonson AM, Lawson MA, Caputo MP, Matt SM, Leyshon BJ, Johnson RW. Maternal viral infection causes global alterations in porcine fetal microglia. *Proc Natl Acad Sci U S A*. 2019;116(40):20190–20200.
68. Herculano-Houzel S, Collins CE, Wong P, Kaas JH. Cellular scaling rules for primate brains. *Proc Natl Acad Sci U S A*. 2007;104(9):3562–3567.

69. Ventura-Antunes L, Mota B, Herculano-Houzel S. Different scaling of white matter volume, cortical connectivity, and gyrification across rodent and primate brains. *Front Neuroanat*. 2013;7:3.
70. Izpisua Belmonte JC, Callaway EM, Caddick SJ, et al. Brains, genes, and primates. *Neuron*. 2015;86(3):617–631.
71. Geirsdottir L, David E, Keren-Shaul H, et al. Cross-species single-cell analysis reveals divergence of the primate microglia program. *Cell*. 2019;179(7):1609–22.e16.
72. Andoh M, Ikegaya Y, Koyama R. Microglia as possible therapeutic targets for autism spectrum disorders. *Prog Mol Biol Transl Sci*. 2019;167:223–245.
73. Iaccarino HF, Singer AC, Martorell AJ, et al. Gamma frequency entrainment attenuates amyloid load and modifies microglia. *Nature*. 2016;540(7632):230–235.
74. Martorell AJ, Paulson AL, Suk HJ, et al. Multi-sensory gamma stimulation ameliorates Alzheimer's-associated pathology and improves cognition. *Cell*. 2019;177(2):256–71.e22.
75. Adaikkan C, Middleton SJ, Marco A, et al. Gamma entrainment binds higher-order brain regions and offers neuroprotection. *Neuron*. 2019;102(5):929–43.e8.

CHAPTER TEN

The early overgrowth theory of autism spectrum disorder: Insight into convergent mechanisms from valproic acid exposure and translational models

Oswald Chen[†], Isabelle Tahmazian[†], Hannah J. Ferrara[†], Bin Hu, Taylor Chomiak*

Division of Translational Neuroscience, Department of Clinical Neurosciences, Hotchkiss Brain Institute, Alberta Children's Hospital Research Institute, Cumming School of Medicine, University of Calgary, Calgary, AB, Canada

*Corresponding author: e-mail address: tgchomia@ucalgary.ca

Contents

Abstract

The development of new approaches for the clinical management of autism spectrum disorder (ASD) can only be realized through a better understanding of the neurobiological changes associated with ASD. One strategy for gaining deeper insight into the neurobiological mechanisms associated with ASD is to identify converging pathogenic processes associated with human idiopathic clinicopathology that are conserved in translational models of ASD. In this chapter, we first present the early overgrowth theory of ASD. Second, we introduce valproic acid (VPA), one of the most robust and well-known environmental risk factors associated with ASD, and we summarize the rapidly

[†] These authors contributed equally.

Progress in Molecular Biology and Translational Science, Volume 173
ISSN 1877-1173
https://doi.org/10.1016/bs.pmbts.2020.04.014

growing body of animal research literature using VPA as an ASD translational model. Lastly, we will detail the mechanisms of action of VPA and its impact on functional neural systems, as well as discuss future research directions that could have a lasting impact on the field.

1. Introduction

Autism spectrum disorder (ASD) is a neurodevelopmental disorder characterized by impairments in communication and behavior that has become a top priority for public and philanthropic support.[1,2] This is partially due to the fact that 20 years ago the prevalence of ASD was around 4 in 10,000, with estimates now closer to 1 in 68, or even 1 in 50.[2–4] The rapid growth in the estimated prevalence of ASD has encouraged further investigation of the disorder to provide the best evidence-based care to this population.[3] Indeed, there has been tremendous support aimed at increasing our understanding of ASD, its causes, and evaluating potential treatments.[1–3] Nevertheless, the majority of children with ASD currently remain within the spectrum as adults, and regardless of their intellectual functioning, continue to experience problems with independent living, employment, social relationships, and mental health as adults.[5–7]

Current approaches to the treatment of ASD are complex, often requiring chronic management.[8–10] Furthermore, there is no standard pharmacological approach for the treatment of ASD symptoms.[8] While behavioral and developmental interventions can be highly effective, they pose a significant economic burden to families.[8–10] In fact, a landmark study has estimated the total individual lifetime cost to be around $1–2 million,[10] which works out to billions of dollars per year in economic costs worldwide. Thus, new and innovative treatment options that provide more cost-effective and clinically beneficial intervention strategies are urgently needed.

The development of new clinically beneficial interventional strategies for ASD can be realized by identifying converging pathogenic processes associated with human idiopathic clinicopathology that are conserved in translational models of ASD. In this chapter, we will build on this idea by first presenting the early overgrowth theory of ASD. We will then introduce valproic acid (VPA), one of the most robust and well-known environmental risk factors associated with ASD, and summarize the rapidly growing body of animal research literature using VPA as a translational animal model of ASD. The last two sections of this chapter will focus on the mechanisms of action of VPA in the context of the early overgrowth theory, and then discuss future directions of research that could have a lasting impact on the field.

2. The early overgrowth theory of ASD

In his now classical paper, Kanner described a group of 11 children with characteristic features of ASD.[11] Since Kanner's paper, the views on the basis of ASD have dramatically changed. The predominance of psychoanalytical theories in post–World War II child psychiatry that led to the misconception that ASD was an infant's response to early disturbances of the mother–child relationship (i.e., "refrigerator mother" theory),[12] gave rise to more modern biologically-based theories that underscore a clear neurobiological, genetic, and epigenetic basis.[12–20]

Over the last three decades, many independent labs have provided empirical evidence supporting the contemporary idea known as the early overgrowth theory.[13–16,21–24] The early overgrowth theory is based on clinical data from head circumference measurements, post-mortem tissue, and magnetic resonance imaging (MRI) studies of children with ASD. The data reveal a pattern of dysregulated and accelerated brain growth in which areas of the brain, particularly temporal and frontal association cortices, undergo rapid overgrowth during early childhood.[13,15,16,24] While it is known that a subset of children with ASD have a small brain size, sometimes associated with extremely rare chromosomal defects, this is most often not the case.[13] In fact, data from over 8000 subjects from 44 MRI and 27 head circumference studies spanning toddlerhood to adulthood, have found significant brain and head circumference overgrowth in ASD compared to controls with the most pronounced increases occurring at earlier ages.[13,21] Furthermore, cutting-edge experimental techniques and experimental data from human-cell *in vitro* models lends further support to the early overgrowth theory of ASD.[25,26] Using ASD and control subject-derived induced pluripotent stem cells (iPSCs), Schafer et al. have recently used transcriptome-wide screening to show an ASD-associated temporal dysregulation within a specific neurodevelopmental gene module that can cause marked changes in the maturational and developmental sequence of cortical neurons and acceleration of their growth.[25] Importantly, this study directly links epigenetic regulation, accelerated neuronal maturation, and ASD. Thus, both clinical data and experimental results, including those from *in vitro* human models, are pointing to a multistage progressive cascade of

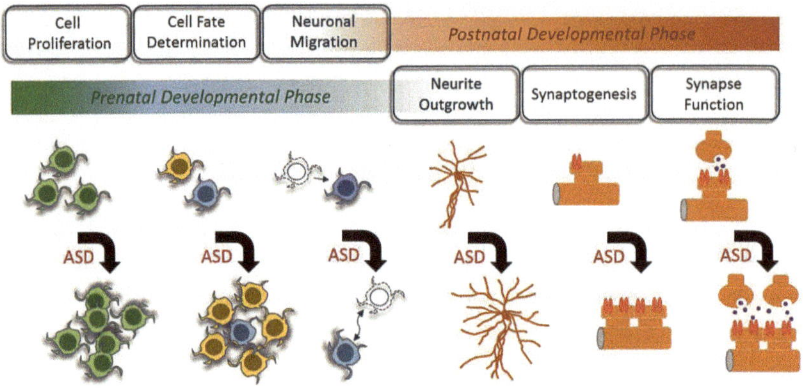

Fig. 1 Multistage events of progressive neural development and maturation. A schematic of the key progressive processes occurring during cortical neuronal maturation at different stages of prenatal (green/blue) and postnatal (orange) development. Arrows indicating "ASD" represent cellular and molecular changes at each stage associated with ASD and ASD translational models. See text for details.

ASD pathogenic processes that involves temporal dysregulation of specific gene networks, morphological growth acceleration, and aberrant neuronal connectivity[13,25–28] (Fig. 1).

3. Valproic acid and ASD

3.1 Epidemiological evidence

While gene-based translational research has dominated the ASD field over the last few decades, accumulating evidence highlights the importance of environmental influences and suggests that a large proportion of variance in ASD liability may indeed be determined by environmental factors.[29,30] The impact of any given environmental exposure likely varies across a population according to individual genetic differences, making it difficult to identify clear mechanisms associated with exposure and ASD.[27] A key strategy for gaining deeper insight into the neurobiological mechanisms associated with ASD, is to identify converging pathogenic processes that are associated with environmental exposure in translational animal models that are conserved in human idiopathic clinicopathology.

One of the most robust, well-studied, and generally-accepted environmental risk factors associated with ASD is *in utero* valproic acid (VPA) exposure.[14,16] VPA is a commonly used antiepileptic and mood stabilizer drug that can interfere with a variety of biological processes.[31] In late 2014, the European Medicines Agency cited accumulating research linking VPA to ASD[14] as a means to strengthen existing restrictions on the use of VPA. The Agency stipulated that in girls (irrespective of age and epilepsy type) and in women of childbearing potential, VPA should not be prescribed unless other treatments are ineffective or not tolerated.[32] This regulatory action was based on accumulating evidence assessing the link between prenatal exposure to antiepileptic drugs (AEDs) and subsequent deficits in cognitive function, with a more recent focus on the relationship between AEDs and ASD. Fetal exposure to VPA has been known to cause congenital malformations and developmental delays, features of what is now known as fetal valproate syndrome (FVS). Many earlier studies have attempted to elucidate a possible link between FVS, fetal exposure to VPA, and ASD. For example, in 1993 Laegreid and Colleagues evaluated seven children all diagnosed with FVS.[33] Of those seven, two exhibited autistic traits, while two others had severe developmental delays.[33] Several studies following this 1993 study have also found links between *in utero* VPA exposure, FVS, and autism.[34–39]

Since these initial studies linking VPA to ASD, there have been several subsequent studies conducted on country-wide scales confirming these earlier reports. For example, Christensen et al. published a study evaluating over half a million children born from 1996 to 2006 in Denmark.[40] The study controlled for known contributors of ASD and reported the absolute risk and hazard ratios for both ASD and childhood autism (defined as the most severe diagnosis). Results showed there was a significantly higher risk of a child developing ASD as well as childhood autism when prenatally exposed to VPA (4.42% and 2.5%, respectively; compared to 1.53% and 0.48% for nonexposed, respectively).[40] In addition, a 2016 study conducted in the UK assessed the risk of prescribed psychotropic medications during pregnancy by integrating data from two separate country-wide medical databases. Similar to the findings of the Danish study, it was found that in pregnancies where psychotropics such as valproate were prescribed during pregnancy,[41] there was a threefold increase in instances of neurodevelopmental and behavioral disorders in children exposed *in utero* to VPA.[41] In 2018, a similar study to the abovementioned studies with respect to scale was completed in the United States. Hisle-Gorman and colleagues (2018) used the Military Health System database to assess 29 previously reported neonatal, perinatal, and prenatal risk factors of ASD, including prenatal exposure to AEDs such as VPA.[42] The study matched 3 control children with 1 child diagnosed with ASD based on sex and date of birth, for a total cohort of 8760 children diagnosed with ASD and 26,280 controls.

ASD was found to be significantly associated with prenatal exposure to AEDs, with VPA being the most commonly prescribed AED.[42] The general conclusions of these and several additional studies are that *in utero* VPA exposure increases the risk of neurodevelopmental disorders such as ASD, with some reports even correlating higher prescribed doses of maternal VPA to increased severity of ASD symptoms and poorer clinical outcome.[43–45] Thus, there is sufficient clinical data to support the view that VPA exposure is a valid and relevant pathological model for understanding the cellular and molecular mechanisms underlying ASD.

3.2 The VPA animal model of ASD

Parallel to these studies in humans, there is a large body of animal research literature using VPA as a translational animal model of ASD (see Refs. 14,16,46–48 for several reviews on the topic). For an animal model to be considered a relevant model of a condition described in humans, it should fit several criteria: construct, face and predictive validity.[49] Construct validity relates to the identity of causation between the animal phenotype and the human disease, while face validity infers that the phenomenological aspect observed in the animal is similar to the one observed in the human condition (e.g., similar neuroanatomical and behavioral features).[49] Remarkably, VPA-induced ASD-like behavioral features appear to be conserved across species, as there is evidence supporting the role of VPA in ASD-like pathogenic processes in nonhuman primates, rat, mouse, voles, fish, and chicks.[14,16,46–48,50–54] Finally, predictive validity refers to the idea that treatments that are effective in the animal model, will translate back to effective treatments in humans.[49]

VPA exposure is one of the most cited animal models of ASD, and several authors have pointed out that this animal model is indeed a relevant translational model as it exhibits all three validity criteria.[14,47,48] For example, rats exposed to VPA *in utero* can exhibit physical malformations (e.g., ear malformations) comparable to similar abnormalities observed in some individuals with ASD (e.g., posterior rotation of the outer ear).[55] Furthermore, numerous other studies have also shown that prenatal VPA exposure can lead to both core (decreased social interactions, repetitive/stereotypic-like activity, and changes in specific types of pup ultrasonic vocalizations) as well as non-core ASD-like abnormalities (sensorimotor abnormalities, increased neuropsychiatric-like features, attention disengagement impairments, and microbiome features).[56–75] Although the effects of

VPA have been tested in rodents for many years, VPA has recently been gaining traction with respect to studying potential treatment interventions (i.e., predictive validity). For instance, Schneider et al. previously reported that environmental enrichment, such as multisensory stimulation, reversed almost all behavioral alterations observed in pups exposed to VPA *in utero*.[58] Woo and Leon, on the other hand, have since published two clinical studies supporting the use of environmental enrichment in the form of multiple sensorimotor stimuli for improving some of the symptoms in ASD.[76,77] Newer experimental data using the VPA animal model have also suggested several other potential therapeutic approaches such as chemical compounds (e.g., histone deacetylase (HDAC) inhibitors, resveratrol, cannabidivarin),[61,78,79] reward-coupled locomotor activity,[56] gut microbes,[80] and molecular hydrogen.[67] However, many of these potential strategies for the clinical management of ASD have yet to be rigorously tested in humans.

Similar to the prenatal model, many ASD-like behavioral and pathological features have been observed with VPA exposure during the early postnatal period[81–88] (equivalent to the third trimester and perinatal period in humans). This similarity suggests that the mechanism by which VPA alters neuronal development and maturation can extend over different developmental time points in rodents.[14] The postnatal VPA model could be crucial for studying certain clinical observations in humans. For example, maternal serum VPA free-fraction increases in the third trimester and is highest at birth.[89] This is consistent with the idea that bioavailability may play an important role in VPA teratogenicity. Understanding the spectrum of prenatal, perinatal, and early postnatal VPA exposure (and possibly even paternal VPA exposure on the offspring[90]), may help better elucidate the neuropathological mechanisms associated with ASD.[14]

4. VPA and the cellular and molecular mechanisms of early overgrowth

VPA is known to have a variety of effects on biological systems.[31] Its classical mechanisms of actions were thought to be mediated primarily by modulating voltage-gated ion channels and synaptic signaling.[31] However, we now know that VPA is a potent HDAC inhibitor (Fig. 2) that can block HDAC activity in neurons.[91–96] As HDAC activity is central to the regulation of gene transcription and phenotype differentiation,[97] it is not

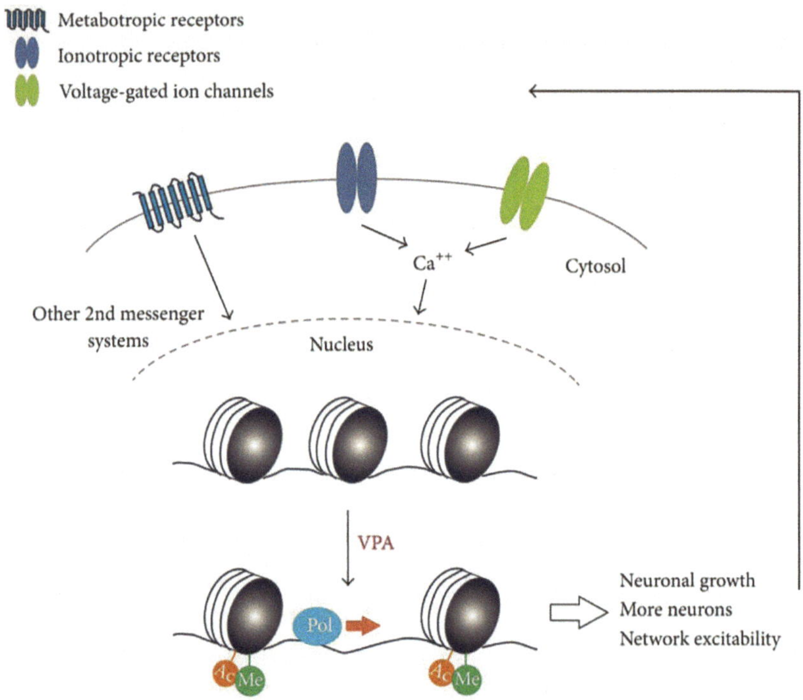

Fig. 2 Key molecular and cellular changes associated with VPA related to ASD neuropathology. Common clinicopathology associated with ASD is accelerated or early brain overgrowth and increased network excitability. Early or accelerated overgrowth may result from more synapses, increased dendritic growth, and/or increased number of cells, while increased network excitability may result from hyper-connectivity and/or hyperplasticity of microcircuits, presumably driven by synaptic mechanisms. All of these aspects are also likely regulated by histone acetylation (orange circles "Ac") and/or histone methylation (green circles "Me"), and possibly DNA methylation (DNA methylation not shown). VPA has been shown to increase histone acetylation and histone methylation that can promote gene activation (symbolized by blue circle "Pol"; RNA polymerase). Increased connectivity and network excitability may further influence this process by activating metabotropic, ionotropic, and voltage-gated ion channels and subsequent intracellular signaling cascades. *Figure and legend from Chomiak T, Turner N, Hu B. What we have learned about autism spectrum disorder from valproic acid.* Pathol Res Int. *2013;2013, an open access article distributed under the Creative Commons Attribution License.*

surprising that many behavioral outcomes are affected by VPA exposure. In fact, the impact of VPA exposure on the developing nervous system is just as diverse, and can lead to alterations in many functional systems as will be summarized next.

4.1 Modulation of functional systems

4.1.1 Gamma-aminobutyric acid (GABA)

Classically, the mechanism of action of VPA had focused on increases in brain concentrations of gamma-aminobutyric acid (GABA), the major inhibitory neurotransmitter of the nervous system. The effect of VPA on the GABAergic system had been postulated in the 1970s, when it was proposed to enhance inhibitory GABAergic activity.[31] The main mechanisms through which VPA increases GABA levels have generally been attributed to both inhibition of GABA catabolism by GABA transaminase, and by increasing the availability of the GABA precursor α-ketoglutarate through inactivation of α-ketoglutarate dehydrogenase.[31] Furthermore, it has also been observed that VPA affects not only the catalytic activity, but also the expression of the key enzyme involved in GABA synthesis (glutamate decarboxylase—GAD) due to its action on histone acetylation.[31] As GABA can exert trophic effects during neuronal development,[98–100] these molecular mechanisms may collectively act to promote network overgrowth.

More recently, work from Ben-Ari's lab has provided evidence of a much more detailed mechanism of VPA action involving alterations in the developmental switch from excitatory-to-inhibitory GABA.[101] It is well known that neuronal intracellular chloride $[Cl^-]_i$ gradients change during development, shifting the reversal potential of $[Cl^-]_i$ more negative, which leads to a switch of GABA-mediated current from excitatory to inhibitory.[102,103] Remarkably, the developmental excitatory-inhibitory GABA switch is not only absent in pyramidal neurons in both VPA-exposed rats and Fragile X mice (a genetic model of ASD), but it can also be prevented at birth in normal rats by an oxytocin receptor antagonist, which also leads to postnatal autistic-like behavioral changes similar to those seen in VPA exposed rats and Fragile X mice.[101,104] This model has been further supported by a 2019 study that found that the abolition of the oxytocin-mediated excitatory-to-inhibitory shift of GABA action during labor and birth, can contribute to the pathogenesis of ASD by stimulating neural growth.[105] This VPA-mediated abnormality in the early developmental regulation of neuronal $[Cl^-]_i$, common to two distinct animal models of ASD,[101] has been observed by several studies.[101,105,106] This implies that persistently elevated $[Cl^-]_i$, which drives GABA's excitatory effects, is likely important in the pathogenesis of ASD.[104] While the idea that oxytocin signaling may play a role in ASD is not new,[107] the experimental results highlighted above suggest that a lack of an oxytocin-mediated excitatory-to-inhibitory shift of GABA may be a fundamental molecular event involved in persistently elevated $[Cl^-]_i$ and aberrant GABA-mediated excitatory network activity.

Finally, impairments in the GABAergic system may contribute to an increased synaptic excitation/inhibition ratio, a previously proposed network model of ASD pathology that is also exhibited in the VPA model.[108,109] Although the increased excitation/inhibition ratio in ASD has recently been challenged,[110] it should be pointed out that changes observed in adult tissue may reflect compensatory changes in ASD circuits rather than mechanistic determinants of network malformation. Nevertheless, when taken altogether, the data appear to support the role of GABAergic dysfunction in ASD.

4.1.2 Glutamate

GABAergic depolarization acts in synergy with *N*-methyl-D-aspartate (NMDA) glutamate receptor-mediated plasticity to enhance the trophic effects on neurite growth, migration, and synapse formation associated with excitatory post-synaptic activity.[103] The role of altered glutamatergic activity in ASD has also been previously proposed.[111–113] For example, several genetic studies have pointed to the involvement of the glutamatergic system in ASD. These include single nucleotide polymorphisms (SNPs) in genes encoding ionotropic and metabotropic glutamate receptors,[114,115] in addition to several glutamatergic gene mutations related to excitatory synapses such as neuroligin[116,117] and SHANK[118] (also see Ref. 112). Neuroligins are particularly interesting, as their interaction with presynaptically expressed neurexins are thought to increase both spine density and neuronal connectivity, which may contribute to early overgrowth in ASD.[119]

VPA has a wide variety of mechanisms of action on the glutamatergic system.[31] For example, VPA has been shown to up-regulate the transcription factor Pax6, which is involved in glutamatergic neuronal cell-fate differentiation.[120] Pax6 up-regulation has been shown to be transiently increased in the embryonic cortex of the VPA animal model of ASD, resulting in increased glutamatergic neuronal differentiation and increased expression of glutamatergic proteins in the postnatal brain of the exposed offspring.[120] This up-regulation appeared to be mediated by increased acetylated histone binding on the Pax6 promoter region, since it was observed with other HDAC inhibitors such as trichostatin A and sodium butyrate, but not with valpromide, which is devoid of HDAC inhibitor activity.[120]

Additional research has shown that VPA can significantly increase glutamate concentrations,[121] and increase glutamatergic synaptic transmission and synaptic plasticity in the neocortex.[72,122] The latter appears to be mediated by a selective over-expression of NR2A and NR2B subunits of NMDA receptors, as well as calcium/calmodulin-dependent protein kinase II (CAMK-II).[72] These cellular and molecular changes effectively led to a boosted NMDA synaptic current component,[72] which has subsequently led to a proposal that ASD may result from "hyperfunctioning" cortical glutamatergic microcircuits.[112,123–126] Similarly, electrophysiological data from amygdaloid nucleus slices has also shown increased reactivity to stimulation, increased long-term potentiation, and impaired inhibition in response to VPA exposure.[68,73] However, this notion of "hyperfunctioning" circuits has been challenged, as abnormalities have been shown to peak soon after birth, but are gradually corrected for by compensatory mechanisms with continued maturation.[127] It is important to point out that NMDA receptor expression is stimulus-dependent. This dependence could explain the high levels observed during brain development as observed in VPA-exposed rat offspring,[72] and the idea that these high levels might relax with age, yet be more readily inducible.[112] Hence, VPA-dependent glutamatergic neuronal differentiation and early developmental glutamatergic activity-dependent synaptic plasticity may act to enhance the trophic effects on synapse formation and neurite outgrowth,[103,128,129] as well as the subsequent emergence of the characteristic behaviors observed in ASD.[14,16]

4.1.3 Serotonin

Serotonin (5-HT) has a long-standing history with ASD. Early clinical work dating back to the 1960s and 1970s has consistently reported a significant elevation in blood 5-HT in autistic children.[130–138] One of the initial models involving the central actions of peripheral 5-HT in ASD proposed that during early development, when the blood brain barrier is not yet fully formed, 5-HT crosses the blood brain barrier to act on the central nervous system.[139] This centrally-acting 5-HT may then lead to a transient hyperinnervation of 5-HT cortical fibers (reviewed in Ref. 16). More recently, clinical evidence suggesting increased numbers of cortical 5-HT axons in ASD and experimental results in ASD animal models have provided additional supporting evidence for increased brain serotonergic activity in ASD.[140–143] Using the rodent prenatal VPA model, several studies have found elevations in 5-HT in different brain regions.[71,144,145] In addition, the development of a separate rodent model of ASD based on the 5-HT agonist

5-methoxytryptamine found that treated animals also exhibited some ASD-like features including repetitive behavior, abnormal sensory processing, and communicative abnormalities.[146] More recently, Wang et al. have linked serotonergic and glutamatergic systems in the VPA animal model.[142] These authors have provided important evidence of functional system interactions in this ASD animal model, as prenatal VPA exposure can profoundly enhance glutamatergic synaptic transmission onto putative 5-HT neurons and increase spontaneous firing in these neurons thought to drive increased 5-HT release.[142] Given 5-HT promotes neuronal differentiation, synaptogenesis, and network excitability,[147,148] while 5-HT depletion can reduce dendritic arborization and complexity,[149] these findings further support a critical role of 5-HT in neurodevelopment[143] and suggest a possible interaction between the serotonergic and glutamatergic systems associated with early overgrowth in ASD.

4.1.4 Dopamine

Much like 5-HT, the monoamine dopamine (DA) has been known for some time to be associated with ASD.[16,150] Early clinical work using biochemical analysis suggested that increased dopaminergic action may play a role in the early presentation of ASD.[16,151,152] Indeed, several additional lines of evidence have supported the idea of increased DA signaling in ASD. First, perinatal risk factors consistently associated with ASD, such as cesarean births and low-Apgar scores, may be associated with elevated DA transmission in the newborn.[153–156] Second, DiCarlo et al. have reported altered DA-dependent behaviors corresponding to behavioral phenotypes observed in ASD based on a *de novo* mutation in the SLC6A3 gene that drives persistent reverse DA transport (i.e., anomalous DA efflux) and synaptic hyperdopaminergia.[157,158] Third, VPA exposure in pregnant dams has been shown to result in increased levels of DA in neocortical networks in their offspring.[145,159] Fourth, alterations in hedonic behaviors have been observed in mice exposed to perinatal L-DOPA (DA precursor),[160] a finding consistent with that reported in the VPA animal model of ASD.[161] This further supports the notion that early elevations in DA can have a direct impact on behavioral development. Finally, as VPA is a potent HDAC inhibitor,[31] increased dopaminergic signaling may be a result of disrupted HDAC-mediated Nurr1-dependent suppression of dopaminergic genes in prenatal DA neurons.[162,163] This finding is particularly interesting because some cortical neurons also express high levels of Nurr1,[164,165] an established nuclear orphan receptor[166] that regulates expression of genes necessary for functional DA signaling (e.g., tyrosine hydroxylase, vesicular monoamine transporter, dopamine transporter, and aromatic L-amino acid decarboxylase).[162] This expression raises the intriguing question of what the impact of VPA exposure is on Nurr1-positive cortical neurons.[164,165]

Increased DA signaling may have both direct and indirect influences on early ASD overgrowth. Direct experimental evidence has shown that DA and DA-receptor (particularly D_2 receptor—D_2R) activation can result in premature overgrowth of DA neurons and cultured neocortical frontal neurons *in vitro*.[167–169] In fact, DA-dependent signaling may be preferentially biased toward D_2R pathway activation during early development, as VPA can differentially potentiate D_2R signaling activity in immature neurons via an HDAC-dependent mechanism.[170] An important prediction from this D_2R molecular mechanism is that cortical neurons with D_2R, which would have been pathologically targeted during early development, should also be involved in the manifestation of abnormal ASD-like behavior. A recent optogenetic study using D_2R-Cre mice to label D_2R-positive neurons has provided evidence that layer 5 subcortically-projecting D_2R prefrontal cortical neurons are indeed involved in disrupted social behavior in the VPA animal model of ASD.[171] It appears that abnormal neural activity specific to D_2R prefrontal cortical neurons impairs social behavior,[171] supporting the idea that early pathogenic D_2R activation may lead to lasting changes in D_2R-neuron function. Finally, Perreault et al. have reported a novel physiological role for the dopamine D_5 receptor in the prefrontal cortex that links DA signals to enhanced brain-derived neurotrophic factor expression and signaling.[172] The mechanism involves an Akt-dependent intracellular signaling cascade,[172] an intracellular pathway commonly associated with ASD.[13] With the accumulating evidence supporting the involvement of brain-derived neurotrophic factor in ASD (as we shall see next), enhanced DA signaling may act synergistically with this trophic factor in ASD pathogenesis.

4.1.5 Brain-derived neurotrophic factor

Brain-derived neurotrophic factor (BDNF) is a potent stimulator of dendritic growth and branching.[173–175] It is therefore not surprising that hyperactive BDNF has been proposed to be a potential mechanism in the pathogenesis of ASD.[176] Indeed, there is accumulating clinical evidence to support the notion that early BDNF signaling may play an etiological role in ASD. For example, serum BDNF levels, which positively correlate to cortical BDNF levels,[177,178] are increased in individuals with ASD compared to controls.[179,180] A recently conducted systematic review and meta-analysis on the role of BDNF in ASD have also concluded that BDNF is significantly higher in ASD compared to controls.[181,182] Furthermore, experimental results have shown that VPA can up-regulate BDNF expression through epigenetic modulation in neural progenitor cells, immature neurons *in vitro*, and in the offspring of pregnant dams exposed to VPA *in vivo*.[93,94,183–186] Thus, both clinical and experimental research findings are starting to point to an important link between BDNF in ASD.

Terminal differentiation is the process of morphological and functional maturation of individual perinatal neurons that continues into the postnatal phase of maturation.[187–189] While the precise cellular mechanisms of BDNF action with respect to the emergence of ASD are still not fully understood, research from our group has suggested that it may disrupt neuronal somato-dendritic decoupling during terminal differentiation.[190,191] Somato-dendritic decoupling represents an important developmental process that regulates cortical neuron cell-soma accessibility to electrical and chemical signals[190,191] (Fig. 3). The mechanism of somatic compartmentalization arises from a complex cytosolic lipid-protein structure that increases somato-dendritic inferfacial resistance and is age-dependent (decreasing with advancing age).[190,191] Given that a multitude of developmental genomic and epigenetic programs are hosted in the soma, and that their activation often requires externally-driven electrical stimulation and somatic accessibility,[193,194] protracted somatic compartmentalization via decoupling may regulate cellular maturation.[190,191] This regulation may take the form of blocking soluble trans-compartment signaling cascades, keeping somatic spike-counting machinery "off-line", and/or preventing developmentally premature bidirectional somato-dendritic communication.[195–197] In further support of the fact that somato-dendritic decoupling might play a role in ASD, it has been shown that BDNF stimulation can mimic the effects of VPA exposure, inducing premature somato-dendritic coupling and early neural overgrowth in ASD translational models.[57,84,191]

Moreover, somato-dendritic decoupling is thought to help minimize recurrent cortical network connectivity during early development to allow for the sequential building of more complex cortical networks.[190] An additional consequence of premature somato-dendritic coupling might therefore be to further enhance pathogenic activity-dependent processes involved in neuronal as well as synaptic network maturation and growth in ASD.[14,16] As it has been shown that activity-induced NR2A activation

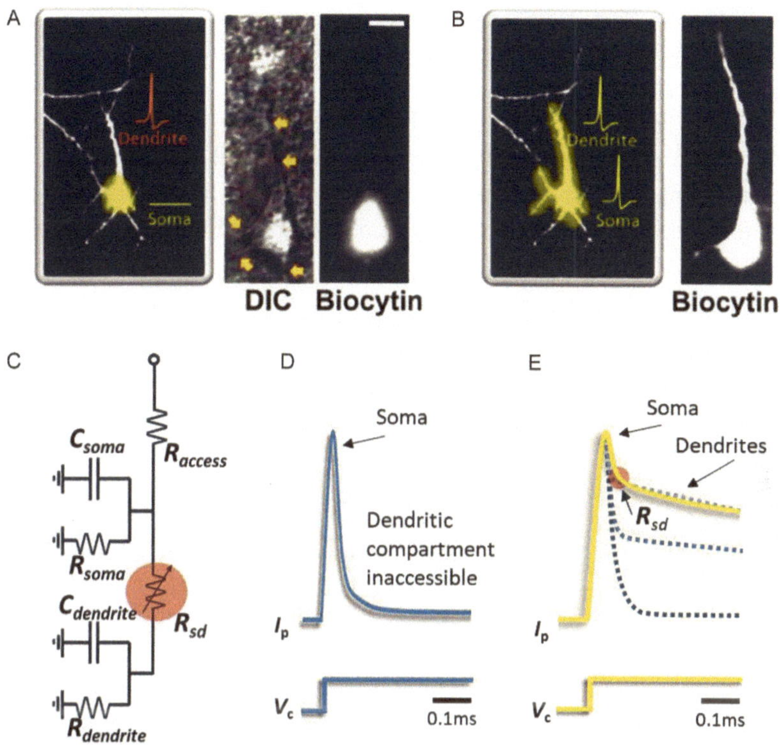

Fig. 3 Somato-dendritic decoupling in cortical neurons. (A) Schematic illustrating somato-dendritic decoupling (left) in a cortical neuron. Here, dendrites can receive afferent inputs and even spike (denoted in red), but this information does not converge at the level of the soma. An experimentally identified high-order temporal lobe neocortical dormant neuron that exhibits somato-dendritic decoupling (right). Yellow arrows indicate observable dendrites that lack diffusible biocytin labeling, despite the labeled somatic compartment. Biocytin was delivered via the patch pipette during patch-clamp recordings. (B) Schematic illustrating somato-dendritic coupling (left), and an experimentally identified somato-dendritically coupled neuron (right). (C) An equivalent circuit of somato-dendritic de/coupling. The red dot represents the interface of the two compartments. (D, E) During patch-clamp experiments, electrophysiological recordings were used to confirm a functionally compartmentalized soma or a dendritically-coupled soma. The low-electrical resistance inflection point (red dot) in the illustrative normalized capacitance current transient in panel E (which is not present in the panel D trace) represents the junction between the two compartments that can be modeled by the equivalent circuit in panel C. Dotted lines in panel E represent PartSim model simulations[192] with increasing R_{sd} (moving from top to bottom). *Figure and legend modified from Chomiak T, Hu B. Mechanisms of hierarchical cortical maturation.* Front Cell Neurosci. *2017;11; Chomiak T, Hung J, Nguyen MD, Hu B. Somato-dendritic decoupling as a novel mechanism for protracted cortical maturation.* BMC Biol. *2016;14. These articles are distributed under the terms of the Creative Commons Attribution 4.0 International License (http://creativecommons.org/licenses/by/4.0/).*

can lead to increased BDNF expression,[198] increased recurrent cortical connectivity and network excitability may be further influenced by BDNF-dependent HDAC modulation[199] and/or activity-dependent signaling cascades (see Fig. 2 for example). Thus, the data to date point to a new cellular developmental model involving a mechanism of somato-dendritic decoupling that may be particularly vulnerable to pathogenic environmental influences, and the premature initiation of activity-dependent developmental programs.

5. Concluding remarks and future directions

In this chapter, we have provided a framework to better understand the neurobiological changes associated with ASD. We described the link between converging pathogenic processes that are associated with human clinicopathology and conserved in ASD translational animal models. By integrating findings from both human and animal research, the data discussed in this chapter reveal cellular and molecular insights into the contemporary view of an early overgrowth model of ASD.

As the field continues to move forward, there are a few key points to consider that may help accelerate discovery. First, since the potential mechanisms discussed in this chapter are likely interconnected, it will be important for future research to consider interactions between functional systems. For instance, while both the dopaminergic and serotonergic systems have had a long-standing association with ASD, emerging evidence suggests that other neuromodulatory transmitter systems such as histamine and acetylcholine may also be important.[8,200] Given that histamine and acetylcholine may interact with the dopaminergic and serotonergic systems (e.g., Ref. 201), a better understanding of these interactions and their potential bi-directional cross-talk in ASD models may lead to deeper insight into the pathogenesis associated with ASD and its clinical management. Second, it will be important for researchers to keep in mind that the mechanism of overgrowth may not be uniform. This may be reflected in temporally and/or regionally-specific expression patterns,[202] or possibly in differences with respect to increases in the numbers of neurons versus neurite overgrowth in different brain regions. Finally, we have seen a growing wave of optogenetic applications that selectively target specific circuits, allowing investigators to achieve circuit-level insight within freely behaving mammals.[203] Despite improved optogenetic tools for use in well-defined neuronal populations, there has been relatively little research devoted to applying these tools to ASD translational models. This is likely related to the fact that traditional approaches such as viral transduction are of limited use to investigate local network interactions during development. That is because most promoters have been shown to undergo qualitative and quantitative transitions during development that can lead to unspecific and unstable expression.[204] In addition, viruses often require 10–14 days before reliable and sufficient expression is achieved,[204] thus also limiting its use during early postnatal development. However, given recent advances for overcoming these limitations via brain area-, layer- and cell type-specific expression of opsins by *in utero* electroporation, optogenetic manipulation of neonatal neuronal networks opens new perspectives for the interrogation of neuronal networks and the elucidation of cellular interactions during normal and pathological brain development.[204]

Acknowledgments

TC would like to thank the CSM Optogenetics Facility, the Hotchkiss Brain Institute, and CaPRI for their continued support. TC is a new investigator funded by the Alberta Children's Hospital Research Institute (ACHRI) Behaviour & The Developing Brain Pilot Research Program, and the Hotchkiss Brain Institute/Department of Clinical Neurosciences/Tourmaline Oil Chair in Parkinson's Disease Pilot Research Fund Program. TC was previously supported by the Sinneave Family Foundation.

References

1. Bishop DV. Which neurodevelopmental disorders get researched and why? *PLoS One*. 2010;5:e15112.
2. Park HY, Harwood RL, Yu SM, Kavanagh L, Lu MC. Autism and other developmental disabilities research programs of the maternal and child health Bureau. *Pediatrics*. 2016;137(Suppl):S61–S66.
3. Murray DS, Fedele A, Shui A, Coury DL. The autism speaks autism treatment network registry data: opportunities for investigators. *Pediatrics*. 2016;137(Suppl):S72–S78.
4. Blumberg SJ, Bramlett MD, Kogan MD, et al. *Changes in Prevalence of Parent-Reported Autism Spectrum Disorder in School-Aged U.S. Children: 2007 to 2011–2012*. Hyattsville, MD: National Center for Health Statistics; 2013. National Health Statistics Report 65.
5. Howlin P, Goode S, Hutton J, Rutter M. Adult outcome for children with autism. *J Child Psychol Psychiatry*. 2004;45:212–229.
6. Magiati I, Tay XW, Howlin P. Cognitive, language, social and behavioural outcomes in adults with autism spectrum disorders: a systematic review of longitudinal follow-up studies in adulthood. *Clin Psychol Rev*. 2014;34:73–86.
7. Gray KM, et al. Adult outcomes in autism: community inclusion and living skills. *J Autism Dev Disord*. 2014;44:3006–3015.
8. Eissa N, et al. Current enlightenment about etiology and pharmacological treatment of autism spectrum disorder. *Front Neurosci*. 2018;12:304.
9. Accordino RE, Kidd C, Politte LC, Henry CA, McDougle CJ. Psychopharmacological interventions in autism spectrum disorder. *Expert Opin Pharmacother*. 2016;17:937–952.

10. Buescher AV, Cidav Z, Knapp M, Mandell DS. Costs of autism spectrum disorders in the United Kingdom and the United States. *JAMA Pediatr.* 2014;168:721–728.
11. Kanner L. Autistic disturbances of affective contact. *Nerv Child.* 1943;2:217–250.
12. Fombonne E. Modern views of autism. *Can J Psychiatr.* 2003;48:503–505.
13. Courchesne E, et al. The ASD living biology: from cell proliferation to clinical phenotype. *Mol Psychiatry.* 2019;24:88–107.
14. Chomiak T, Turner N, Hu B. What we have learned about autism spectrum disorder from valproic acid. *Pathol Res Int.* 2013;2013:712758.
15. Courchesne E, et al. Mapping early brain development in autism. *Neuron.* 2007; 56:399–413.
16. Chomiak T, Hu B. Alterations of neocortical development and maturation in autism: insight from valproic acid exposure and animal models of autism. *Neurotoxicol Teratol.* 2013;36:57–66.
17. Wing L. The history of ideas on autism: legends, myths and reality. *Autism.* 1997; 1:13–23.
18. Wiśniowiecka-Kowalnik B, Nowakowska BA. Genetics and epigenetics of autism spectrum disorder-current evidence in the field. *J Appl Genet.* 2019;60:37–47.
19. Sun W, et al. Histone acetylome-wide association study of autism spectrum disorder. *Cell.* 2016;167:1385–1397.e11.
20. Loke YJ, Hannan AJ, Craig JM. The role of epigenetic change in autism spectrum disorders. *Front Neurol.* 2015;6:107.
21. Sacco R, Gabriele S, Persico AM. Head circumference and brain size in autism spectrum disorder: a systematic review and meta-analysis. *Psychiatry Res.* 2015;234:239–251.
22. Amaral DG, Schumann CM, Nordahl CW. Neuroanatomy of autism. *Trends Neurosci.* 2008;31:137–145.
23. Courchesne E, Carper R, Akshoomoff N. Evidence of brain overgrowth in the first year of life in autism. *JAMA.* 2003;290:337–344.
24. Schumann CM, et al. Longitudinal magnetic resonance imaging study of cortical development through early childhood in autism. *J Neurosci.* 2010;30:4419–4427.
25. Schafer ST, et al. Pathological priming causes developmental gene network heterochronicity in autistic subject-derived neurons. *Nat Neurosci.* 2019;22:243–255.
26. Marchetto MC, et al. Altered proliferation and networks in neural cells derived from idiopathic autistic individuals. *Mol Psychiatry.* 2017;22:820–835.
27. Stamou M, Streifel KM, Goines PE, Lein PJ. Neuronal connectivity as a convergent target of gene x environment interactions that confer risk for Autism Spectrum Disorders. *Neurotoxicol Teratol.* 2013;36:3–16.
28. Deneault E, et al. CNTN5-/+ or EHMT2-/+ human iPSC-derived neurons from individuals with autism develop hyperactive neuronal networks. *elife.* 2019;8:e40092.
29. Carter CJ, Blizard RA. Autism genes are selectively targeted by environmental pollutants including pesticides, heavy metals, bisphenol A, phthalates and many others in food, cosmetics or household products. *Neurochem Int.* 2016;101:83–109.
30. Modabbernia A, Velthorst E, Reichenberg A. Environmental risk factors for autism: an evidence-based review of systematic reviews and meta-analyses. *Mol Autism.* 2017;8:13.
31. Monti B, Polazzi E, Contestabile A. Biochemical, molecular and epigenetic mechanisms of valproic acid neuroprotection. *Curr Mol Pharmacol.* 2009;2:95–109.
32. Tomson T, Battino D, Perucca E. Valproic acid after five decades of use in epilepsy: time to reconsider the indications of a time-honoured drug. *Lancet Neurol.* 2016; 15:210–218.
33. Laegreid L, Kyllerman M, Hedner T, Hagberg B, Viggedahl G. Benzodiazepine amplification of valproate teratogenic effects in children of mothers with absence epilepsy. *Neuropediatrics.* 1993;24:88–92.

34. Christianson AL, Chesler N, Kromberg JG. Fetal valproate syndrome: clinical and neuro-developmental features in two sibling pairs. *Dev Med Child Neurol.* 1994; 36:361–369.
35. Williams G, et al. Fetal valproate syndrome and autism: additional evidence of an association. *Dev Med Child Neurol.* 2001;43:202–206.
36. Williams PG, Hersh JH. A male with fetal valproate syndrome and autism. *Dev Med Child Neurol.* 1997;39:632–634.
37. Moore SJ, et al. A clinical study of 57 children with fetal anticonvulsant syndromes. *J Med Genet.* 2000;37:489–497.
38. Rasalam AD, et al. Characteristics of fetal anticonvulsant syndrome associated autistic disorder. *Dev Med Child Neurol.* 2005;47:551–555.
39. Bromley RL, et al. The prevalence of neurodevelopmental disorders in children prenatally exposed to antiepileptic drugs. *J Neurol Neurosurg Psychiatry.* 2013; 84:637–643.
40. Christensen J, et al. Prenatal valproate exposure and risk of autism spectrum disorders and childhood autism. *JAMA.* 2013;309:1696–1703.
41. Petersen I, et al. Risks and benefits of psychotropic medication in pregnancy: cohort studies based on UK electronic primary care health records. *Health Technol Assess.* 2016;20:1–176.
42. Hisle-Gorman E, et al. Prenatal, perinatal, and neonatal risk factors of autism spectrum disorder. *Pediatr Res.* 2018;84:190–198.
43. Vajda FJE, et al. The Australian register of antiepileptic drugs in pregnancy: the first 1002 pregnancies. *Aust N Z J Obstet Gynaecol.* 2007;47:468–474.
44. Wood AG, et al. Prospective assessment of autism traits in children exposed to antiepileptic drugs during pregnancy. *Epilepsia.* 2015;56:1047–1055.
45. Deshmukh U, et al. Behavioral outcomes in children exposed prenatally to lamotrigine, valproate, or carbamazepine. *Neurotoxicol Teratol.* 2016;54:5–14.
46. Roullet FI, Lai JK, Foster JA. In utero exposure to valproic acid and autism—a current review of clinical and animal studies. *Neurotoxicol Teratol.* 2013;36:47–56.
47. Mabunga DF, Gonzales EL, Kim JW, Kim KC, Shin CY. Exploring the validity of valproic acid animal model of autism. *Exp Neurol.* 2015;24:285–300.
48. Nicolini C, Fahnestock M. The valproic acid-induced rodent model of autism. *Exp Neurol.* 2017;299(Pt A):217–227.
49. Belzung C, Leman S, Vourc'h P, Andres C. Rodent models for autism: a critcal review. *Drug Discov Today Dis Model.* 2005;2:93–101.
50. Chomiak T, Cihal A, Clark T, Ranawaya R, Hu B. Concurrent arm swing-stepping (CASS) test for dual task-related movement incoordination and hesitation in Parkinson's disease. *J Parkinsons Dis.* 2013;3:P12.08.
51. Zachar G, et al. Valproate exposure in ovo attenuates the acquisition of social preferences of young post-hatch domestic chicks. *Front Physiol.* 2019;10:881.
52. Sailer L, Duclot F, Wang Z, Kabbaj M. Consequences of prenatal exposure to valproic acid in the socially monogamous prairie voles. *Sci Rep.* 2019;9:2453.
53. Dwivedi S, et al. Larval zebrafish model for studying the effects of valproic acid on neurodevelopment: an approach towards modeling autism. *J Pharmacol Toxicol Methods.* 2019;95:56–65.
54. Zhao H, et al. Maternal valproic acid exposure leads to neurogenesis defects and autism-like behaviors in non human primates. *Transl Psychiatry.* 2019;9:267.
55. Dufour-Rainfray D, et al. Fetal exposure to teratogens: evidence of genes involved in autism. *Neurosci Biobehav Rev.* 2011;35:1254–1265.
56. Chomiak T, Brown AR, Teskey GC, Hu B. Motivational wheel running reverses cueing behavioural inflexibility in rodents. *J Neural Transm.* 2017;124:1635–1640.

57. Chomiak T, et al. Auditory-cued sensorimotor task reveals disengagement deficits in rats exposed to the autism-associated teratogen valproic acid. *Neuroscience*. 2014; 268:212–220.
58. Schneider T, Turczak J, Przewlocki R. Environmental enrichment reverses behavioral alterations in rats prenatally exposed to valproic acid: issues for a therapeutic approach in autism. *Neuropsychopharmacology*. 2006;31:36–46.
59. Schneider T, et al. Gender-specific behavioral and immunological alterations in an animal model of autism induced by prenatal exposure to valproic acid. *Psychoneuroendocrinology*. 2008;33:728–740.
60. Vorhees CV. Behavioral teratogenicity of valproic acid: selective effects on behavior after prenatal exposure to rats. *Psychopharmacology*. 1987;92:173–179.
61. Zamberletti E, et al. Cannabidivarin treatment ameliorates autism-like behaviors and restores hippocampal endocannabinoid system and glia alterations induced by prenatal valproic acid exposure in rats. *Front Cell Neurosci*. 2019;13:367.
62. Guimarães-Souza EM, Joselevitch C, Britto LRG, Chiavegatto S. Retinal alterations in a pre-clinical model of an autism spectrum disorder. *Mol Autism*. 2019;10:19.
63. Kotajima-Murakami H, et al. Effects of rapamycin on social interaction deficits and gene expression in mice exposed to valproic acid in utero. *Mol Brain*. 2019;12:3.
64. Liu F, Horton-Sparks K, Hull V, Li RW, Martínez-Cerdeño V. The valproic acid rat model of autism presents with gut bacterial dysbiosis similar to that in human autism. *Mol Autism*. 2018;9:61.
65. Mirza R, Sharma B. Benefits of Fenofibrate in prenatal valproic acid-induced autism spectrum disorder related phenotype in rats. *Brain Res Bull*. 2019;147:36–46.
66. Hajisoltani R, et al. Hyperexcitability of hippocampal CA1 pyramidal neurons in male offspring of a rat model of autism spectrum disorder (ASD) induced by prenatal exposure to valproic acid: a possible involvement of Ih channel current. *Brain Res*. 2019;1708:188–199.
67. Guo Q, et al. Hydrogen-rich water ameliorates autistic-like behavioral abnormalities in valproic acid-treated adolescent mice offspring. *Front Behav Neurosci*. 2018;12:170.
68. Lin HC, Gean PW, Wang CC, Chan YH, Chen PS. The amygdala excitatory/inhibitory balance in a valproate-induced rat autism model. *PLoS One*. 2013;8:e55248.
69. Bambini-Junior V, et al. Animal model of autism induced by prenatal exposure to valproate: behavioral changes and liver parameters. *Brain Res*. 2011;1408:8–16.
70. Binkerd PE, Rowland JM, Nau H, Hendrickx AG. Evaluation of valproic acid (VPA) developmental toxicity and pharmacokinetics in Sprague-Dawley rats. *Fundam Appl Toxicol*. 1988;11:485–493.
71. Dufour-Rainfray D, et al. Behavior and serotonergic disorders in rats exposed prenatally to valproate: a model for autism. *Neurosci Lett*. 2010;470:55–59.
72. Rinaldi T, Kulangara K, Antoniello K, Markram H. Elevated NMDA receptor levels and enhanced postsynaptic long-term potentiation induced by prenatal exposure to valproic acid. *Proc Natl Acad Sci USA*. 2007;104:13501–13506.
73. Markram K, Rinaldi T, La Mendola D, Sandi C, Markram H. Abnormal fear conditioning and amygdala processing in an animal model of autism. *Neuropsychopharmacology*. 2008;33:901–912.
74. Roullet FI, Wollaston L, Decatanzaro D, Foster JA. Behavioral and molecular changes in the mouse in response to prenatal exposure to the anti-epileptic drug valproic acid. *Neuroscience*. 2010;170:514–522.
75. Schneider T, Przewlocki R. Behavioral alterations in rats prenatally exposed to valproic acid: animal model of autism. *Neuropsychopharmacology*. 2005;30:80–89.
76. Woo CC, Leon M. Environmental enrichment as an effective treatment for autism: a randomized controlled trial. *Behav Neurosci*. 2013;127:487–497.

77. Woo CC, Donnelly JH, Steinberg-Epstein R, Leon M. Environmental enrichment as a therapy for autism: a clinical trial replication and extension. *Behav Neurosci*. 2015; 129:412–422.
78. Foley AG, et al. Class I histone deacetylase inhibition ameliorates social cognition and cell adhesion molecule plasticity deficits in a rodent model of autism spectrum disorder. *Neuropharmacology*. 2012;63:750–760.
79. Bambini-Junior V, et al. Resveratrol prevents social deficits in animal model of autism induced by valproic acid. *Neurosci Lett*. 2014;583:176–181.
80. Sgritta M, et al. Mechanisms underlying microbial-mediated changes in social behavior in mouse models of autism spectrum disorder. *Neuron*. 2019;101:246–259.e6.
81. Oguchi-Katayama A, Monma A, Sekino Y, Moriguchi T, Sato K. Comparative gene expression analysis of the amygdala in autistic rat models produced by pre- and post-natal exposures to valproic acid. *J Toxicol Sci*. 2013;38:391–402.
82. Reynolds S, Millette A, Devine DP. Sensory and motor characterization in the post-natal valproate rat model of autism. *Dev Neurosci*. 2012;34:258–267.
83. Yochum CL, Dowling P, Reuhl KR, Wagner GC, Ming X. VPA-induced apoptosis and behavioral deficits in neonatal mice. *Brain Res*. 2008;1203:126–132.
84. Chomiak T, Karnik V, Block E, Hu B. Altering the trajectory of early postnatal cortical development can lead to structural and behavioural features of autism. *BMC Neurosci*. 2010;11:102.
85. Mony TJ, Lee JW, Dreyfus C, DiCicco-Bloom E, Lee HJ. Valproic acid exposure during early postnatal gliogenesis leads to autistic-like behaviors in rats. *Clin Psychopharmacol Neurosci*. 2016;14:338–344.
86. Ornoy A, et al. S-adenosyl methionine prevents ASD like behaviors triggered by early postnatal valproic acid exposure in very young mice. *Neurotoxicol Teratol*. 2019; 71:64–74.
87. Morakotsriwan N, Wattanathorn J, Kirisattayakul W, Chaisiwamongkol K. Autistic-like behaviors, oxidative stress status, and histopathological changes in cerebellum of valproic acid rat model of autism are improved by the combined extract of purple rice and silkworm pupae. *Oxidative Med Cell Longev*. 2016;2016:3206561.
88. Kim J-E, et al. Treadmill exercise ameliorates motor disturbance through inhibition of apoptosis in the cerebellum of valproic acid-induced autistic rat pups. *Mol Med Rep*. 2013;8:327–334.
89. Jager-Roman E, et al. Fetal growth, major malformations, and minor anomalies in infants born to women receiving valproic acid. *J Pediatr*. 1986;108:997–1004.
90. Ibi D, et al. Paternal valproic acid exposure in mice triggers behavioral alterations in offspring. *Neurotoxicol Teratol*. 2019;76:106837.
91. Jessberger S, et al. Epigenetic modulation of seizure-induced neurogenesis and cognitive decline. *J Neurosci*. 2007;27:5967–5975.
92. Akhtar MW, et al. Histone deacetylases 1 and 2 form a developmental switch that controls excitatory synapse maturation and function. *J Neurosci*. 2009;29: 8288–8297.
93. Fukuchi M, et al. Valproic acid induces up- or down-regulation of gene expression responsible for the neuronal excitation and inhibition in rat cortical neurons through its epigenetic actions. *Neurosci Res*. 2009;65:35–43.
94. Yasuda S, Liang MH, Marinova Z, Yahyavi A, Chuang DM. The mood stabilizers lithium and valproate selectively activate the promoter IV of brain-derived neurotrophic factor in neurons. *Mol Psychiatry*. 2009;14:51–59.
95. Lee RS, et al. Search for common targets of lithium and valproic acid identifies novel epigenetic effects of lithium on the rat leptin receptor gene. *Transl Psychiatry*. 2015;5:e600.

96. Zhang X, et al. PI3K/AKT/mTOR signaling mediates valproic acid-induced neuronal differentiation of neural stem cells through epigenetic modifications. *Stem Cell Rep.* 2017;8:1256–1269.
97. Cho Y, Cavalli V. HDAC signaling in neuronal development and axon regeneration. *Curr Opin Neurobiol.* 2014;27:118–126.
98. Cancedda L, Fiumelli H, Chen K, Poo MM. Excitatory GABA action is essential for morphological maturation of cortical neurons in vivo. *J Neurosci.* 2007;27: 5224–5235.
99. Owens DF, Kriegstein AR. Is there more to GABA than synaptic inhibition? *Nat Rev Neurosci.* 2002;3:715–727.
100. Represa A, Ben-Ari Y. Trophic actions of GABA on neuronal development. *Trends Neurosci.* 2005;28:278–283.
101. Tyzio R, et al. Oxytocin-mediated GABA inhibition during delivery attenuates autism pathogenesis in rodent offspring. *Science.* 2014;343:675–679.
102. Ben-Ari Y. Excitatory actions of gaba during development: the nature of the nurture. *Nat Rev Neurosci.* 2002;3:728–739.
103. Ben-Ari Y. The GABA excitatory/inhibitory developmental sequence: a personal journey. *Neuroscience.* 2014;279:187–219.
104. Zimmerman AW, Connors SL. Could autism be treated prenatally? *Science.* 2014; 343:620–621.
105. Cloarec R, et al. Pyramidal neuron growth and increased hippocampal volume during labor and birth in autism. *Sci Adv.* 2019;5:eaav0394.
106. Roux S, Lohof A, Ben-Ari Y, Poulain B, Bossu J-L. Maturation of GABAergic transmission in cerebellar purkinje cells is sex dependent and altered in the valproate model of autism. *Front Cell Neurosci.* 2018;12:232.
107. Gregory SG, et al. Genomic and epigenetic evidence for oxytocin receptor deficiency in autism. *BMC Med.* 2009;7:62.
108. Rubenstein JL, Merzenich MM. Model of autism: increased ratio of excitation/inhibition in key neural systems. *Genes Brain Behav.* 2003;2:255–267.
109. Banerjee A, et al. Impairment of cortical GABAergic synaptic transmission in an environmental rat model of autism. *Int J Neuropsychopharmacol.* 2013;16:1309–1318.
110. Horder J, et al. Glutamate and GABA in autism spectrum disorder-a translational magnetic resonance spectroscopy study in man and rodent models. *Transl Psychiatry.* 2018;8:106.
111. Rojas DC. The role of glutamate and its receptors in autism and the use of glutamate receptor antagonists in treatment. *J Neural Transm.* 2014;121:891–905.
112. Markram K, Markram H. The intense world theory—a unifying theory of the neurobiology of autism. *Front Hum Neurosci.* 2010;4:224.
113. Chiocchetti AG, Bour HS, Freitag CM. Glutamatergic candidate genes in autism spectrum disorder: an overview. *J Neural Transm.* 2014;121:1081–1106.
114. Shuang M, et al. Family-based association study between autism and glutamate receptor 6 gene in Chinese Han trios. *Am J Med Genet.* 2004;131B:48–50.
115. Serajee FJ, Zhong H, Nabi R, Huq AH. The metabotropic glutamate receptor 8 gene at 7q31: partial duplication and possible association with autism. *J Med Genet.* 2003;40:e42.
116. Jamain S, et al. Mutations of the X-linked genes encoding neuroligins NLGN3 and NLGN4 are associated with autism. *Nat Genet.* 2003;34:27–29.
117. Laumonnier F, et al. X-linked mental retardation and autism are associated with a mutation in the NLGN4 gene, a member of the neuroligin family. *Am J Hum Genet.* 2004;74:552–557.
118. Monteiro P, Feng G. SHANK proteins: roles at the synapse and in autism spectrum disorder. *Nat Rev Neurosci.* 2017;18:147–157.

119. Penzes P, Cahill ME, Jones KA, VanLeeuwen JE, Woolfrey KM. Dendritic spine pathology in neuropsychiatric disorders. *Nat Neurosci*. 2011;14:285–293.
120. Kim KC, et al. Pax6-dependent cortical glutamatergic neuronal differentiation regulates autism-like behavior in prenatally valproic acid-exposed rat offspring. *Mol Neurobiol*. 2014;49:512–528.
121. Zieminska E, et al. Glutamate, glutamine and GABA levels in rat brain measured using MRS, HPLC and NMR methods in study of two models of autism. *Front Mol Neurosci*. 2018;11:418.
122. Iijima Y, et al. Distinct defects in synaptic differentiation of neocortical neurons in response to prenatal valproate exposure. *Sci Rep*. 2016;6:27400.
123. Markram H, Rinaldi T, Markram K. The intense world syndrome—an alternative hypothesis for autism. *Front Neurosci*. 2007;1:77–96.
124. Rinaldi T, Perrodin C, Markram H. Hyper-connectivity and hyper-plasticity in the medial prefrontal cortex in the valproic acid animal model of autism. *Front Neural Circuits*. 2008;2:4.
125. Silvia GT, et al. Enhanced long term microcircuit plasticity in the valproic acid animal model of autism. *Front Synaptic Neurosci*. 2009;1:1.
126. Mohammadi S, et al. Improvement of autistic-like behaviors in adult rats prenatally exposed to valproic acid through early suppression of NMDA receptor function. *Psychopharmacology*. 2019;237:199–208. https://doi.org/10.1007/s00213-019-05357-2.
127. Walcott EC, Higgins EA, Desai NS. Synaptic and intrinsic balancing during postnatal development in rat pups exposed to valproic acid in utero. *J Neurosci*. 2011;31: 13097–13109.
128. Finsterwald C, Fiumelli H, Cardinaux J-R, Martin J-L. Regulation of dendritic development by BDNF requires activation of CRTC1 by glutamate. *J Biol Chem*. 2010;285: 28587–28595.
129. Wong ROL, Ghosh A. Activity-dependent regulation of dendritic growth and patterning. *Nat Rev Neurosci*. 2002;3:803–812.
130. Muller CL, Anacker AMJ, Veenstra-VanderWeele J. The serotonin system in autism spectrum disorder: from biomarker to animal models. *Neuroscience*. 2016;321:24–41.
131. Gabriele S, Sacco R, Persico AM. Blood serotonin levels in autism spectrum disorder: a systematic review and meta-analysis. *Eur Neuropsychopharmacol*. 2014;24:919–929.
132. Abramson RK, et al. Elevated blood serotonin in autistic probands and their first-degree relatives. *J Autism Dev Disord*. 1989;19:397–407.
133. Anderson GM, et al. Whole blood serotonin in autistic and normal subjects. *J Child Psychol Psychiatry*. 1987;28:885–900.
134. Leboyer M, et al. Whole blood serotonin and plasma beta-endorphin in autistic probands and their first-degree relatives. *Biol Psychiatry*. 1999;45:158–163.
135. McBride PA, et al. Effects of diagnosis, race, and puberty on platelet serotonin levels in autism and mental retardation. *J Am Acad Child Adolesc Psychiatry*. 1998;37:767–776.
136. Ritvo ER, et al. Increased blood serotonin and platelets in early infantile autism. *Arch Gen Psychiatry*. 1970;23:566–572.
137. Schain RJ, Freedman DX. Studies on 5-hydroxyindole metabolism in autistic and other mentally retarded children. *J Pediatr*. 1961;58:315–320.
138. Takahashi S, Kanai H, Miyamoto Y. Reassessment of elevated serotonin levels in blood platelets in early infantile autism. *J Autism Child Schizophr*. 1976;6:317–326.
139. Whitaker-Azmitia PM. Behavioral and cellular consequences of increasing serotonergic activity during brain development: a role in autism? *Int J Dev Neurosci*. 2005; 23:75–83.
140. Azmitia EC, Singh JS, Whitaker-Azmitia PM. Increased serotonin axons (immunoreactive to 5-HT transporter) in postmortem brains from young autism donors. *Neuropharmacology*. 2011;60:1347–1354.

141. Tanaka M, et al. Brain hyperserotonemia causes autism-relevant social deficits in mice. *Mol Autism*. 2018;9:60.
142. Wang R, Hausknecht K, Shen R-Y, Haj-Dahmane S. Potentiation of glutamatergic synaptic transmission onto dorsal raphe serotonergic neurons in the valproic acid model of autism. *Front Pharmacol*. 2018;9:1185.
143. Kepser L-J, Homberg JR. The neurodevelopmental effects of serotonin: a behavioural perspective. *Behav Brain Res*. 2015;277:3–13.
144. Tsujino N, et al. Abnormality of circadian rhythm accompanied by an increase in frontal cortex serotonin in animal model of autism. *Neurosci Res*. 2007;57:289–295.
145. Narita N, et al. Increased monoamine concentration in the brain and blood of fetal thalidomide- and valproic acid-exposed rat: putative animal models for autism. *Pediatr Res*. 2002;52:576–579.
146. Kahne D, et al. Behavioral and magnetic resonance spectroscopic studies in the rat hyperserotonemic model of autism. *Physiol Behav*. 2002;75:403–410.
147. Chubakov AR, Gromova EA, Konovalov GV, Sarkisova EF, Chumasov EI. The effects of serotonin on the morpho-functional development of rat cerebral neocortex in tissue culture. *Brain Res*. 1986;369:285–297.
148. Fricker AD, Rios C, Devi LA, Gomes I. Serotonin receptor activation leads to neurite outgrowth and neuronal survival. *Brain Res Mol Brain Res*. 2005;138:228–235.
149. Vitalis T, Cases O, Passemard S, Callebert J, Parnavelas JG. Embryonic depletion of serotonin affects cortical development. *Eur J Neurosci*. 2007;26:331–344.
150. Pavăl D. A dopamine hypothesis of autism spectrum disorder. *Dev Neurosci*. 2017; 39:355–360.
151. Gillberg C, Svennerholm L. CSF monoamines in autistic syndromes and other pervasive developmental disorders of early childhood. *Br J Psychiatry*. 1987;151:89–94.
152. Previc FH. Prenatal influences on brain dopamine and their relevance to the rising incidence of autism. *Med Hypotheses*. 2007;68:46–60.
153. Decker MJ, Rye DB. Neonatal intermittent hypoxia impairs dopamine signaling and executive functioning. *Sleep Breath*. 2002;6:205–210.
154. Glasson EJ, et al. Perinatal factors and the development of autism: a population study. *Arch Gen Psychiatry*. 2004;61:618–627.
155. Hultman CM, Sparen P, Cnattingius S. Perinatal risk factors for infantile autism. *Epidemiology*. 2002;13:417–423.
156. Seeman P, et al. Dopamine supersensitivity correlates with D2High states, implying many paths to psychosis. *Proc Natl Acad Sci USA*. 2005;102:3513–3518.
157. DiCarlo GE, et al. Autism-linked dopamine transporter mutation alters striatal dopamine neurotransmission and dopamine-dependent behaviors. *J Clin Invest*. 2019;129: 3407–3419.
158. Hamilton PJ, et al. De novo mutation in the dopamine transporter gene associates dopamine dysfunction with autism spectrum disorder. *Mol Psychiatry*. 2013;18:1315–1323.
159. Yochum CL, Bhattacharya P, Patti L, Mirochnitchenko O, Wagner GC. Animal model of autism using GSTM1 knockout mice and early post-natal sodium valproate treatment. *Behav Brain Res*. 2010;210:202–210.
160. de Matos LO, et al. L-Dopa treatment during perinatal development leads to different behavioral alterations in female vs male juvenile Swiss mice. *Pharmacol Biochem Behav*. 2018;173:1–14.
161. Schneider T, Ziolkowska B, Gieryk A, Tyminska A, Przewlocki R. Prenatal exposure to valproic acid disturbs the enkephalinergic system functioning, basal hedonic tone, and emotional responses in an animal model of autism. *Psychopharmacology*. 2007; 193:547–555.
162. Jacobs FM, et al. Pitx3 potentiates Nurr1 in dopamine neuron terminal differentiation through release of SMRT-mediated repression. *Development*. 2009;136:531–540.

163. Green AL, et al. Valproate increases dopamine transporter expression through histone acetylation and enhanced promoter binding of Nurr1. *Neuropharmacology*. 2017; 125:189–196.
164. Xing G, et al. Rat nurr1 is prominently expressed in perirhinal cortex, and differentially induced in the hippocampal dentate gyrus by electroconvulsive vs. kindled seizures. *Brain Res Mol Brain Res*. 1997;47:251–261.
165. Arimatsu Y, Ishida M, Kaneko T, Ichinose S, Omori A. Organization and development of corticocortical associative neurons expressing the orphan nuclear receptor Nurr1. *J Comp Neurol*. 2003;466:180–196.
166. Shi Y. Orphan nuclear receptors in drug discovery. *Drug Discov Today*. 2007; 12:440–445.
167. Swarzenski BC, Tang L, Oh YJ, O'Malley KL, Todd RD. Morphogenic potentials of D2, D3, and D4 dopamine receptors revealed in transfected neuronal cell lines. *Proc Natl Acad Sci USA*. 1994;91:649–653.
168. Reinoso BS, Undie AS, Levitt P. Dopamine receptors mediate differential morphological effects on cerebral cortical neurons in vitro. *J Neurosci Res*. 1996;43:439–453.
169. Todd RD. Neural development is regulated by classical neurotransmitters: dopamine D2 receptor stimulation enhances neurite outgrowth. *Biol Psychiatry*. 1992;31: 794–807.
170. Lee S, et al. Valproate alters dopamine signaling in association with induction of Par-4 protein expression. *PLoS One*. 2012;7:e45618.
171. Brumback AC, et al. Identifying specific prefrontal neurons that contribute to autism-associated abnormalities in physiology and social behavior. *Mol Psychiatry*. 2018; 23:2078–2089.
172. Perreault ML, Jones-Tabah J, O'Dowd BF, George SR. A physiological role for the dopamine D5 receptor as a regulator of BDNF and Akt signalling in rodent prefrontal cortex. *Int J Neuropsychopharmacol*. 2013;16:477–483.
173. Horch HW, Katz LC. BDNF release from single cells elicits local dendritic growth in nearby neurons. *Nat Neurosci*. 2002;5:1177–1184.
174. Maisonpierre PC, et al. NT-3, BDNF, and NGF in the developing rat nervous system: parallel as well as reciprocal patterns of expression. *Neuron*. 1990;5:501–509.
175. McAllister AK, Lo DC, Katz LC. Neurotrophins regulate dendritic growth in developing visual cortex. *Neuron*. 1995;15:791–803.
176. Tsai SJ. Is autism caused by early hyperactivity of brain-derived neurotrophic factor? *Med Hypotheses*. 2005;65:79–82.
177. Karege F, Schwald M, Cisse M. Postnatal developmental profile of brain-derived neurotrophic factor in rat brain and platelets. *Neurosci Lett*. 2002;328:261–264.
178. Klein AB, et al. Blood BDNF concentrations reflect brain-tissue BDNF levels across species. *Int J Neuropsychopharmacol*. 2011;14:347–353.
179. Miyazaki K, et al. Serum neurotrophin concentrations in autism and mental retardation: a pilot study. *Brain Dev*. 2004;26:292–295.
180. Nelson KB, et al. Neuropeptides and neurotrophins in neonatal blood of children with autism or mental retardation. *Ann Neurol*. 2001;49:597–606.
181. Saghazadeh A, Rezaei N. Brain-derived neurotrophic factor levels in autism: a systematic review and meta-analysis. *J Autism Dev Disord*. 2017;47:1018–1029.
182. Qin X-Y, et al. Association of peripheral blood levels of brain-derived neurotrophic factor with autism spectrum disorder in children. *JAMA Pediatr*. 2016;170:1079.
183. Konopko MA, Densmore AL, Krueger BK. Sexually dimorphic epigenetic regulation of brain-derived neurotrophic factor in fetal brain in the valproic acid model of autism spectrum disorder. *Dev Neurosci*. 2017;39:507–518.
184. Almeida LEF, Roby CD, Krueger BK. Increased BDNF expression in fetal brain in the valproic acid model of autism. *Mol Cell Neurosci*. 2014;59:57–62.

185. Bennett GD, Wlodarczyk B, Calvin JA, Craig JC, Finnell RH. Valproic acid-induced alterations in growth and neurotrophic factor gene expression in murine embryos [corrected]. *Reprod Toxicol.* 2000;14:1–11.
186. Ko HM, et al. Dual mechanisms for the regulation of brain-derived neurotrophic factor by valproic acid in neural progenitor cells. *Korean J Physiol Pharmacol.* 2018;22:679.
187. Abrous DN, Koehl M, Le Moal M. Adult neurogenesis: from precursors to network and physiology. *Physiol Rev.* 2005;85:523–569.
188. Hobert O. Regulation of terminal differentiation programs in the nervous system. *Annu Rev Cell Dev Biol.* 2011;27:681–696.
189. Picken Bahrey HL, Moody WJ. Early development of voltage-gated ion currents and firing properties in neurons of the mouse cerebral cortex. *J Neurophysiol.* 2003; 89:1761–1773.
190. Chomiak T, Hu B. Mechanisms of hierarchical cortical maturation. *Front Cell Neurosci.* 2017;11:272.
191. Chomiak T, Hung J, Nguyen MD, Hu B. Somato-dendritic decoupling as a novel mechanism for protracted cortical maturation. *BMC Biol.* 2016;14:48.
192. PartSim. 2019. Circuit Simulation Made Easy. Available at: http://partsim.com.
193. Zhang LI, Poo MM. Electrical activity and development of neural circuits. *Nat Neurosci.* 2001;4(Suppl):1207–1214.
194. Spitzer NC. Electrical activity in early neuronal development. *Nature.* 2006; 444:707–712.
195. Herbst WA, Martin KC. Regulated transport of signaling proteins from synapse to nucleus. *Curr Opin Neurobiol.* 2017;45:78–84.
196. Wild AR, et al. Synapse-to-nucleus communication through NFAT is mediated by L-type Ca^{2+} channel Ca^{2+} spike propagation to the soma. *Cell Rep.* 2019; 26:3537–3550.e4.
197. Nonaka M, et al. Untangling the two-way signalling route from synapses to the nucleus, and from the nucleus back to the synapses. *Philos Trans R Soc Lond Ser B Biol Sci.* 2014;369:20130150.
198. Chen Q, et al. Differential roles of NR2A- and NR2B-containing NMDA receptors in activity-dependent brain-derived neurotrophic factor gene regulation and limbic epileptogenesis. *J Neurosci.* 2007;27:542–552.
199. Nott A, Watson PM, Robinson JD, Crepaldi L, Riccio A. S-nitrosylation of histone deacetylase 2 induces chromatin remodelling in neurons. *Nature.* 2008;455:411–415.
200. Wright C, et al. Altered expression of histamine signaling genes in autism spectrum disorder. *Transl Psychiatry.* 2017;7:e1126.
201. Calabresi P, Di Filippo M. ACh/dopamine crosstalk in motor control and reward: a crucial role for alpha 6-containing nicotinic receptors? *Neuron.* 2008;60:4–7.
202. Hou Q, et al. A developmental study of abnormal behaviors and altered GABAergic signaling in the VPA-treated rat model of autism. *Front Behav Neurosci.* 2018;12:182.
203. Kim CK, Adhikari A, Deisseroth K. Integration of optogenetics with complementary methodologies in systems neuroscience. *Nat Rev Neurosci.* 2017;18:222–235.
204. Bitzenhofer SH, Ahlbeck J, Hanganu-Opatz IL. Methodological approach for optogenetic manipulation of neonatal neuronal networks. *Front Cell Neurosci.* 2017;11:239.

CHAPTER ELEVEN

The role of neuroglia in autism spectrum disorders

Caterina Scuderi[a,*], Alexei Verkhratsky[b,c]
[a]Department of Physiology and Pharmacology "Vittorio Erspamer", SAPIENZA University of Rome, Rome, Italy
[b]Faculty of Biology, Medicine and Health, The University of Manchester, Manchester, United Kingdom
[c]Achucarro Center for Neuroscience, IKERBASQUE, Basque Foundation for Science, Bilbao, Spain
[*]Corresponding author: e-mail address: caterina.scuderi@uniroma1.it

Contents

Abstract

Neuroglia are a large class of neural cells of ectodermal (astroglia, oligodendroglia, and peripheral glial cells) and mesodermal (microglia) origin. Neuroglial cells provide homeostatic support, protection, and defense to the nervous tissue. Pathological potential of neuroglia has been acknowledged since their discovery. Research of the recent decade has shown the key role of all classes of glial cells in autism spectrum disorders (ASD), although molecular mechanisms defining glial contribution to ASD are yet to be fully characterized. This narrative conceptualizes recent findings of the broader roles of glial cells, including their active participation in the control of cerebral environment and regulation of synaptic development and scaling, highlighting their putative involvement in the etiopathogenesis of ASD.

1. Autism spectrum disorders

Autism is grouped under the hypernum of autism spectrum disorders (ASD) based on the definition presented in 2013 update of DSM-5.[1] Both terms (autism and ASD) are indistinctively used to designate a group of

Progress in Molecular Biology and Translational Science, Volume 173
ISSN 1877-1173
https://doi.org/10.1016/bs.pmbts.2020.04.011

polyetiological neurodevelopmental disorders that, despite arising from different pathological mechanisms, result in aberrant social communication and interaction, impaired language skills, as well as restrictive and repetitive behaviors. Autism as a distinct nosological form was described by the psychiatrist Leo Kanner in 1943.[2] The word "autism" comes from the Greek word αυτός, which literally translates as "being alone with yourself". In modern days, ASD embrace numerous conditions previously thought to belong to other mental illnesses, including Rett's syndrome, Asperger's syndrome, pervasive developmental disorder not otherwise specified, and childhood disintegrative disorder.[3,4] Clinical presentation of ASD are exceedingly heterogeneous and include lack of attention, hyperactivity, irritability, self-injuring, and many other behavioral modifications.[5] In particular, autism is associated with compromised production and understanding of verbal language (the autistic children do not speak or speak little and with little connection to reality), malfunctional nonverbal communication (the autistic subject is unable to communicate through facial expressions, tones of voice and physical and visual contact) and compromised sociability (autistic children are unable to fit into the social and family context).[6–11] Finally, subjects with autism are unable to tune in to the emotions of others. Another frequent behavioral manifestation is the excessive susceptibility to environmental stimuli, such as noises, food, people or moving objects. At the same time some subjects affected by ASD demonstrate highly developed cognitive abilities, memory, or musical and mathematical skills.[12,13]

The interest in ASD has grown substantially in recent years, which reflects an almost exponential increase in the number of ASD diagnoses that placed ASD among the most frequent developmental disorders (Fig. 1).[14] Epidemiological studies highlight that ASD affect one in 160 children worldwide, with males diagnosed 4.5 times more than females, and with global increase in prevalence.[15] There is no credible explanation for this progressive increase in the prevalence of ASD, even if the improvement of the diagnostic criteria and the environmental influences, still under study, are considered valid reasons. Clinical signs emerge in the childhood and tend to persist into adolescence and adulthood. While some people with ASD can live autonomously, others are severely disabled and require special support. Several co-morbidities have been found in patients with ASD, including epilepsy, anxiety, depression, bipolar disorder, schizophrenia, attention deficit and hyperactivity disorder, sleep perturbations, gastrointestinal distress, and eating disorders.[16] These conditions profoundly affect the quality of life of the patients. Worldwide, people with ASD are often stigmatized

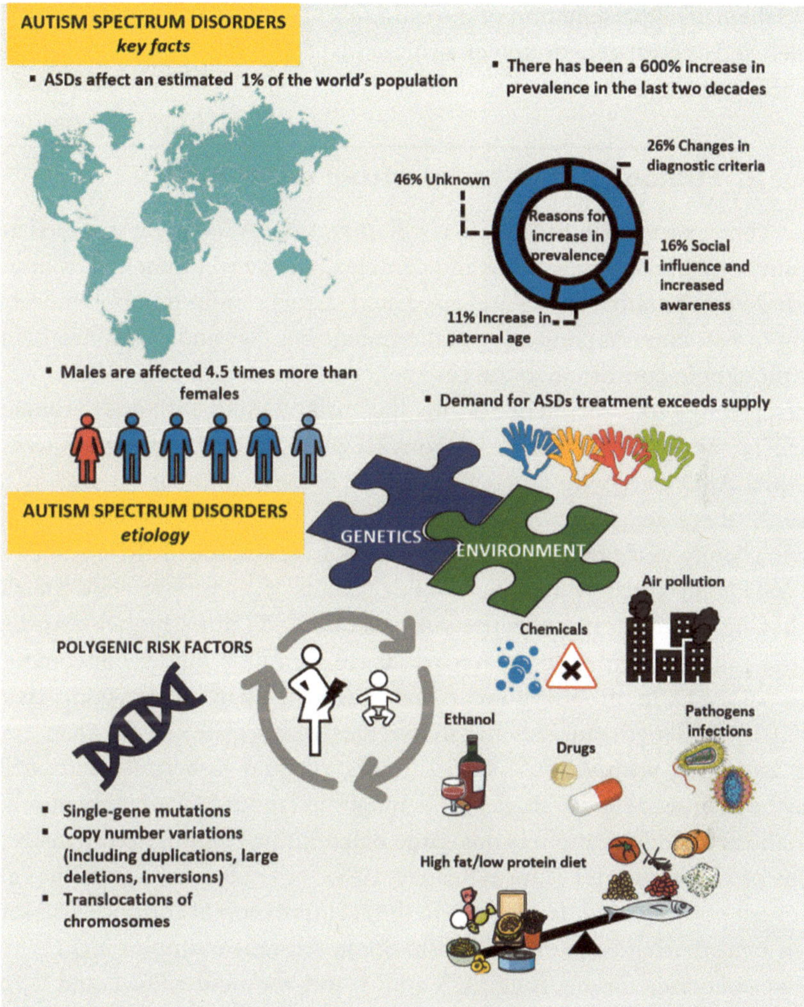

Fig. 1 Autism spectrum disorders: key facts and etiology.

and discriminated. Early diagnosis is essential for a better prognosis. ASD treatment requires integrated approaches for controlling mental illness that comorbid with ASD, including pedagogical and psychotherapeutic interventions, as well as pharmacological treatment. No drugs are currently available to cure autism. However, some treatments are employed for specific maladaptive behavioral symptoms, such as self- and hetero-aggression, hyperactivity, stereotyped behavior, and insomnia.

Schematic representation of the most recent epidemiological data (upper panel) and of both genetic and environmental factors (lower panel) involved in ASD etiology.

2. Etiology of autism spectrum disorders

The exact etiology of ASD is still unknown; ASD being stratified as a group of highly heterogeneous and complex neurodevelopmental disorders with diverse putative causes and associated factors resulting in polymorphic symptomatology. At present, both the genetic heritage and the environment are thought to contribute to the emergence and evolution of these disorders. The precise role of genetics in the pathogenesis of ASD remains uncharacterized and no single responsible gene has been identified so far. Despite ASD cases with a genetic etiology account only for 10–20% of all cases,[17] there are several hints that genetics does play a defining role.[18] Twin, family and population-based studies have estimated the heritability of ASD approximately at 50%.[19,20] For parents of an autistic child the risk of having another one with the same disease is 20 times higher than that in the general population. Moreover, there are much higher concordance rates of ASD in monozygotic twins (70–90%) than in dizygotic twins (0–10%).[17] Genetic analysis has shown that many genes (more than 100) are associated with ASD.[21] Genetic abnormalities associated with ASD may be a consequence of single-gene mutations (about 5%), or copy number variations including duplications, large deletions, inversions, and translocations of chromosomes (approximately 10%).[22–25] These abnormalities are frequent polygenic risk factors due to the accumulation of common variants, each contributing to part of the risk. Some relevant examples include the genes encoding for neuroligins 3 and 4 and for neurexins 1 and 3, all involved in the formation and operation of synapses; for SHANK3, involved in the development of dendrites; and for reelin, essential for the stabilization of the basic laminar organization of the cerebral cortex.[26–30] Unfortunately, the usefulness of this information remains elusive. Manipulations of ASD-associated genes in animal models and subsequent functional analyses may provide information on their physiological functions, highlighting complex phenotypic associations with these genes.

Environmental factors and neurotoxic agents add to the genetic background as etiological components of ASD.[31] Conceptually ASD originate from altered course of development of the central nervous system (CNS), which stipulates significant damage of physiology and behavior in adulthood.[32] The aforementioned evidence on the concordance rates of ASD in twins is an indirect proof of the existence of other nongenetic etiological factors. In particular, the perinatal environment has attracted much attention for its implications on brain development and functions.[33] Environmental risk factors that affect neurodevelopment and result in long-term alterations in brain physiology include in utero exposure to viral infections (e.g., rubella virus and cytomegalovirus), toxins, stress, alcohol, various drugs (for example thalidomide or valproic acid), as well as maternal care or diet.[34–39] Additional conceivable risk factors, not linked to genetic mutations, are represented by advanced age of the parents at conception, premature birth and low birth body weight, and even a short (less than 1 year) interval between pregnancies.

In summary, genetic and environmental risk factors are not to be considered separately. They can act either alone or by interacting with each other. It is probable that a particular combination of genes confers an individual susceptibility, that is, a latent risk; the concomitant presence of one or more environmental factors promotes the emergence of typical ASD symptoms. Although heterogeneous, many of the variants found in ASD-associated genes converge in interrelated functional modules. A proper understanding of the relationships between genotype and phenotype may elucidate the pathobiological mechanisms of specific genetic variations, hopefully revealing new diagnostic and therapeutic targets.

3. Molecular aspects of autism spectrum disorders

The relationship between genetic factors, environmental factors and clinical and subclinical manifestations in ASD is rather intricate, especially given the complex multifactorial etiology of this pathology. Ongoing research is focused on understanding how these elements interact in the developing brain. Postmortem studies, deep phenotype analyses of patients sharing highly penetrant mutations, functional neuroimaging studies, and functional tests in model organisms represent to date the best instruments for the comprehension of the biology of ASD.[40,41]

Brain areas involved in the control of emotions and social interaction as well as in motor coordination, which all are compromised in subjects with ASD, have been studied extensively. Neuropathological studies conducted in post-mortem samples from young subjects diagnosed with ADS show abnormalities in several brain regions, in particular in the cerebellum, hippocampus, archicortex, brainstem, and other subcortical regions, with a regional specificity that correlates with the severity of the disease.[22,42,43] Moreover, increased brain size and head circumference compared to control subjects is frequently observed in a subset of patients with ASD.[44] Disorganization of gray and white matter, cortical dysplasia, and heterotopia (nodules of misplaced neurons) have been also identified in subjects with ASD.[45–47]

These abnormalities are probably related to alterations in both maturation and migration of neurons in ASD patients. Many other modifications have been reported in ASD subjects; these include (i) significant increase in the neuropil (comprising dendrites, nonmyelinated axons, synapses, microvasculature,[22] and processes of glial cell[22,48]) in the frontopolar cortex and the anterior cingulate gyrus, but not in the primary visual, motor, and somatosensory cortices[49]; (ii) reduced dendritic branching in the hippocampus[50]; (iii) fewer dendrites in the prefrontal cortex[51]; (iv) decrease of pyramidal neurons in the inferior frontal cortex[52]; (v) reduced number and size of Purkinje cells in the cerebellum.[42,53–55] Moreover, in several brain regions (including superior temporal cortex, fusiform cortex, midbrain, and cerebellum) of ASD patients higher level of two molecular markers of angiogenesis (which normally decline in postnatal development), nestin in pericytes and CD-34 in endothelial cells has been detected, suggesting that vascular remodeling persists beyond childhood.[56]

The ASD research is mainly focused on neurons, with much less attention dedicated to pathology of neuroglia. In human ASD samples, increased microglial density was sometimes observed in cortices[57–59]; no changes in the number of microglia, astrocytes or oligodendrocytes was found in the amygdala,[60] whereas increased number of oligodendrocytes and glial fibrillary acidic protein (GFAP)-immunoreactive ependymal cells as well as activated microglia were observed in the striatum.[60,61] In addition, altered microglial spatial organization has been characterized in the dorsolateral prefrontal cortex of ASD patients, highlighting increased intimate microglia–neuron interaction, including encirclement of neurons by microglial processes.[62]

Several animal models of ASD have been developed in recent years. These models are based on both genetic and environmental alterations, and these models replicate some behavioral phenotypes of ASD, including social deficits and stereotypies.[22] Nevertheless, translation from animals to humans remains questionable.[63,64] In particular, anatomical modifications idiosyncratic to ASD patients were not observed in autistic-like animals. For example, decreased brain weight or volume were seen in several genetically modified rodent models in comparison to controls,[22,65–68] while in humans a generalized increased brain size has been observed.[44] The animal models of idiopathic ASD are based on the ablation of a single ASD-related gene and, for this reason, cannot represent the pathology in its entirety, because multiple factors are implicated in the etiopathogenesis of the disease.

Keeping in mind the high heterogeneity of ASD, no single model can recapitulate the full spectrum of symptoms. Notwithstanding these limitations, preclinical ASD models may be useful for characterization of molecular pathways likely to be relevant for the future development of new therapeutics. Accumulated evidence converges in highlighting the presence of an altered density of the dendritic spines and modifications in synapses, as well as the presence of inflammation. At the cellular and molecular level, these processes are fundamentally controlled by glial cells, which contribute to the neuropathologic changes observed in ASD.[69] In the following sections, we will provide a brief introduction on the basic function of different neuroglia in the human brain.

4. Neuroglia in autism spectrum disorders

ASD arise from a complex combination of genetic and environmental factors which causes an erratic execution of perinatal programming with long-lasting consequences. Glial cells play a key role in the regulation and fine tuning of many neurodevelopmental processes, and in particular in shaping synaptic connectivity. Therefore, in the last 10 years, the interest in the involvement of these cells in the etiopathogenesis of ASD and associated comorbidities has grown remarkably.

4.1 Neuroglia: Homeostatic cells of the CNS

Neuroglia are homeostatic and defensive cells of the CNS represented by heterogeneous cell populations, which are classified into macroglia and microglia.[70–72] The macroglia includes astrocytes, oligodendrocytes and oligodendroglial precursors also known as NG-2 glia; all macroglial cells sustain homeostasis of the CNS, while oligodendrocytes create connectome through myelination of CNS axons. Oligodendrocytes are mainly involved in the formation of the myelin sheath, thus regulating the transmission of action potentials through myelinated axons and profoundly affecting the functional connectome integrating neurons[73]; oligodendrocytes also provide homeostatic support to axons.[74] The NG-2-glia (also known as oligodendroglial precursor cells or synantocytes) present highly branched morphology and are widely disseminated throughout the brain. Their functions are still enigmatic, even if it is known that NG-2-glia participate in the control of myelination and in the brain plasticity they express a plethora of ion channels and receptors and receive direct synaptic input from neurons.[75] All types of glial cells support neuronal networks, sustain nervous tissue and maintain the function of CNS as an organ; neuroglia contribute to neuronal development and synaptic connectivity and are fundamental for CNS protection.[70–72,76,77]

4.1.1 Microglia

Microglial cells represent the immunocompetent portion of the CNS, executing numerous defensive functions.[76] Microglia originate from myeloid progenitors that enter the CNS early in the embryonic development.[78,79] Microglia represent the only type of glial cells populating early embryonic brain, since astrogenesis and oligodendrogenesis occur later in the perinatal period.[69] Microglia are potentially phagocytic cells, capable of incorporating cellular debris, waste products and pathogens.[80] In the healthy brain microglia exist in surveillant state scanning the surrounding environment by regular extension and retraction of their processes[81] and are capable of morphological and functional plasticity in order to restore the impairments in homeostasis.[82] At the same time, microglial cells contribute to physiological functions particularly important for the formation, shaping, and functioning of the synapses.[83] Highly dynamic microglial processes constantly check presynaptic boutons and postsynaptic spines.[84] This surveillance is crucial for normal brain development during pre- and postnatal periods when synaptic pruning operated by the microglia is critical for shaping neuronal ensembles.[85,86] Wrong execution of the early postnatal process of synapses elimination causes long-lasting severe modifications as demonstrated by preclinical studies, which have highlighted the presence of aberrant functional connectivity, aberrant behavior and deficient hippocampal long-term potentiation (LTP).[87]

4.1.2 Astrocytes

Astrocytes are principal homeostatic cells of the CNS and they are represented by many morphologically and functionally distinct subtypes. Major subtypes of astroglial cells are (i) protoplasmic astrocytes of the gray matter, (ii) fibrous astrocytes of the white matter, (iii) velate astrocytes of the cerebellum and olfactory bulb, (iv) radial astrocytes represented by (iva) Müller retinal glial cells, (ivb) cerebellar Bergmann glial cells, (ivc) tanycytes of the hypothalamus and parts of the spinal cord, and (ivd) stem radial astrocytes of neurogenic niches, (v) pituicytes in the neuro-hypophysis, (vi) perivascular and marginal astrocytes, (vii) Gomori-positive astrocytes localized in arcuate nucleus of the hypothalamus and in the hippocampus; these cells are rich in iron (viii) surface-associated astrocytes; (ix) ependymocytes, choroid plexus cells and retinal pigment epithelial cells. The brain of humans and higher primates contain specific astroglial cells classified as interlaminar, polarized and varicose projection astrocytes.[76]

Astrocytes and oligodendrocytes are intercellularly connected into syncytia by gap junctions. Gap junctions are formed by hundreds of intercellular channels or connexons which penetrate two very closely (2–3 nm) adjacent membranes of neighboring cells.[88] Connexons are assembled from 12 individual subunits, known as connexins, which form large channels permeable to molecules with M.W. up to 1000 Da, including ions, second messengers, energy substrates etc. On average there are 230 gap junctions connecting astrocytic pairs in the gray matter; while every single astrocyte is typically connected with 50–100 astrocytes in its close vicinity.[68] Astrocytes are generally coupled in segregated networks specific for distinct anatomical portions of the brain. For example, astroglial syncytia are formed within and confined to individual barrels of somatosensory cortex or to individual glomeruli in the olfactory bulb.[89–91] Astrocytes are also coupled to oligodendrocytes, with gap junctions localized on oligodendroglial somata, processes and outer (abaxonal) layer of the myelin sheath.[92,93] Pathological impairment of astroglia-oligodendroglia coupling results in severe neurological conditions: for example, mutations in GJB1 gene encoding connexin 32 results in Charcot-Marie-Tooth disease,[94,95] whereas mutations in GJC2 gene that encodes connexin 43 are associated with leucoencephalopathies[96] and hypomyelinating leukodystrophies known as Pelizaeus-Merzbacher disease.[97]

Astrocytes maintain CNS homoeostasis at all levels of organization (molecular, cellular, network, organ and system levels).[76] Astrocytes, for example, regulate CNS microenvironment by tuning extracellular ions and pH, releasing vasoactive substances to regulate blood flow, and buffering reactive oxygen species.[72,89,98,99] Astroglial cells are an important component of the blood-brain barrier, thus controlling the communication between the CNS and the periphery. Indeed, astrocytes provide neurons with energy substrates (most notable lactate) and convert neuroactive substances transported by the bloodstream (for example converting triiodothyronine to thyroxine), promoting thus neuronal survival and the maintenance of CNS homeostasis.[100]

Astrocytes, which form "synaptic cradle"[101] participate in brain development through regulating synaptogenesis, synaptic maturation, synaptic maintenance and synaptic extinction. Astroglia are bona fide secretory cells being thus a part of "gliocrine" system; astrocytes secrete ~200 molecules such as neurotrophic factors or energy substrates.[102–106] Astrocytes support and regulate synaptic transmission by regulating neurotransmitter levels in the synaptic cleft, by catabolizing major neurotransmitters such as glutamate, noradrenaline and adenosine, and by supplying neurons with neurotransmitter precursor glutamine or L-serine. Astrocytes employ specific form of excitability, mediated by transient spatio-temporally organized fluctuations of intracellular ion concentration; of which Ca^{2+} and Na^{+} signaling are the most explored, although other ions (for example K^{+}, Cl^{-} and Mg^{2+}) may also have astroglia-specific signaling role.[68,107–111]

4.2 The role of neuroglia in autism spectrum disorders

4.2.1 Neuroglia and ASD genes

Over the last 15 years, growing evidence indicates that glial cells exert a fundamental role in pathogenesis of ASD (Fig. 2).[69,112,113] Studies on animal models are consistent with clinical observations in demonstrating glial abnormalities in ASD. Some of the genes which confer susceptibility to ASD have a key role in brain development and they are highly expressed in glial cells.[114–121] For instance, mutations in the genes that encode cell adhesion molecules neuroligins (NLs) have been linked to ASD in humans.[122,123] These NLs have important roles in the formation, maturation and function of synaptic structures.[124,125] Mutations in the genes that encode for NLs have been linked to a number of neurodevelopmental disorders including autism.[126] Indeed, genetic analyses of rare variants of ASD revealed a potential association with NL3 and NL4 mutations in humans.[122,123] Studies on transgenic mice mimicking the mutations observed in ASD patients revealed impaired social[127] as well as stereotyping[128] behaviors highlighting a conserved role of NLs in defining ASD phenotype among species. Astroglial involvement has been hypothesized since NLs are expressed specifically in astrocytes and oligodendrocytes,[27,129–131] while growing evidence indicates a key role for astrocytic NLs in regulating synaptogenesis and synaptic transmission that are impaired in ASD.[131,132]

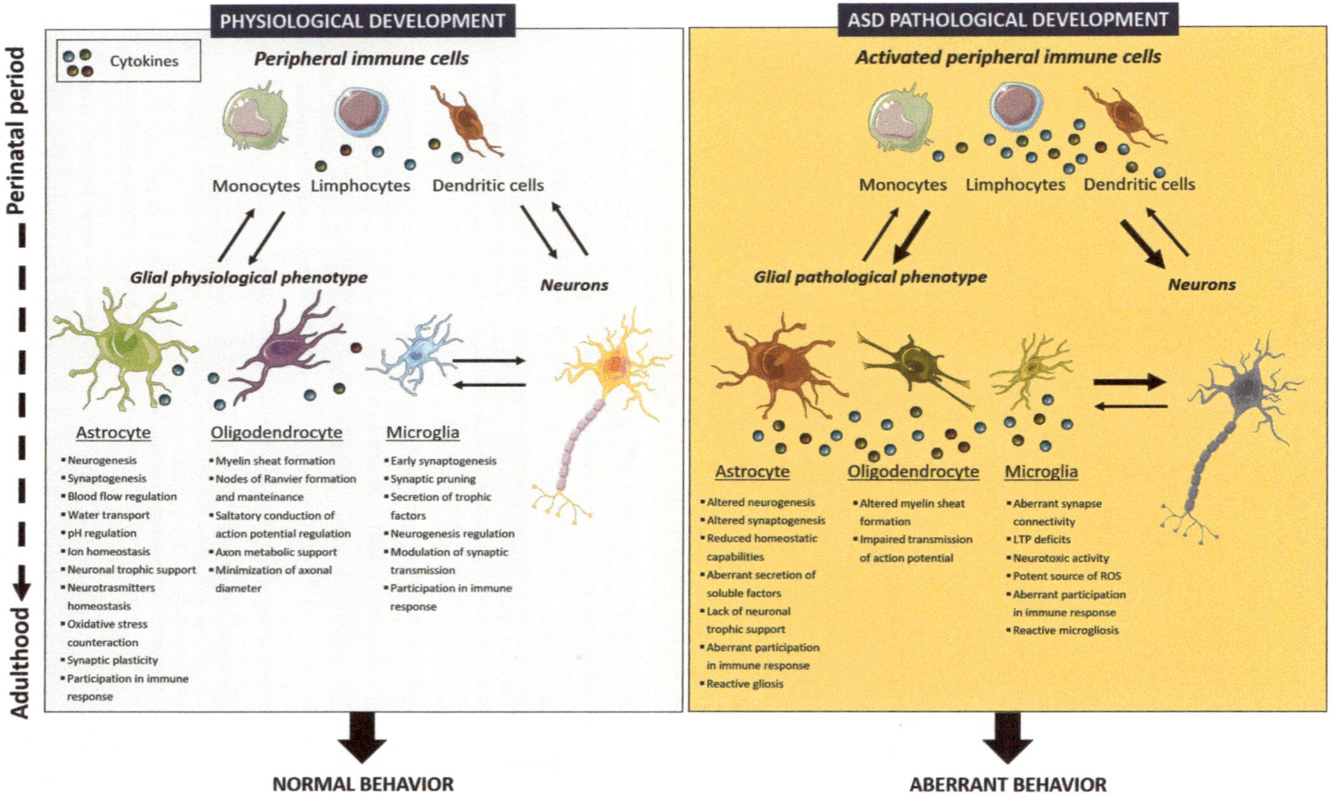

Fig. 2 Functions of glial cells in developing brain and their interaction with immune system. Schematic representation of the functions exerted by glial cells during brain development in physiological (left) and pathological (right) conditions from the perinatal period to adulthood. The interaction between immune cells, glial cells, and neurons allows the physiological development of brain structure and function, thus resulting in normal brain function and behaviors. Environmental stressors (e.g., viral infections, toxins, stress, alcohol, various drugs, as well as maternal care or diet) can prompt an inflammatory response in the perinatal period, causing modifications to the normal communication among immune cells, glial cells, and neurons. Then the function of these cells results perturbated along with the interaction with the others, resulting in abnormal brain function and behavior.

A detailed gene coexpression network analysis of autistic brains further corroborates the contribution of glial cells to the pathophysiology of ASD.[133] In particular, a significant downregulation of genes linked to synaptic function and an up-regulation of genes implicated both in reactive astrogliosis and in the immune and inflammatory response have been revealed.[114] These observations are in line with data obtained from immunohistochemical experiments, performed in human post-mortem brain samples, and a PET scan study demonstrating the presence of microglial activation and proliferation in several brain regions of young ASD patients.[57,58,62,134–137] Activated microglia in ASD patients subjects were characterized by an increased soma, short processes, extended filopodia as well as increased microglial cell density in the gray matter.[59] The neuroinflammatory status was further corroborated by higher levels of proinflammatory cytokines/chemokine, including IL-1β, IL-6, IL-8, INF-γ and TNF-α, detected in both brain specimens and blood of autistic patients compared with typically developing controls.[58,138–142]

4.2.2 Neuroglia reactivity and immune response in ASD

Conceptually, ASD can be considered as a failure of the brain development in response to different environmental challenges (e.g., chemicals, toxins, trauma, ischemia, endogenous factors) with, as the above-mentioned data suggest, an activation of the immune response. Neuroimmune alterations, which are intrinsically diverse, are probably responsible for the phenotypic heterogeneity observed in patients with ASD. Indeed, it is now recognized that abnormal immune response and inflammation are implicated in the etiology of several neuropathologies, including those with an onset at the early phase of the brain development, which applies to ASD.[133,143,144] Perinatal brain development is particularly susceptible to the consequences of abnormal immune activation.[33,145] The context in which the brain development program occurs assumes crucial importance since, potentially, any stimulus might cause its erratic execution provoking irreversible and long-lasting repercussions.

Glial cells, being the fundamental component of the brain homeostasis, are implicated in the pathophysiology of all brain disorders, including ASD. Besides maintaining homoeostasis of the brain, neuroglia represents the CNS defensive system. Glial cells rapidly respond to brain injuries, by undergoing important changes in their morphology, biochemistry and function. Pathological changes of astroglia are complex which are classified into

(i) reactive astrogliosis; (ii) astroglial atrophy with loss of function and (iii) pathological remodeling of astrocytes.[146,147] Different types of astroglial pathological states may coexist or appear in sequence in disease evolution.[146] Conceptually, reactive astrogliosis is an evolutionary conserved, intricate, multistage, disease-specific defensive response, which, however, may acquire maladaptive features and exacerbate nervous tissue damage.[147] Microglial cells similarly undergo activation in response to brain injury, and similarly this activation produces multiple context-specific pathological phenotypes.[83,148] Activated microglia acting in concert with reactive astrocytes, often foster a neuroinflammatory response, manifested by the synthesis of cytokines and proinflammatory mediators. It has been proposed that reactive astrogliosis and activation of microglia may exacerbate the inflammatory state caused by immune activation, thus contributing to the pathogenesis of ASD.[112] However, in depth molecular analyses explaining the mechanisms correlating glial reactivity to ASD pathogenesis are yet to be performed. The development of neural networks is highly susceptible to insults affecting glial pathways associated with synaptic plasticity; therefore, it is possible that the glial reactive remodeling may disturb the ability of microglia and astrocytes to modulate genesis, maturation, pruning, elimination or function of developing synapses.[112,149] For example, animals with genetic deletion of the CX3C chemokine receptor 1 (CX3CR1) display several ASD-related behaviors and deficits in hippocampal LTP, as well as impairments in synaptic pruning and functional connectivity.[87] In addition, CX3CR1 receptors are highly expressed in microglial cells that are responsible for the early postnatal synaptic pruning through the phagocytosis of pre- and postsynaptic elements.[85,86]

Loss-of-function mutation in Hoxb8, the gene exclusively expressed in bone marrow-derived microglia, causes pathological behavior in mice; this behavior is manifested by compulsive grooming and hair removal, resembling stereotyped behavior in humans suffering from trichotillomania.[150] Restoring the normal microglia through bone marrow transplantation eliminated pathological behaviors in mice.[151] It is universally appreciated that glial cells are the principal contributors to the neuroinflammatory response regulated by IL-1 signaling.[152–154] Mutations in the IL-1 cytokine receptor family are associated with ASD. Genetic studies revealed the presence in ASD patients of a single-nucleotide polymorphism in the gene encoding for IL-1 receptor type 2 (IL-1R2)[155,156] that is highly enriched in microglial cells,[120] and a rare mutation of the gene encoding for IL-1 receptor accessory protein-like 1,[157] which is highly expressed in astrocytes.[120]

4.2.3 Neuroglia and glutamate excitotoxicity in ASD

Preclinical data further corroborate that dysfunctions or alterations in the physiological duties of glial cells affect the brain development. For instance, loss-of-functions mutations in the X-linked Mecp2 gene encoding methyl-CpG-binding protein 2 (MeCP2) are associated with the human Rett syndrome, one of the neurodevelopmental disorders in ASD characterized by profound intellectual and motor disability.[158] It has been demonstrated that Mecp2-null microglial cells produce and release high levels of glutamate, causing excitotoxicity,[159] while Mecp2-null astrocytes, when co-cultured with hippocampal neurons suppress dendritic ramification arguably because of aberrant secretion of neurotrophic factors.[160] Specific re-expression of healthy Mecp2 in astrocytes of Mecp2-null mice ameliorated animal behaviors and prolonged their lifespan compared to globally deficient mice[161]; the expression of wild-type MECP2 protein in the microglia of Mecp2-null animals normalized mice phenotype.[162] Accordingly, changes in the synaptic levels of glutamate have been detected in a rodent model of autism[163] as well as in patients with ASD,[134] and correlated with the altered expression of the glutamate transporters expressed in astrocytes.[164]

4.2.4 Neuroglia and environmental factors in ASD

Besides genetic predisposition, the environment is emerging as a key etiological factor of ASD. The most direct evidence correlating the exposure to environmental factors to disorders of brain development, like ASD, derives from epidemiological and in vivo studies. Exposure in utero or during the first weeks of neonatal life to numerous agents, including air pollution, endocrine disruptors, chemicals, drugs, or pathogens has detrimental consequences on brain development and, subsequently, fundamental importance in ASD.[165–172] Both morphology and functions of glial cells seem to be significantly influenced by these factors.[163,173–181] For instance, clinical data indicate a higher occurrence of ASD-like behaviors in subjects prenatally exposed to pathogens like cytomegalovirus, herpes simplex, clostridia, and rubella.[182–188] Rubella virus selectively targets astrocytes and oligodendrocytes in the developing brain and modifies their functions, thus resulting in neurologic deficits.[189] A preclinical study demonstrated that influenza viral infection in mid-pregnancy leads to alterations in connexin-43 and aquaporin-4 expression in astrocytes resulting in abnormal glial-neuronal communication and aberrant neuronal migration.[190]

Evidence from preclinical studies identified a link between maternal diet during pregnancy and lactation and changes in the offspring behavior.[191,192] Neuroglia most likely contribute to this mechanism since such alterations appear to be related to modifications of the inflammatory response. High-fat diet fed dams generate offspring with increased anxiety- and depression-related behaviors, and decreased cognitive functions,[193,194] correlated with microglial activation and increased expression of pro-inflammatory cytokines in amygdala and hippocampus.[195,196]

Additionally, an exacerbated peripheral response to an inflammatory challenge and increased anxiety-related behaviors has been found in mice perinatally exposed to a high-fat diet,[195,196] whereas hyperactivity, impairments in working memory, and depressive-like behaviors have been detected in animal models of perinatal malnutrition.[197–201] Dysbalanced perinatal diet (high fat, low protein) has been suggested to alter cerebral expression of cytokines under physiological conditions and after inflammatory and psychological stressors in the adult animals.[202]

Recent investigations have demonstrated that prenatal exposure to the anticonvulsant and mood stabilizer valproic acid, an ASD model of acknowledged clinical relevance, impairs gliogenesis, alters brain cholesterol/isoprenoid metabolism, as well as morphology and functions of astrocytes, microglia, and oligodendrocytes, with signs of chronic neuroinflammation and modifications in local connectivity.[203–206] These data are difficult to interpret because they are markedly different in different brain areas and are affected by the postnatal age in which the investigations were carried out. However, these observations confirm that glial cells are highly dynamic elements of the CNS and, by virtue of their complex physiology, they are able to modulate mammalian behavior opening a new field of research in the context of ASD.

4.3 Mapping neuroglia in a preclinical rodent model of ASD

We have recently performed an in-depth analysis of the three major types of neuroglia, astrocytes, oligodendrocytes, and microglia in the brain of rats exposed prenatally to valproic acid (VPA), an animal model of ASD being of acknowledged clinical relevance (Fig. 3).[203] Our decision to use this environmental model raised from two key considerations: (i) the limitations of the genetic models of ASD (discussed above), and (ii) the results of several epidemiological studies that have linked valproic acid usage during pregnancy to higher risks of ASD.[207–209] Indeed, when administered during

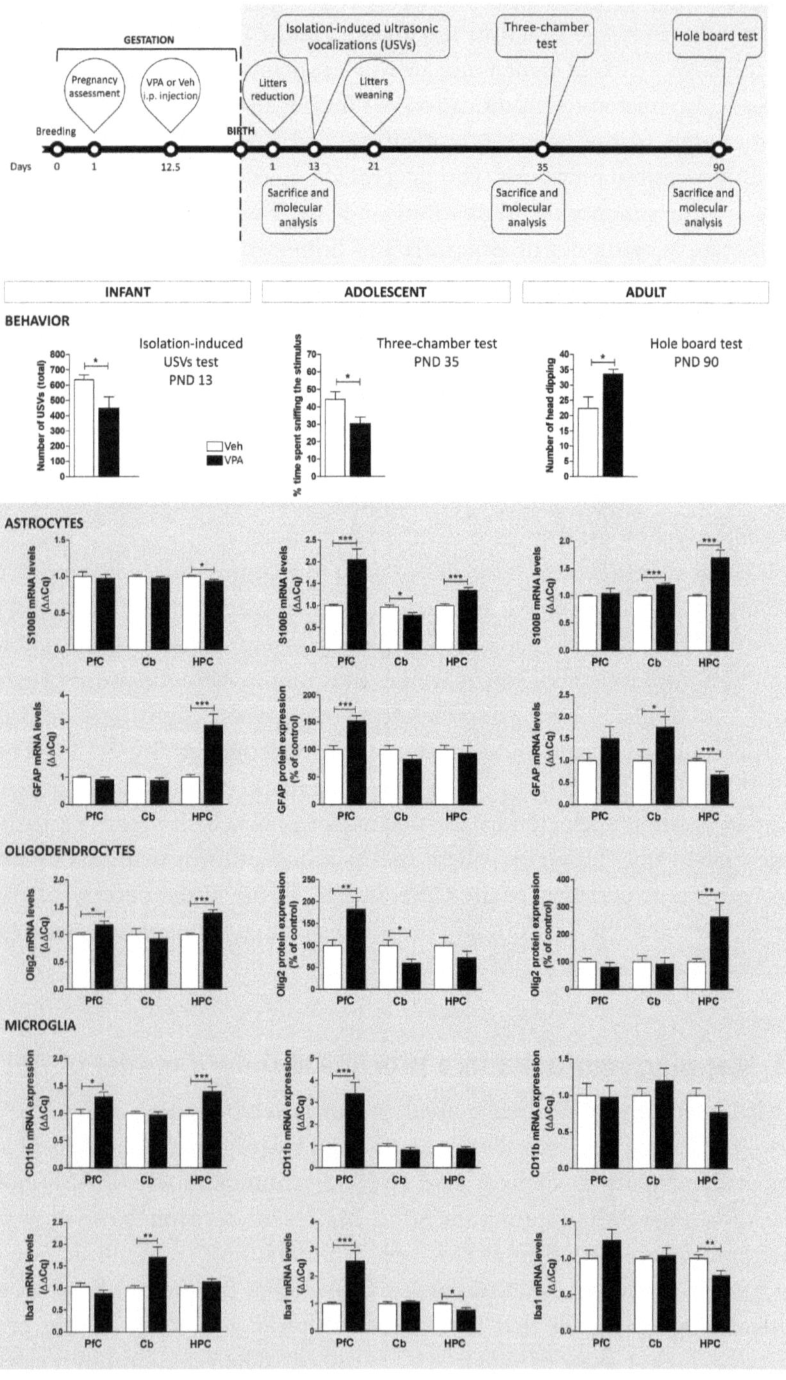

Fig. 3 See legend on opposite page.

pregnancy, this antiepileptic and mood stabilizer drug was reported to induce autistic-like features in the children who were exposed to this drug in utero; these ASD symptoms included reduced communication and sociability,[209] stereotyped behaviors[210] as well as many congenital malformations.[211] These clinical observations have stipulated the development of VPA preclinical model, which has become widely used for studies of ASD with face and construct validity.[212–214] Briefly, pregnant rats, on gestational day 12.5, received an intraperitoneal injection of either VPA (500 mg/kg in saline) or saline. This dose of VPA as well as the choice of the administration times are known to induce autistic-like characteristics in the exposed rat offsprings.[215] According to the epidemiological data that report a higher incidence of ASD in boys than in girls,[15] and the evidence showing that the autistic-like behaviors displayed by rats prenatally exposed to VPA are more pronounced in the male than in the female offspring[216,217] we used only the male offspring born from VPA-exposed dams. Furthermore, it was our intention to carry out a study that gave information on the role of glial cells

Fig. 3 Mapping neuroglia in a preclinical rodent model of ASD. This figure shows results of the in-depth analysis of the three major types of neuroglia, astrocytes, oligodendrocytes, and microglia in the brain of rats prenatally exposed to valproic acid (VPA).[203] A schematic representation of the experimental design is presented in the upper panel. Behavioral and molecular analyses were carried out in animals at three different ages, postnatal day (PND) 13, PND 35, and PND 90, to account for human infancy, adolescence, and adulthood, respectively. The autistic-like phenotype was assessed through reduced isolation-induced ultrasonic vocalization (USV) emission at PND 13, sociability in the three-chamber test at PND 35 and induced stereotypic behavior in the hole-board test at PND 90 (control rat (Veh) $n = 5$, VPA $n = 5$). Gene and protein expression of specific markers of neuroglia were investigated in hippocampus (HPC), prefrontal cortex (PfC), and cerebellum (Cb) of healthy- (open bars, Veh) and autistic-like (black bars, VPA) infant rats. *Astrocytes*: Representation of the relative expression of S100B and GFAP in VPA animals compared to control (Veh), normalized to both TATA box binding protein (TBP) and hypoxanthine guanine phosphoribosyl transferase (HPRT) ($\Delta\Delta$Cq; $N = 3$, in triplicate). Representative western blots for GFAP, densitometric analyses are normalized to β-actin used as loading control. Results are expressed as percentage of control (Veh) ($N = 3$, in triplicate). *Oligodendrocytes*: Representation of the relative expression of Olig2 in VPA animals compared to control (Veh), normalized to both TBP and HPRT ($\Delta\Delta$Cq; $N = 3$, in triplicate). Representative western blots for Olig2 protein and densitometric analyses normalized to β-actin used as loading control. Results are expressed as percentage of control (Veh) ($N = 3$, in triplicate). *Microglia*: Representation of the relative expression of CD11b and Iba1 in VPA animals compared to control (Veh), normalized to both TBP and HPRT ($\Delta\Delta$Cq; $N = 3$, in triplicate). All data are presented as means ± SEM. Statistical analysis was performed by *t*-test ($*P < 0.05$; $**P < 0.01$; $***P < 0.001$ vs Veh group). For details, please refer to Bronzuoli et al.[203]

throughout life. For this reason, we studied animals at three different ages, 13, 35, and 90 days old, to account for human infancy, adolescence, and adulthood, respectively. Considering the different skills of the animals at these three ages, we used three distinct, appropriate behavioral tests to confirm the presence of ASD-like phenotypes. In particular, we deployed the isolation-induced ultrasonic vocalizations to test pups, the three-chamber test to analyze adolescent rats, and the hole board test to examine adult animals. At all ages, animals prenatally exposed to VPA showed impairments in the three core symptoms of ASD, confirming the manifestation of ASD-like behaviors that persist throughout life (Fig. 3). Indeed, when separated from the dams, infant VPA-exposed pups vocalized significantly less compared to control pups; adolescent VPA-exposed rats showed decreased sociability compared to control animals since they spent less time sniffing the stimulus animal; lastly, adult valproic acid-exposed rats made more head dipping than control animals in the hole board test, thus showing stereotypic behaviors.

Subsequently, we examined the effects of prenatal exposure to VPA on gene and protein expression of specific markers of neuroglia in brain regions associated with ASD pathogenesis (namely, hippocampus, prefrontal cortex, cerebellum, and amygdala).[218–220] We observed that prenatal exposure to VPA heterogeneously impacts on astrocytes in different brain regions, causing transcriptional modifications of two important astrocytic markers, the neurotrophin/Ca^{2+} binding protein S100B and the cytoskeletal glial fibrillary acidic protein (GFAP).[221–223] These modifications were particularly evident in adolescent and adult rats. By testing Olig2, a transcriptional factor essential for oligodendrocyte development,[224] we examined the effects of prenatal VPA exposure on oligodendrocytes. Both transcriptional and translational modifications of Olig2 occurred mainly in the prefrontal cortex and in the hippocampus. Of note, these alterations are particularly evident during adolescence but seem to be compensated in adulthood. Lastly, to study microglia, we analyzed CD11b, a marker of microglial activation, and Iba1, a Ca^{2+}-binding protein constitutively expressed by both surveillant and activated microglia.[83,225,226] We found that prenatal exposure to VPA switches microglial phenotypes from surveillant to activated. This phenomenon occurs mainly in the prefrontal cortex and in the hippocampus and is particularly evident during infancy and adolescence, while is somewhat diminished in adulthood.

To summarize, molecular analyses indicate that prenatal VPA exposure affects all types of neuroglia, mainly causing transcriptional modifications. These alterations are brain region- and age-dependent. We noticed the most prominent changes in the prefrontal cortex and in the hippocampus of autistic-like animals while the amygdala seems to be spared. Moreover, these modifications are particularly evident during infancy and adolescence, while they appear to be mitigated in adulthood.

Collected data indicate that the pathological phenotype of neuroglia in rats exposed to VPA seems to be rather mild with scant signs of widespread and chronic neuroinflammation. The intricate heterogeneity of neuroglia response that we observed in this preclinical model is hard to explain, it requires further investigations and opens new scenarios for the better comprehension and characterizations of ASD.

5. Conclusions

Despite the significant impact of neuroglia on several key processes associated with ASD pathogenesis, such as synaptic stripping, brain immune function, the epigenetic interface for environmental stimuli, stem cell development, synaptic transmission, and others, ASD research until recently has ignored neuroglia. As presented in this chapter exciting new data highlighting the involvement of astrocytes, microglia, and oligodendrocytes in the mechanisms involved in the onset and progression of ASD are emerging from both preclinical and clinical studies. These findings are changing the traditional neurocentric view of the pathology and hence ASD should be considered to be due, at least in part, to disorders affecting neuroglial cells or neuron-glia interactions. This opens a new and, in many aspects, unexplored field of research on ASD. Indeed, investigations carried hitherto have been focused mainly on the involvement of neuroglia in the immune response and inflammation, focusing scientists' attention primarily on microglia. However, accumulating evidence has made clear that macroglia dysfunctions contribute to the pathogenesis of ASD. Unfortunately, the results of these investigations are often convoluted and heterogeneous, reflecting the complexity of the molecular and cellular mechanisms underlying ASD. Conceptually. ASD are multifactorial diseases of brain development arising from the interaction of genetic and environmental factors. The awareness that genetic susceptibility is highly involved in ASD has led to many studies that have identified hundreds of genes whose variants are related to ASD. Unfortunately, the usefulness of this information remains elusive until the mechanistic roles of single genes are clarified. Besides genetic background, environmental stressors are definitively recognized as key players in ASD etiology. During the last two decades, the number of presumed environmental risk factors has grown. However, mechanistic studies providing exhaustive causative links are yet to be performed. Lastly, it remains to be elucidated if genetic and environmental risk factors converge to a single pathophysiological mechanism or do they act on several pathways.

Adding to the confusion the period on which such risk factors exert their action on brain development assumes a crucial role. According to the triple hit hypothesis of autism, suggested by Casanova in 2007, clinical manifestations happen when three factors impinge to various degrees upon a subject[227] This extremely variable combination of interactions may explain the high variability of clinical signs among patients. Unfortunately, there is still a lack of detailed molecular studies explaining the mechanisms of this interaction. Although the role of glial cells in these three factors is increasingly demonstrated, it remains to be clarified whether it is the observed alterations in both the structure and functionality of glial cells that cause the ASD or whether they are a consequence of the diseases.

References

1. American Psychiatric Association. Neurodevelopmental disorders. In: *Diagnostic and Statistical Manual of Mental Disorders*. 5th ed. Washington, DC: American Psychiatric Association; 2013.
2. Kanner L. Autistic disturbances of affective contact. *Nerv Child*. 1943;2:217–250.
3. Volkmar FR, Rutter M. Childhood disintegrative disorder: results of the DSM-IV autism field trial. *J Am Acad Child Adolesc Psychiatry*. 1995;34(8):1092–1095.
4. Volkmar FR, Pauls D. Autism. *Lancet*. 2003;362(9390):1133–1141.
5. Volkman F, Klin A, Cohen DJ. Diagnosis and classification of autism and related conditions. In: Cohen D, Volkman FR, eds. *Handbook of Autism and Pervasive Developmental Disorders*. 2nd ed. John Wiley & Sons Inc.; 1997:5–40
6. Johnson CP, Myers SM, American Academy of Pediatrics Council on Children With Disabilities. Identification and evaluation of children with autism spectrum disorders. *Pediatrics*. 2007;120(5):1183–1215.
7. Mostert-Kerckhoffs MA, Staal WG, Houben RH, de Jonge MV. Stop and change: inhibition and flexibility skills are related to repetitive behavior in children and young adults with autism spectrum disorders. *J Autism Dev Disord*. 2015;45(10):3148–3158.
8. Leung RC, Vogan VM, Powell TL, Anagnostou E, Taylor MJ. The role of executive functions in social impairment in autism spectrum disorder. *Child Neuropsychol*. 2016;22(3):336–344.
9. de Vries M, Geurts H. Influence of autism traits and executive functioning on quality of life in children with an autism spectrum disorder. *J Autism Dev Disord*. 2015;45(9): 2734–2743.
10. American Psychiatric Association. *The Diagnostic and Statistical Manual of Mental Disorders*. 5th ed. Arlington, VA: American Psychiatric Publishing; 2013.
11. Johnson CP. Early clinical characteristics of children with autism. In: Gupta VB, ed. *Autistic Spectrum Disorders in Children*. New York: Marcel Dekker, Inc.; 2004:85–123
12. Hughes JEA, Ward J, Gruffydd E, et al. Savant syndrome has a distinct psychological profile in autism. *Mol Autism*. 2018;9:53.

13. Treffert DA. The savant syndrome: an extraordinary condition. A synopsis: past, present, future. *Philos Trans R Soc Lond B Biol Sci*. 2009;364(1522):1351–1357.
14. Baio J, Wiggins L, Christensen DL, et al. Prevalence of autism spectrum disorder among children aged 8 years—autism and developmental disabilities monitoring network, 11 sites, United States, 2014. *MMWR Surveill Summ*. 2018;67(6):1–23.
15. Elsabbagh M, Divan G, Koh YJ, et al. Global prevalence of autism and other pervasive developmental disorders. *Autism Res*. 2012;5(3):160–179.
16. Croen LA, Zerbo O, Qian Y, et al. The health status of adults on the autism spectrum. *Autism*. 2015;19(7):814–823.
17. Abrahams BS, Geschwind DH. Advances in autism genetics: on the threshold of a new neurobiology. *Nat Rev Genet*. 2008;9(5):341–355.
18. Wang K, Zhang H, Ma D, et al. Common genetic variants on 5p14.1 associate with autism spectrum disorders. *Nature*. 2009;459(7246):528–533.
19. Gaugler T, Klei L, Sanders SJ, et al. Most genetic risk for autism resides with common variation. *Nat Genet*. 2014;46(8):881–885.
20. Sandin S, Lichtenstein P, Kuja-Halkola R, Larsson H, Hultman CM, Reichenberg A. The familial risk of autism. *JAMA*. 2014;311(17):1770–1777.
21. Yuen RK, Thiruvahindrapuram B, Merico D, et al. Whole-genome sequencing of quartet families with autism spectrum disorder. *Nat Med*. 2015;21(2):185–191.
22. Varghese M, Keshav N, Jacot-Descombes S, et al. Autism spectrum disorder: neuropathology and animal models. *Acta Neuropathol*. 2017;134(4):537–566.
23. Freitag CM. The genetics of autistic disorders and its clinical relevance: a review of the literature. *Mol Psychiatry*. 2007;12(1):2–22.
24. Iossifov I, O'Roak BJ, Sanders SJ, et al. The contribution of de novo coding mutations to autism spectrum disorder. *Nature*. 2014;515(7526):216–221.
25. Nisar S, Hashem S, Bhat AA, et al. Association of genes with phenotype in autism spectrum disorder. *Aging (Albany NY)*. 2019;11(22):10742–10770.
26. Autism Genome Project C, Szatmari P, Paterson AD, et al. Mapping autism risk loci using genetic linkage and chromosomal rearrangements. *Nat Genet*. 2007;39(3):319–328.
27. Chubykin AA, Liu X, Comoletti D, Tsigelny I, Taylor P, Sudhof TC. Dissection of synapse induction by neuroligins: effect of a neuroligin mutation associated with autism. *J Biol Chem*. 2005;280(23):22365–22374.
28. Durand CM, Betancur C, Boeckers TM, et al. Mutations in the gene encoding the synaptic scaffolding protein SHANK3 are associated with autism spectrum disorders. *Nat Genet*. 2007;39(1):25–27.
29. Feng J, Schroer R, Yan J, et al. High frequency of neurexin 1beta signal peptide structural variants in patients with autism. *Neurosci Lett*. 2006;409(1):10–13.
30. Folsom TD, Fatemi SH. The involvement of Reelin in neurodevelopmental disorders. *Neuropharmacology*. 2013;68:122–135.
31. Lichtenstein P, Carlstrom E, Rastam M, Gillberg C, Anckarsater H. The genetics of autism spectrum disorders and related neuropsychiatric disorders in childhood. *Am J Psychiatry*. 2010;167(11):1357–1363.
32. Bennet L, Gunn AJ. The fetal origins of adult mental illness. In: Wintour-Coghlan M, Owens J, eds. *Early Life Origins of Health and Disease*. Springer: New York; 2006:204–218. Advances in Experimental Medicine and Biology.
33. Hallmayer J, Cleveland S, Torres A, et al. Genetic heritability and shared environmental factors among twin pairs with autism. *Arch Gen Psychiatry*. 2011;68(11):1095–1102.
34. Krakowiak P, Walker CK, Bremer AA, et al. Maternal metabolic conditions and risk for autism and other neurodevelopmental disorders. *Pediatrics*. 2012;129(5):e1121–e1128.
35. Riley EP, McGee CL. Fetal alcohol spectrum disorders: an overview with emphasis on changes in brain and behavior. *Exp Biol Med (Maywood)*. 2005;230(6):357–365.

36. Ross EJ, Graham DL, Money KM, Stanwood GD. Developmental consequences of fetal exposure to drugs: what we know and what we still must learn. *Neuropsychopharmacology*. 2015;40(1):61–87.
37. Brown AS. Epidemiologic studies of exposure to prenatal infection and risk of schizophrenia and autism. *Dev Neurobiol*. 2012;72(10):1272–1276.
38. Lupien SJ, McEwen BS, Gunnar MR, Heim C. Effects of stress throughout the lifespan on the brain, behaviour and cognition. *Nat Rev Neurosci*. 2009;10(6):434–445.
39. Grandjean P, Landrigan PJ. Developmental neurotoxicity of industrial chemicals. *Lancet*. 2006;368(9553):2167–2178.
40. Chen JA, Penagarikano O, Belgard TG, Swarup V, Geschwind DH. The emerging picture of autism spectrum disorder: genetics and pathology. *Annu Rev Pathol*. 2015;10:111–144.
41. de la Torre-Ubieta L, Won H, Stein JL, Geschwind DH. Advancing the understanding of autism disease mechanisms through genetics. *Nat Med*. 2016;22(4):345–361.
42. Wegiel J, Flory M, Kuchna I, et al. Brain-region-specific alterations of the trajectories of neuronal volume growth throughout the lifespan in autism. *Acta Neuropathol Commun*. 2014;2:28.
43. Wegiel J, Flory M, Kuchna I, et al. Neuronal nucleus and cytoplasm volume deficit in children with autism and volume increase in adolescents and adults. *Acta Neuropathol Commun*. 2015;3:2.
44. Sacco R, Gabriele S, Persico AM. Head circumference and brain size in autism spectrum disorder: a systematic review and meta-analysis. *Psychiatry Res*. 2015;234(2):239–251.
45. Wegiel J, Kuchna I, Nowicki K, et al. The neuropathology of autism: defects of neurogenesis and neuronal migration, and dysplastic changes. *Acta Neuropathol*. 2010;119(6):755–770.
46. Stoner R, Chow ML, Boyle MP, et al. Patches of disorganization in the neocortex of children with autism. *N Engl J Med*. 2014;370(13):1209–1219.
47. Casanova MF, El-Baz AS, Kamat SS, et al. Focal cortical dysplasias in autism spectrum disorders. *Acta Neuropathol Commun*. 2013;1:67.
48. Spocter MA, Hopkins WD, Barks SK, et al. Neuropil distribution in the cerebral cortex differs between humans and chimpanzees. *J Comp Neurol*. 2012;520(13):2917–2929.
49. Carper RA, Moses P, Tigue ZD, Courchesne E. Cerebral lobes in autism: early hyperplasia and abnormal age effects. *Neuroimage*. 2002;16(4):1038–1051.
50. Raymond GV, Bauman ML, Kemper TL. Hippocampus in autism: a Golgi analysis. *Acta Neuropathol*. 1996;91(1):117–119.
51. Mukaetova-Ladinska EB, Arnold H, Jaros E, Perry R, Perry E. Depletion of MAP2 expression and laminar cytoarchitectonic changes in dorsolateral prefrontal cortex in adult autistic individuals. *Neuropathol Appl Neurobiol*. 2004;30(6):615–623.
52. Jacot-Descombes S, Uppal N, Wicinski B, et al. Decreased pyramidal neuron size in Brodmann areas 44 and 45 in patients with autism. *Acta Neuropathol*. 2012;124(1):67–79.
53. Wegiel J, Kuchna I, Nowicki K, et al. Contribution of olivofloccular circuitry developmental defects to atypical gaze in autism. *Brain Res*. 2013;1512:106–122.
54. Skefos J, Cummings C, Enzer K, et al. Regional alterations in purkinje cell density in patients with autism. *PLoS One*. 2014;9(2):e81255.
55. Fatemi SH, Halt AR, Realmuto G, et al. Purkinje cell size is reduced in cerebellum of patients with autism. *Cell Mol Neurobiol*. 2002;22(2):171–175.
56. Azmitia EC, Saccomano ZT, Alzoobaee MF, Boldrini M, Whitaker-Azmitia PM. Persistent angiogenesis in the autism brain: an immunocytochemical study of postmortem cortex, brainstem and cerebellum. *J Autism Dev Disord*. 2016;46(4):1307–1318.
57. Tetreault NA, Hakeem AY, Jiang S, et al. Microglia in the cerebral cortex in autism. *J Autism Dev Disord*. 2012;42(12):2569–2584.

58. Vargas DL, Nascimbene C, Krishnan C, Zimmerman AW, Pardo CA. Neuroglial activation and neuroinflammation in the brain of patients with autism. *Ann Neurol.* 2005;57(1):67–81.
59. Morgan JT, Chana G, Pardo CA, et al. Microglial activation and increased microglial density observed in the dorsolateral prefrontal cortex in autism. *Biol Psychiatry.* 2010;68(4):368–376.
60. Morgan JT, Barger N, Amaral DG, Schumann CM. Stereological study of amygdala glial populations in adolescents and adults with autism spectrum disorder. *PLoS One.* 2014;9(10):e110356.
61. Kotagiri P, Chance SA, Szele FG, Esiri MM. Subventricular zone cytoarchitecture changes in autism. *Dev Neurobiol.* 2014;74(1):25–41.
62. Morgan JT, Chana G, Abramson I, Semendeferi K, Courchesne E, Everall IP. Abnormal microglial-neuronal spatial organization in the dorsolateral prefrontal cortex in autism. *Brain Res.* 2012;1456:72–81.
63. Godlee F. How predictive and productive is animal research? *BMJ.* 2014;348:g3719.
64. Leenaars CHC, Kouwenaar C, Stafleu FR, et al. Animal to human translation: a systematic scoping review of reported concordance rates. *J Transl Med.* 2019;17(1):223.
65. Dodero L, Damiano M, Galbusera A, et al. Neuroimaging evidence of major morpho-anatomical and functional abnormalities in the BTBR T+TF/J mouse model of autism. *PLoS One.* 2013;8(10):e76655.
66. Miller VM, Gupta D, Neu N, Cotroneo A, Boulay CB, Seegal RF. Novel inter-hemispheric white matter connectivity in the BTBR mouse model of autism. *Brain Res.* 2013;1513:26–33.
67. Penzes P, Cahill ME, Jones KA, VanLeeuwen JE, Woolfrey KM. Dendritic spine pathology in neuropsychiatric disorders. *Nat Neurosci.* 2011;14(3):285–293.
68. Wahlsten D, Metten P, Crabbe JC. Survey of 21 inbred mouse strains in two laboratories reveals that BTBR T/+ tf/tf has severely reduced hippocampal commissure and absent corpus callosum. *Brain Res.* 2003;971(1):47–54.
69. Zeidan-Chulia F, Salmina AB, Malinovskaya NA, Noda M, Verkhratsky A, Moreira JC. The glial perspective of autism spectrum disorders. *Neurosci Biobehav Rev.* 2014;38:160–172.
70. Kettenmann H, Verkhratsky A. Neuroglia: the 150 years after. *Trends Neurosci.* 2008;31(12):653–659.
71. Kettenmann H, Ransom BR. *Neuroglia.* 3rd ed. Oxford: Oxford University Press; 2013.
72. Verkhratsky A, Butt A. *Glial Physiology and Pathophysiology.* 1st ed. Wiley-Blackwell; 2013.
73. Fields RD, Woo DH, Basser PJ. Glial regulation of the neuronal connectome through local and long-distant communication. *Neuron.* 2015;86(2):374–386.
74. Simons M, Nave KA. Oligodendrocytes: myelination and axonal support. *Cold Spring Harb Perspect Biol.* 2015;8(1):a020479.
75. Eugenin-von Bernhardi J, Dimou L. NG2-glia, more than progenitor cells. *Adv Exp Med Biol.* 2016;949:27–45.
76. Verkhratsky A, Nedergaard M. Physiology of astroglia. *Physiol Rev.* 2018;98(1):239–389.
77. Verkhratsky A, Ho MS, Zorec R, Parpura V. The concept of neuroglia. *Adv Exp Med Biol.* 2019;1175:1–13.
78. Ginhoux F, Greter M, Leboeuf M, et al. Fate mapping analysis reveals that adult microglia derive from primitive macrophages. *Science.* 2010;330(6005):841–845.
79. Kershman J. Genesis of microglia in the human brain. *Arch Neurol Psychiatr.* 1939;41(1):24–50.
80. Nayak D, Roth TL, McGavern DB. Microglia development and function. *Annu Rev Immunol.* 2014;32:367–402.

81. Nimmerjahn A, Kirchhoff F, Helmchen F. Resting microglial cells are highly dynamic surveillants of brain parenchyma in vivo. *Science*. 2005;308(5726):1314–1318.
82. Cronk JC, Kipnis J. Microglia—the brain's busy bees. *F1000Prime Rep*. 2013;5:53.
83. Kettenmann H, Hanisch U-K, Noda M, Verkhratsky A. Physiology of microglia. *Physiol Rev*. 2011;91(2):461–553.
84. Tremblay ME, Lowery RL, Majewska AK. Microglial interactions with synapses are modulated by visual experience. *PLoS Biol*. 2010;8(11):e1000527.
85. Paolicelli RC, Bolasco G, Pagani F, et al. Synaptic pruning by microglia is necessary for normal brain development. *Science*. 2011;333(6048):1456–1458.
86. Schafer DP, Lehrman EK, Kautzman AG, et al. Microglia sculpt postnatal neural circuits in an activity and complement-dependent manner. *Neuron*. 2012;74(4):691–705.
87. Zhan Y, Paolicelli RC, Sforazzini F, et al. Deficient neuron-microglia signaling results in impaired functional brain connectivity and social behavior. *Nat Neurosci*. 2014;17(3):400–406.
88. Evans WH, Martin PE. Gap junctions: structure and function (review). *Mol Membr Biol*. 2002;19(2):121–136.
89. Giaume C, Koulakoff A, Roux L, Holcman D, Rouach N. Astroglial networks: a step further in neuroglial and gliovascular interactions. *Nat Rev Neurosci*. 2010;11(2):87–99.
90. Houades V, Koulakoff A, Ezan P, Seif I, Giaume C. Gap junction-mediated astrocytic networks in the mouse barrel cortex. *J Neurosci*. 2008;28(20):5207–5217.
91. Houades V, Rouach N, Ezan P, Kirchhoff F, Koulakoff A, Giaume C. Shapes of astrocyte networks in the juvenile brain. *Neuron Glia Biol*. 2006;2(1):3–14.
92. Massa PT, Mugnaini E. Cell junctions and intramembrane particles of astrocytes and oligodendrocytes: a freeze-fracture study. *Neuroscience*. 1982;7(2):523–538.
93. Rash JE, Duffy HS, Dudek FE, Bilhartz BL, Whalen LR, Yasumura T. Grid-mapped freeze-fracture analysis of gap junctions in gray and white matter of adult rat central nervous system, with evidence for a "panglial syncytium" that is not coupled to neurons. *J Comp Neurol*. 1997;388(2):265–292.
94. Lee MJ, Nelson I, Houlden H, et al. Six novel connexin32 (GJB1) mutations in X-linked Charcot-Marie-tooth disease. *J Neurol Neurosurg Psychiatry*. 2002;73(3):304–306.
95. Taylor RA, Simon EM, Marks HG, Scherer SS. The CNS phenotype of X-linked Charcot-Marie-tooth disease: more than a peripheral problem. *Neurology*. 2003;61(11):1475–1478.
96. Bugiani M, Al Shahwan S, Lamantea E, et al. GJA12 mutations in children with recessive hypomyelinating leukoencephalopathy. *Neurology*. 2006;67(2):273–279.
97. Uhlenberg B, Schuelke M, Ruschendorf F, et al. Mutations in the gene encoding gap junction protein alpha 12 (connexin 46.6) cause Pelizaeus-Merzbacher-like disease. *Am J Hum Genet*. 2004;75(2):251–260.
98. Wang DD, Bordey A. The astrocyte odyssey. *Prog Neurobiol*. 2008;86(4):342–367.
99. Parpura V, Verkhratsky A. Neuroglia at the crossroads of homoeostasis, metabolism and signalling: evolution of the concept. *ASN Neuro*. 2012;4(4):201–205.
100. Keaney J, Campbell M. The dynamic blood-brain barrier. *FEBS J*. 2015;282(21):4067–4079.
101. Verkhratsky A, Nedergaard M. Astroglial cradle in the life of the synapse. *Philos Trans R Soc Lond B Biol Sci*. 2014;369(1654):20130595.
102. Araque A, Carmignoto G, Haydon PG, Oliet SH, Robitaille R, Volterra A. Gliotransmitters travel in time and space. *Neuron*. 2014;81(4):728–739.
103. Butt A, Verkhratsky A. Neuroglia: realising their true potential. *Brain Neurosci Adv*. 2018;2239821281881 7495.
104. Petrelli F, Bezzi P. Novel insights into gliotransmitters. *Curr Opin Pharmacol*. 2016;26:138–145.

105. Verkhratsky A, Matteoli M, Parpura V, Mothet JP, Zorec R. Astrocytes as secretory cells of the central nervous system: idiosyncrasies of vesicular secretion. *EMBO J.* 2016;35(3):239–257.
106. Parpura V, Zorec R. Gliotransmission: exocytotic release from astrocytes. *Brain Res Rev.* 2010;63(1–2):83–92.
107. Rose CR, Verkhratsky A. Principles of sodium homeostasis and sodium signalling in astroglia. *Glia.* 2016;64(10):1611–1627.
108. Kirischuk S, Parpura V, Verkhratsky A. Sodium dynamics: another key to astroglial excitability? *Trends Neurosci.* 2012;35(8):497–506.
109. Verkhratsky A, Orkand RK, Kettenmann H. Glial calcium: homeostasis and signaling function. *Physiol Rev.* 1998;78(1):99–141.
110. Verkhratsky A, Rose CR. Na(+)-dependent transporters: the backbone of astroglial homeostatic function. *Cell Calcium.* 2020;85:102136.
111. Verkhratsky A, Untiet V, Rose CR. Ionic signalling in astroglia beyond calcium. *J Physiol.* 2019. https://doi.org/10.1113/JP277478.
112. Petrelli F, Pucci L, Bezzi P. Astrocytes and microglia and their potential link with autism spectrum disorders. *Front Cell Neurosci.* 2016;10:21.
113. Sloan SA, Barres BA. Mechanisms of astrocyte development and their contributions to neurodevelopmental disorders. *Curr Opin Neurobiol.* 2014;27:75–81.
114. Johnson MB, Kawasawa YI, Mason CE, et al. Functional and evolutionary insights into human brain development through global transcriptome analysis. *Neuron.* 2009;62(4):494–509.
115. Willsey AJ, State MW. Autism spectrum disorders: from genes to neurobiology. *Curr Opin Neurobiol.* 2015;30:92–99.
116. Ming X, Johnson WG, Stenroos ES, Mars A, Lambert GH, Buyske S. Genetic variant of glutathione peroxidase 1 in autism. *Brain Dev.* 2010;32(2):105–109.
117. Ronesi JA, Collins KA, Hays SA, et al. Disrupted Homer scaffolds mediate abnormal mGluR5 function in a mouse model of fragile X syndrome. *Nat Neurosci.* 2012;15(3):431–440:[S431].
118. Barnby G, Abbott A, Sykes N, et al. Candidate-gene screening and association analysis at the autism-susceptibility locus on chromosome 16p: evidence of association at GRIN2A and ABAT. *Am J Hum Genet.* 2005;76(6):950–966.
119. Maekawa M, Iwayama Y, Arai R, et al. Polymorphism screening of brain-expressed FABP7, 5 and 3 genes and association studies in autism and schizophrenia in Japanese subjects. *J Hum Genet.* 2010;55(2):127–130.
120. Cahoy JD, Emery B, Kaushal A, et al. A transcriptome database for astrocytes, neurons, and oligodendrocytes: a new resource for understanding brain development and function. *J Neurosci.* 2008;28(1):264–278.
121. Zhang Y, Sloan SA, Clarke LE, et al. Purification and characterization of progenitor and mature human astrocytes reveals transcriptional and functional differences with mouse. *Neuron.* 2016;89(1):37–53.
122. Jamain S, Quach H, Betancur C, et al. Mutations of the X-linked genes encoding neuroligins NLGN3 and NLGN4 are associated with autism. *Nat Genet.* 2003;34(1):27–29.
123. Ylisaukko-oja T, Rehnstrom K, Auranen M, et al. Analysis of four neuroligin genes as candidates for autism. *Eur J Hum Genet.* 2005;13(12):1285–1292.
124. Dalva MB, McClelland AC, Kayser MS. Cell adhesion molecules: signalling functions at the synapse. *Nat Rev Neurosci.* 2007;8(3):206–220.
125. Varoqueaux F, Aramuni G, Rawson RL, et al. Neuroligins determine synapse maturation and function. *Neuron.* 2006;51(6):741–754.
126. Singh SK, Eroglu C. Neuroligins provide molecular links between syndromic and nonsyndromic autism. *Sci Signal.* 2013;6(283):re4.

127. Jaramillo TC, Liu S, Pettersen A, Birnbaum SG, Powell CM. Autism-related neuroligin-3 mutation alters social behavior and spatial learning. *Autism Res.* 2014;7(2):264–272.
128. Burrows EL, Laskaris L, Koyama L, et al. A neuroligin-3 mutation implicated in autism causes abnormal aggression and increases repetitive behavior in mice. *Mol Autism.* 2015;6:62.
129. Scheiffele P, Fan J, Choih J, Fetter R, Serafini T. Neuroligin expressed in nonneuronal cells triggers presynaptic development in contacting axons. *Cell.* 2000;101(6): 657–669.
130. Fu Z, Washbourne P, Ortinski P, Vicini S. Functional excitatory synapses in HEK293 cells expressing neuroligin and glutamate receptors. *J Neurophysiol.* 2003;90(6):3950–3957.
131. Sakers K, Eroglu C. Control of neural development and function by glial neuroligins. *Curr Opin Neurobiol.* 2019;57:163–170.
132. Stogsdill JA, Ramirez J, Liu D, et al. Astrocytic neuroligins control astrocyte morphogenesis and synaptogenesis. *Nature.* 2017;551(7679):192–197.
133. Voineagu I, Wang X, Johnston P, et al. Transcriptomic analysis of autistic brain reveals convergent molecular pathology. *Nature.* 2011;474(7351):380–384.
134. Purcell AE, Jeon OH, Zimmerman AW, Blue ME, Pevsner J. Postmortem brain abnormalities of the glutamate neurotransmitter system in autism. *Neurology.* 2001;57(9):1618–1628.
135. Edmonson C, Ziats MN, Rennert OM. Altered glial marker expression in autistic postmortem prefrontal cortex and cerebellum. *Mol Autism.* 2014;5(1):3.
136. Fatemi SH, Folsom TD, Reutiman TJ, Lee S. Expression of astrocytic markers aquaporin 4 and connexin 43 is altered in brains of subjects with autism. *Synapse.* 2008;62(7):501–507.
137. Suzuki K, Sugihara G, Ouchi Y, et al. Microglial activation in young adults with autism spectrum disorder. *JAMA Psychiat.* 2013;70(1):49–58.
138. Watkins LR, Milligan ED, Maier SF. Glial activation: a driving force for pathological pain. *Trends Neurosci.* 2001;24(8):450–455.
139. Wei H, Zou H, Sheikh AM, et al. IL-6 is increased in the cerebellum of autistic brain and alters neural cell adhesion, migration and synaptic formation. *J Neuroinflammation.* 2011;8:52.
140. Li X, Chauhan A, Sheikh AM, et al. Elevated immune response in the brain of autistic patients. *J Neuroimmunol.* 2009;207(1–2):111–116.
141. Jyonouchi H, Sun S, Le H. Proinflammatory and regulatory cytokine production associated with innate and adaptive immune responses in children with autism spectrum disorders and developmental regression. *J Neuroimmunol.* 2001;120(1):170–179.
142. Hellendall R, Matsushima GK, Knapp GW, Suzuki K, Sherman P, Ting JPY. Nitric oxide exacerbates the course of pathology in a mouse model of demyelinating disesase. *J Neuroimmunol.* 1998;90(1):56.
143. Estes ML, McAllister AK. Immune mediators in the brain and peripheral tissues in autism spectrum disorder. *Nat Rev Neurosci.* 2015;16(8):469–486.
144. Gupta S, Ellis SE, Ashar FN, et al. Transcriptome analysis reveals dysregulation of innate immune response genes and neuronal activity-dependent genes in autism. *Nat Commun.* 2014;5:5748.
145. Depino AM. Perinatal inflammation and adult psychopathology: from preclinical models to humans. *Semin Cell Dev Biol.* 2018;77:104–114.
146. Verkhratsky A, Zorec R, Parpura V. Stratification of astrocytes in healthy and diseased brain. *Brain Pathol.* 2017;27(5):629–644.
147. Pekny M, Pekna M, Messing A, et al. Astrocytes: a central element in neurological diseases. *Acta Neuropathol.* 2016;131(3):323–345.

148. Prinz M, Jung S, Priller J. Microglia biology: one century of evolving concepts. *Cell.* 2019;179(2):292–311.
149. Chung W, Choi SY, Lee E, et al. Social deficits in IRSp53 mutant mice improved by NMDAR and mGluR5 suppression. *Nat Neurosci.* 2015;18(3):435–443.
150. Greer JM, Capecchi MR. Hoxb8 is required for normal grooming behavior in mice. *Neuron.* 2002;33(1):23–34.
151. Chen SK, Tvrdik P, Peden E, et al. Hematopoietic origin of pathological grooming in Hoxb8 mutant mice. *Cell.* 2010;141(5):775–785.
152. Zhang P, Miller BS, Rosenzweig SA, Bhat NR. Activation of C-Jun N-terminal kinase/stress-activated protein kinase in primary glial cultures. *J Neurosci Res.* 1996;46(1):114–121.
153. Molina-Holgado F, Toulmond S, Rothwell NJ. Involvement of interleukin-1 in glial responses to lipopolysaccharide: endogenous versus exogenous interleukin-1 actions. *J Neuroimmunol.* 2000;111(1):1–9.
154. Moynagh PN, Williams DC, O'Neill LA. Interleukin-1 activates transcription factor NF kappa B in glial cells. *Biochem J.* 1993;294(pt 2):343–347.
155. O'Roak BJ, Deriziotis P, Lee C, et al. Exome sequencing in sporadic autism spectrum disorders identifies severe de novo mutations. *Nat Genet.* 2011;43(6):585–589.
156. Sanders SJ, Murtha MT, Gupta AR, et al. De novo mutations revealed by whole-exome sequencing are strongly associated with autism. *Nature.* 2012;485(7397): 237–241.
157. Bhat SS, Ladd S, Grass F, et al. Disruption of the IL1RAPL1 gene associated with a pericentromeric inversion of the X chromosome in a patient with mental retardation and autism. *Clin Genet.* 2008;73(1):94–96.
158. Pohodich AE, Zoghbi HY. Rett syndrome: disruption of epigenetic control of post-natal neurological functions. *Hum Mol Genet.* 2015;24(R1):R10–R16.
159. Maezawa I, Jin LW. Rett syndrome microglia damage dendrites and synapses by the elevated release of glutamate. *J Neurosci.* 2010;30(15):5346–5356.
160. Ballas N, Lioy DT, Grunseich C, Mandel G. Non-cell autonomous influence of MeCP2-deficient glia on neuronal dendritic morphology. *Nat Neurosci.* 2009;12(3): 311–317.
161. Lioy DT, Garg SK, Monaghan CE, et al. A role for glia in the progression of Rett's syndrome. *Nature.* 2011;475(7357):497–500.
162. Derecki NC, Cronk JC, Lu Z, et al. Wild-type microglia arrest pathology in a mouse model of Rett syndrome. *Nature.* 2012;484(7392):105–109.
163. Smith SE, Zhou YD, Zhang G, Jin Z, Stoppel DC, Anderson MP. Increased gene dosage of Ube3a results in autism traits and decreased glutamate synaptic transmission in mice. *Sci Transl Med.* 2011;3(103):103ra197.
164. Pajarillo E, Rizor A, Lee J, Aschner M, Lee E. The role of astrocytic glutamate transporters GLT-1 and GLAST in neurological disorders: potential targets for neurotherapeutics. *Neuropharmacology.* 2019;161:107559.
165. Aronson M, Hagberg B, Gillberg C. Attention deficits and autistic spectrum problems in children exposed to alcohol during gestation: a follow-up study. *Dev Med Child Neurol.* 1997;39(9):583–587.
166. Christianson AL, Chesler N, Kromberg JG. Fetal valproate syndrome: clinical and neurodevelopmental features in two sibling pairs. *Dev Med Child Neurol.* 1994;36(4):361–369.
167. Kern JK, Geier DA, Audhya T, King PG, Sykes LK, Geier MR. Evidence of parallels between mercury intoxication and the brain pathology in autism. *Acta Neurobiol Exp (Wars).* 2012;72(2):113–153.
168. Satomoto M, Satoh Y, Terui K, et al. Neonatal exposure to sevoflurane induces abnormal social behaviors and deficits in fear conditioning in mice. *Anesthesiology.* 2009;110(3):628–637.

169. Windham GC, Zhang L, Gunier R, Croen LA, Grether JK. Autism spectrum disorders in relation to distribution of hazardous air pollutants in the San Francisco bay area. *Environ Health Perspect*. 2006;114(9):1438–1444.
170. Yorbik O, Kurt I, Hasimi A, Ozturk O. Chromium, cadmium, and lead levels in urine of children with autism and typically developing controls. *Biol Trace Elem Res*. 2010;135(1–3):10–15.
171. Block ML, Calderon-Garciduenas L. Air pollution: mechanisms of neuroinflammation and CNS disease. *Trends Neurosci*. 2009;32(9):506–516.
172. Calderon-Garciduenas L, Mora-Tiscareno A, Ontiveros E, et al. Air pollution, cognitive deficits and brain abnormalities: a pilot study with children and dogs. *Brain Cogn*. 2008;68(2):117–127.
173. Wang C, Luan Z, Yang Y, Wang Z, Cui Y, Gu G. Valproic acid induces apoptosis in differentiating hippocampal neurons by the release of tumor necrosis factor-alpha from activated astrocytes. *Neurosci Lett*. 2011;497(2):122–127.
174. Yang Z, Yang S, Qian SY, et al. Cadmium-induced toxicity in rat primary mid-brain neuroglia cultures: role of oxidative stress from microglia. *Toxicol Sci*. 2007;98(2):488–494.
175. Telgkamp P, Backus KH, Deitmer JW. Blockade of AMPA receptors by nickel in cultured rat astrocytes. *Glia*. 1996;16(2):140–146.
176. Sama P, Long TC, Hester S, et al. The cellular and genomic response of an immortalized microglia cell line (BV2) to concentrated ambient particulate matter. *Inhal Toxicol*. 2007;19(13):1079–1087.
177. Liu YP, Yang CS, Tzeng SF. Inhibitory regulation of glutamate aspartate transporter (GLAST) expression in astrocytes by cadmium-induced calcium influx. *J Neurochem*. 2008;105(1):137–150.
178. Guizzetti M, Moore NH, Giordano G, VanDeMark KL, Costa LG. Ethanol inhibits neuritogenesis induced by astrocyte muscarinic receptors. *Glia*. 2010;58(12):1395–1406.
179. Gibbons HM, Smith AM, Teoh HH, et al. Valproic acid induces microglial dysfunction, not apoptosis, in human glial cultures. *Neurobiol Dis*. 2011;41(1):96–103.
180. Fernandez-Lizarbe S, Pascual M, Guerri C. Critical role of TLR4 response in the activation of microglia induced by ethanol. *J Immunol*. 2009;183(7):4733–4744.
181. Eskes C, Honegger P, Juillerat-Jeanneret L, Monnet-Tschudi F. Microglial reaction induced by noncytotoxic methylmercury treatment leads to neuroprotection via interactions with astrocytes and IL-6 release. *Glia*. 2002;37(1):43–52.
182. Hornig M, Lipkin WI. Infectious and immune factors in the pathogenesis of neurodevelopmental disorders: epidemiology, hypotheses, and animal models. *Ment Retard Dev Disabil Res Rev*. 2001;7(3):200–210.
183. Chess S. Autism in children with congenital rubella. *J Autism Child Schizophr*. 1971;1(1):33–47.
184. Chess S, Fernandez P, Korn S. Behavioral consequences of congenital rubella. *J Pediatr*. 1978;93(4):699–703.
185. Bolte ER. Autism and Clostridium tetani. *Med Hypotheses*. 1998;51(2):133–144.
186. Finegold SM, Downes J, Summanen PH. Microbiology of regressive autism. *Anaerobe*. 2012;18(2):260–262.
187. Lipkin WI, Hornig M. Microbiology and immunology of autism spectrum disorders. *Novartis Found Symp*. 2003;251:129–143 discussion 144–128, 281–197.
188. Shi L, Tu N, Patterson PH. Maternal influenza infection is likely to alter fetal brain development indirectly: the virus is not detected in the fetus. *Int J Dev Neurosci*. 2005;23(2–3):299–305.
189. Okada Y, Naka K, Kawamura K, et al. Localization of matrix metalloproteinase 9 (92-kilodalton gelatinase/type IV collagenase = gelatinase B) in osteoclasts: implications for bone resorption. *Lab Invest*. 1995;72(3):311–322.

190. Fatemi SH, Folsom TD, Reutiman TJ, Sidwell RW. Viral regulation of aquaporin 4, connexin 43, microcephalin and nucleolin. *Schizophr Res*. 2008;98(1):163–177.
191. Bolton JL, Bilbo SD. Developmental programming of brain and behavior by perinatal diet: focus on inflammatory mechanisms. *Dialogues Clin Neurosci*. 2014;16(3):307–320.
192. Sullivan EL, Nousen EK, Chamlou KA. Maternal high fat diet consumption during the perinatal period programs offspring behavior. *Physiol Behav*. 2014;123:236–242.
193. Peleg-Raibstein D, Luca E, Wolfrum C. Maternal high-fat diet in mice programs emotional behavior in adulthood. *Behav Brain Res*. 2012;233(2):398–404.
194. Yu H, Bi Y, Ma W, et al. Long-term effects of high lipid and high energy diet on serum lipid, brain fatty acid composition, and memory and learning ability in mice. *Int J Dev Neurosci*. 2010;28(3):271–276.
195. Bilbo SD, Tsang V. Enduring consequences of maternal obesity for brain inflammation and behavior of offspring. *FASEB J*. 2010;24(6):2104–2115.
196. Sasaki A, de Vega WC, St-Cyr S, Pan P, McGowan PO. Perinatal high fat diet alters glucocorticoid signaling and anxiety behavior in adulthood. *Neuroscience*. 2013;240:1–12.
197. Belluscio LM, Alberca CD, Pregi N, Canepa ET. Altered gene expression in hippocampus and depressive-like behavior in young adult female mice by early protein malnutrition. *Genes Brain Behav*. 2016;15(8):741–749.
198. Naik AA, Patro IK, Patro N. Slow physical growth, delayed reflex ontogeny, and permanent behavioral as well as cognitive impairments in rats following intra-generational protein malnutrition. *Front Neurosci*. 2015;9:446.
199. Ranade SC, Rose A, Rao M, Gallego J, Gressens P, Mani S. Different types of nutritional deficiencies affect different domains of spatial memory function checked in a radial arm maze. *Neuroscience*. 2008;152(4):859–866.
200. Reyes-Castro LA, Rodriguez JS, Charco R, et al. Maternal protein restriction in the rat during pregnancy and/or lactation alters cognitive and anxiety behaviors of female offspring. *Int J Dev Neurosci*. 2012;30(1):39–45.
201. Vucetic Z, Totoki K, Schoch H, et al. Early life protein restriction alters dopamine circuitry. *Neuroscience*. 2010;168(2):359–370.
202. Grissom NM, George R, Reyes TM. Suboptimal nutrition in early life affects the inflammatory gene expression profile and behavioral responses to stressors. *Brain Behav Immun*. 2017;63:115–126.
203. Bronzuoli MR, Facchinetti R, Ingrassia D, et al. Neuroglia in the autistic brain: evidence from a preclinical model. *Mol Autism*. 2018;9:66.
204. Cartocci V, Catallo M, Tempestilli M, et al. Altered brain cholesterol/isoprenoid metabolism in a rat model of autism spectrum disorders. *Neuroscience*. 2018;372:27–37.
205. Codagnone MG, Podesta MF, Uccelli NA, Reines A. Differential local connectivity and neuroinflammation profiles in the medial prefrontal cortex and hippocampus in the valproic acid rat model of autism. *Dev Neurosci*. 2015;37(3):215–231.
206. Lee HJ, Dreyfus C, DiCicco-Bloom E. Valproic acid stimulates proliferation of glial precursors during cortical gliogenesis in developing rat. *Dev Neurobiol*. 2016;76(7):780–798.
207. Christensen J, Gronborg TK, Sorensen MJ, et al. Prenatal valproate exposure and risk of autism spectrum disorders and childhood autism. *JAMA*. 2013;309(16):1696–1703.
208. Bromley RL, Mawer GE, Briggs M, et al. The prevalence of neurodevelopmental disorders in children prenatally exposed to antiepileptic drugs. *J Neurol Neurosurg Psychiatry*. 2013;84(6):637–643.
209. Williams G, King J, Cunningham M, Stephan M, Kerr B, Hersh JH. Fetal valproate syndrome and autism: additional evidence of an association. *Dev Med Child Neurol*. 2001;43(3):202–206.

210. Williams PG, Hersh JH. A male with fetal valproate syndrome and autism. *Dev Med Child Neurol.* 1997;39(9):632–634.
211. Kini U, Adab N, Vinten J, et al. Dysmorphic features: an important clue to the diagnosis and severity of fetal anticonvulsant syndromes. *Arch Dis Child Fetal Neonatal Ed.* 2006;91(2):F90–F95.
212. Nicolini C, Fahnestock M. The valproic acid-induced rodent model of autism. *Exp Neurol.* 2018;299(pt A):217–227.
213. Servadio M, Vanderschuren LJ, Trezza V. Modeling autism-relevant behavioral phenotypes in rats and mice: do 'autistic' rodents exist? *Behav Pharmacol.* 2015;26(6):522–540.
214. Roullet FI, Lai JK, Foster JA. In utero exposure to valproic acid and autism—a current review of clinical and animal studies. *Neurotoxicol Teratol.* 2013;36:47–56.
215. Servadio M, Manduca A, Melancia F, et al. Impaired repair of DNA damage is associated with autistic-like traits in rats prenatally exposed to valproic acid. *Eur Neuropsychopharmacol.* 2018;28(1):85–96.
216. Melancia F, Schiavi S, Servadio M, et al. Sex-specific autistic endophenotypes induced by prenatal exposure to valproic acid involve anandamide signalling. *Br J Pharmacol.* 2018;175(18):3699–3712.
217. Kim KC, Kim P, Go HS, et al. Male-specific alteration in excitatory post-synaptic development and social interaction in pre-natal valproic acid exposure model of autism spectrum disorder. *J Neurochem.* 2013;124(6):832–843.
218. Dichter GS, Felder JN, Green SR, Rittenberg AM, Sasson NJ, Bodfish JW. Reward circuitry function in autism spectrum disorders. *Soc Cogn Affect Neurosci.* 2012;7(2):160–172.
219. Donovan AP, Basson MA. The neuroanatomy of autism—a developmental perspective. *J Anat.* 2017;230(1):4–15.
220. Reim D, Distler U, Halbedl S, et al. Proteomic analysis of post-synaptic density fractions from Shank3 mutant mice reveals brain region specific changes relevant to autism spectrum disorder. *Front Mol Neurosci.* 2017;10:26.
221. Bronzuoli MR, Facchinetti R, Steardo L, Scuderi C. Astrocyte: an innovative approach for Alzheimer's disease therapy. *Curr Pharm Des.* 2017;23(33):4979–4989.
222. Scuderi C, Stecca C, Iacomino A, Steardo L. Role of astrocytes in major neurological disorders: the evidence and implications. *IUBMB Life.* 2013;65(12):957–961.
223. Zhang S, Wu M, Peng C, Zhao G, Gu R. GFAP expression in injured astrocytes in rats. *Exp Ther Med.* 2017;14(3):1905–1908.
224. Cantone M, Kuspert M, Reiprich S, et al. A gene regulatory architecture that controls region-independent dynamics of oligodendrocyte differentiation. *Glia.* 2019;67(5):825–843.
225. Chun H, Marriott I, Lee CJ, Cho H. Elucidating the interactive roles of glia in Alzheimer's disease using established and newly developed experimental models. *Front Neurol.* 2018;9:797.
226. Ito D, Imai Y, Ohsawa K, Nakajima K, Fukuuchi Y, Kohsaka S. Microglia-specific localisation of a novel calcium binding protein, Iba1. *Brain Res Mol Brain Res.* 1998;57(1):1–9.
227. Casanova MF. The neuropathology of autism. *Brain Pathol.* 2007;17(4):422–433.

CHAPTER TWELVE

Oxidative stress, metabolic and mitochondrial abnormalities associated with autism spectrum disorder

Morten Thorsen*
Department of Child and Adolescent Psychiatry, Aalborg, Denmark
*Corresponding author: e-mail address: mbt90@live.dk

Contents

Abstract

Autism spectrum disorder is a neurodevelopmental disorder characterized by impaired development and by abnormal function in regards to social interaction, communication and restricted, repetitive behavior. It affects approximately 1% of the worldwide population. Like other psychiatric disorders the diagnosis is based on observation of, and interview with the patient and next of kin, and diagnostic tests. Many genes have been associated with autism, but only few highly penetrant. Some researchers have instead focused on oxidative stress, metabolic abnormalities and mitochondrial dysfunction as an explanation of the disorder. Currently no cure exists for the disorder, making these abnormalities interesting as they are possibly correctable with supplements or treatment. These various processes cannot be seen independently as they are influencing and interacting with each other. Furthermore many of the metabolic changes seen

Progress in Molecular Biology and Translational Science, Volume 173
ISSN 1877-1173
https://doi.org/10.1016/bs.pmbts.2020.04.018

in autism have also been shown in other psychiatric disorders such as attention deficit hyperactivity disorder, schizophrenia and bipolar disorder along with often comorbid disorders like epilepsy and intellectual disability. As such some of these abnormalities are not specific, however, could indicate a similar mechanism for the development of these disorders, with symptomatology and severity varying according to the location and the amount of damage done to proteins, cells and DNA. Clinical studies trying to treat these abnormalities, have widely been successful in correcting the metabolic abnormalities seen, but only some studies have also shown bettering of autistic symptoms. Hopefully with increased knowledge of the pathophysiology of the disorder, future preventive measures or treatment can be developed.

1. Etiology of autism spectrum disorder

Autism spectrum disorder (ASD) is a cluster of pervasive disorder characterized by abnormalities in reciprocal social interactions and in communication, and by a restricted, stereotyped, repetitive repertoire of interests and activities.[1] It affects approximately 1% of the population.[2] The disorder is considered a trait disorder, indicating a lifelong disorder. When measured in disability-adjusted life-years the total burden of ASD is higher than that of conduct disorder and attention deficit hyperactivity disorder (ADHD) together.[2] As with many other psychiatric disorders, the phenotype of patients with ASD can be very heterogeneous, making diagnosing the disorder difficult.[3] Currently the diagnosis is based on interviews with the patient and next of kin, clinical observation, and cognitive and diagnostic tests such as the Autism Diagnostic Observation Schedule, which is currently the gold standard instrument for diagnosing autism.[4] Although the disorder is regarded as a neuropsychiatric disorder with high inheritance, thereby implying a neurobiological basis, no biomarker or candidate gene has of yet been found. The thought of biomarkers in the psychiatric discipline is not a new radical idea, but has been investigated at least as far back as in the first half of the 20th century, where increased levels of lactic acid was found in patients with schizophrenia.[5]

The etiology of autism is not completely understood yet, however, different explanations have been proposed. Twin studies have revealed a high genetic component with concordance rates of 90% among monozygotic twins and 30% among dizygotic twins.[6] The genetic basis is further emphasized with more than 100 disease genes having been reported in subjects with ASD or autistic-like behavior.[7,8] Several loci have been consistently identified in studies as having a strong role in the development of ASD.[9]

However, only few cases of ASD are explained by defects in a single gene or chromosome, and as such it is considered having multifactorial inheritance with multiple genes at work.[10–12] In approximately 25% of cases a genetic cause can be identified, with causative genes being involved in various processes such as chromatin remodeling, metabolism, mRNA translation, and synaptic function.[13] Abnormal cell proliferation, differentiation and migration has been emphasized as key mechanisms in the development of ASD, supported by data from post-mortem brain, imaging and gene studies.[14] Unusual brain growth, with an increased number of neurons in the prefrontal cortex and increased head size, with a later normalization of the latter after the first year of life, could relate to the abnormal proliferation and differentiation seen.[15–19] A larger head size even correlate with autism severity and regression.[19] However, especially in the last decade, research focusing on a more systemic approach to the development of ASD have increasingly emerged.[20] Studies from many different fields of medicine have found autism to be associated with various physiological abnormalities in other organs than the brain, indicating a more systemic abnormality as a cause for the development of ASD among some patients.[20] The risk for development of ASD in a patient is likely heightened by a genetic predisposition, consisting of several genetic risk factors, and exposure to environmental factors.[20] The process is thought to include toxicant exposure, disturbance in the redox state of cells, inflammation, epigenetic factors and energy production in the mitochondria.[20] This has been of increasing interesting with fast growth in publications since 1986.[20] A hypothesis of several potential causative processes, and an underlying genetic predisposition, would be in line with the heterogeneity of autistic symptoms, their severity and progression of the disorder.

2. Oxidative stress and ASD

Oxidative stress exists when there is an imbalance between prooxidative and antioxidative substances, producing free radicals, such as reactive oxygen species (ROS). ROS are produced through natural aerobic metabolism, and removed by various antioxidative substances and enzymes. Examples of ROS are the free radicals superoxide (O_2^-) and hydroxyl radicals ($^{\bullet}OH$) and the non-free radical hydrogen peroxide (H_2O_2).[21]

Free radicals are readily reactive and can be involved in various intracellular processes including transcription, apoptosis and differentiation.[21] At higher levels they can cause damage to cells, through reaction with its membrane, deoxyribonucleic acid (DNA) and other intracellular components.[22] ROS can react with the cell lipid membrane, through lipid peroxidation, thereby reducing the integrity and fluidity of the cell wall, ultimately leading to cell death.[23] Another mechanism is through actual DNA strand breaks ultimately also resulting cell death.[24]

When kept at low levels, however, ROS can also take a more positive role, being involved in the eradication of pathogens, intracellular and cell to cell signaling and cell division. The problem exists when the regulatory systems cannot cope with the production of ROS, resulting in adverse effects as described above, either through impairment of the systems neutralizing ROS or through increased ROS production.[21]

2.1 Where are reactive oxygen species produced?

Reactive oxygen species is produced throughout the cells and various tissues, making up the human body. Most of the production of ROS happens in the mitochondria, the peroxisomes and the endoplasmatic reticulum of the cells. Most abundant is in the mitochondria, in the energy production of adenosine triphosphate (ATP) by means of the electron transport chain (ETC).[25] The ETC consists of five complexes, where especially complex I and III have been found to "leak" superoxide (O_2^-) in the amount of 0.2–2% of the oxygen consumed by mitochondria (see Fig. 1).[26] Which of the two complexes, that should be considered the major contributor of ROS is still not clear, and in fact various other mitochondrial enzyme complexes associated with, but not part of, the ETC are also thought to be contributors of ROS.[27,28]

Both endo- and exogenous processes can result in increased production of free radicals and ROS. The endogenous processes include inflammatory processes, immune cell activation, ischemia, cancer, exercise, mental stress and aging.[21] Exogenous processes include exposure to radiation, heavy metals and pollutants, certain drugs and smoking, which when processed in the body, produce free radicals among other products.[21] The mitochondria have their own enzyme systems for neutralizing ROS, although it is not sufficient to bring it down to the low levels needed in the cell.[29,30] As such, the cells have their own antioxidative systems, mainly made up of various enzymes.[31]

2.2 Antioxidative systems

Various antioxidative enzymes exists, including superoxide dismutase (SOD), glutathione peroxidase (GPX), catalase (CAT) and ceruloplasmin.[31] Along with these, there are substances such as glutathione, its precursors

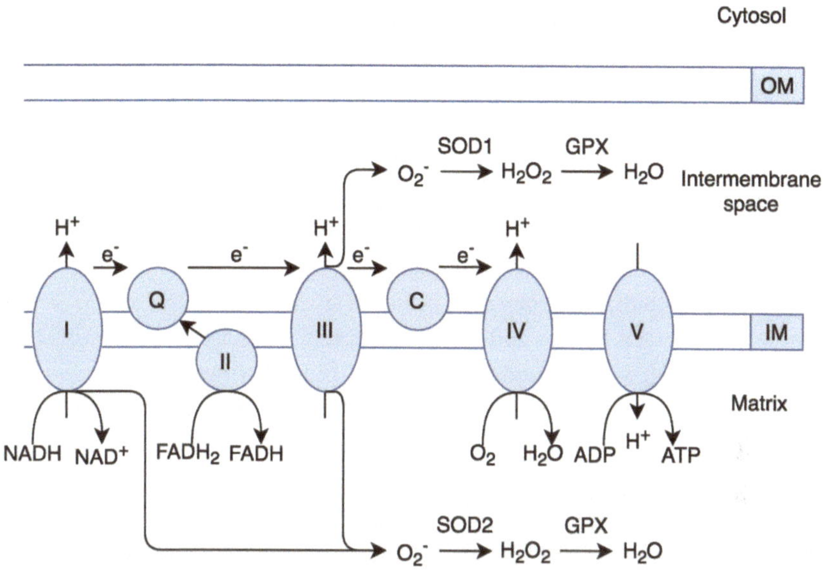

Fig. 1 Production of reactive oxygen species in the mitochondrial electron transport chain (ECT). Nicotine adenine dinucleotide (NADH) and Flavin adenine dinucleotide ($FADH_2$) is provided by the citric acid cycle. The electrons donated therefrom is shuttled through the ECT and ultimately transferred to oxygen (O_2), generating water (H_2O). As depicted, the main generators of ROS are complex I and III. Both complexes leak electrons to form superoxide (O_2^-) in the matrix, while only complex III does in the intermembrane space. In the intermembrane space, superoxide is dismutated by superoxide dismutase 1 (SOD1), to form hydrogen peroxide (H_2O_2). In the matrix superoxide dismutase 2 (SOD2), do the same. In both locations glutathione peroxidase (GPX) further reduced hydrogen peroxide to water. Q, quinone; C, cytochrome C; IM, inner membrane; OM, outer membrane.

(cysteine plus cysteinylglycine, cystathionine, and homocysteine in the transsulfuration pathway, and methionine in the transmethylation cycle). Glutathione exists in a reduced state (GSH) and an oxidized state (GSSG). The latter is produced when neutralizing reactive oxygen species. Various nutrients also have antioxidative capabilities including folate along with vitamins A, B6, C and E.

2.2.1 Superoxide dismutase

In the human body, three types of superoxide dismutase exist, a copper and zinc containing superoxide (SOD1), a manganese containing superoxide dismutase (SOD2) and extracellular superoxide dismutase (SOD3).[22,26] SOD1 is mainly found in the cytosol of the cell and in the intermembrane space of the mitochondria.[32] SOD2 is mainly found in the mitochondrial matrix and inner membrane.[32] Lastly, SOD3 is found extracellular.[26] In the central nervous system (CNS) of rats, SOD1 is mainly found in the cytosol of astrocytes, while SOD2 is mainly found in the mitochondria of neurons.[33] The enzymes acts by catalyzing the dismutation of the free radical superoxide into oxygen and hydrogen peroxide, the latter also being a free radical.[22] The enzyme CAT and various peroxidases catalyze the further decomposition of hydrogen peroxide to water and byproducts.[22]

SOD have been shown with increased activity in erythrocytes in patients with ASD from peripheral blood in multiple studies.[34–37] Only few have differentiated between SOD1 and SOD2, and have found that only SOD1 are elevated in platelets compared to controls.[31,37] Rare single nucleotide polymorphisms in the noncoding region of the SOD1 gene, are also associated with ASD.[38]

2.2.2 Catalase

Usually located in the peroxisome, it catalyzes the decomposition of the free radical hydrogen peroxide (H_2O_2) to water and oxygen.[22,39] It is mainly found in peroxisomes.[40] In autism, reduced activity has been shown in several studies.[34–36]

2.2.3 Glutathione peroxidase

A general name for a whole enzyme family, consisting of several isozymes, each with their own preferred cell and subcellular location including the cytosol and mitochondria.[41] In the mitochondria it reduces H_2O_2 to water in both the intermembrane space and in the matrix.[26] It's function is based on oxidizing two molecules of GSH, forming a disulfide bond, creating the form GSSG. Reduced activity have been shown in erythrocytes from patients with ASD,[34,37,42–44] while studies done on plasma have been more heterogenous.[35,44–46] There are few studies which have used platelets, which have found higher activity in patients with ASD.[37,47]

2.2.4 Glutathione

As stated above reduced GSH can be oxidized into GSSG, and in the process neutralizing free radicals.[48] Glutathione is formed from its precursor cysteine and homocysteine, by a vitamin-B6 dependent reaction.[49] The reduction of GSSG back to GSH, uses nicotinamide adenine dinucleotide phosphate (NADPH) as a substrate.[48] For further elaboration, please see "the transsulfuration pathway."

2.3 Oxidative stress and the brain

Redox regulation have been found to have a critical role in oligodendrocyte-type-2 astrocyte progenitor cells, where growth factors promoting self-renewal result in more reduced cell progenitors, while signaling molecules promoting differentiation result in a more oxidized state.[50] It could be theorized that the developing brain, when exposed to high levels of oxidative stress, overwhelming the antioxidative defense, could result in an abnormal development. Evidence of oxidative damage have been found in the post-mortem brain of patients with ASD, in the cerebellum and Brodmann area 22 in the temporal cortex.[51] The study found increased levels of 3-nitrotyrosine (3-NT), a marker of oxidative protein damage and 8-oxo-deoxyguanosine (8-oxo-dG), a marker of oxidative DNA damage in the aforementioned locations, and furthermore found 8-oxo-dG to inversely correlate with GSH/GSSH ratio in the cerebellum, suggesting a redox system not capable of neutralizing the generation of ROS.[51] In the brain SOD1 is mainly found in astrocytes, while SOD2 is mainly found in the mitochondria of neurons. Markers of astrocytes have also been found to be elevated in post-mortem brains of patients with ASD.[52–54] The study found markers of astrocytes to be significantly higher in the prefrontal cortex and the cerebellum, and showed lower levels of neuron-specific markers in the aforementioned places.[52] Neuronal induced pluripotent stem cells (iPSC) from patients with Rett syndrome, who show autistic-like, show a preference of differentiation toward astrocytes, with a further increment in formation and differentiation and creating a potential harmful environment for the surrounding neurons, disturbing their morphology and function. The correlation between biomarker levels in the CNS and in peripheral blood is not well investigated. However, studies have found peripheral biomarkers, like the plasma level of neurotrophic factor brain-derived neurotrophic factor (BDNF) to correlate with the level in cerebrospinal fluid. Furthermore, changes in gene expression related to the integrity of the blood-brain barrier, have been shown in post-mortem ASD brains.[55] When investigating signs of damage caused by oxidative stress, studies have revealed higher levels of 3-nitrotyrosine and 3-chlorotyrosine, which are signs of oxidative protein damage, in ASD when compared to unaffected children.[56] Signs of oxidative damage to DNA where also shown in the study, by higher levels of 8-oxo-deoxyguanosine.[56]

3. Metabolic abnormalities and ASD

Various abnormalities have been found in patients with ASD. In dries blood spots from neonates, reduced levels of neurotrophic factors have been found in patients later developing autism.[57] Neurotrophic factors are involved in the neurodevelopment, and as such, abnormalities in these could give rise to abnormal development, later manifesting as a developmental disorder.[58] Studies have also shown reduced levels of cytokines in dried blood spots in neonates later developing ASD, perhaps indicating a hypoactive immune system in the neonatal period.[59] With a reduced inflammatory response, they could be more susceptible to toxic exposure from various infections and toxicants. Likewise, fetal exposure to preeclampsia increases the risk of later developing ASD.[60] The lack of sufficient nutrient triggers an increase in oxidative stress in the fetus, which further stimulate the maternal inflammatory response.[61] Some of these inflammatory can cross from mother to fetus through the placenta, and potentially influence the survival and functions of the developing brain.[60]

Another interesting area is the rather complex folate-dependent one-carbon metabolism including the methylation cycle, the transsulfuration pathway and the folate cycle.[56] These pathways and cycles are involved in the DNA synthesis and repair, cell proliferation, methylation of DNA, RNA and proteins, regulation of oxidative stress, apoptosis, and many more functions. As such, this is a vital part in many aspects of the cell life and function.[56] Research so far has mainly focused on the methylation cycle and the transsulfuration pathway in relation to the pathophysiology of ASD.

3.1 The methylation cycle

Starting with the methylation cycle, a global state of hypomethylation can lead to genomic instability and abnormal change of gene expression.[62,63] The process of methylation is caused by methionine first being "activated" into *S*-adenosylmethionine (SAM), which is the main methyl donor for several intracellular processes through methyltransferase. In the methylation process SAM is converted to *S*-adenosylhomocysteine (SAH).[56] To fulfill the cycle SAH is hydrolyzed into adenosine and homocysteine, and the latter gets added a methyl group from 5-methyl-tetrafolate, through a B12 vitamin dependent transfer, facilitated by methionine synthase, thereby regenerating methionine (see Fig. 2).[56]

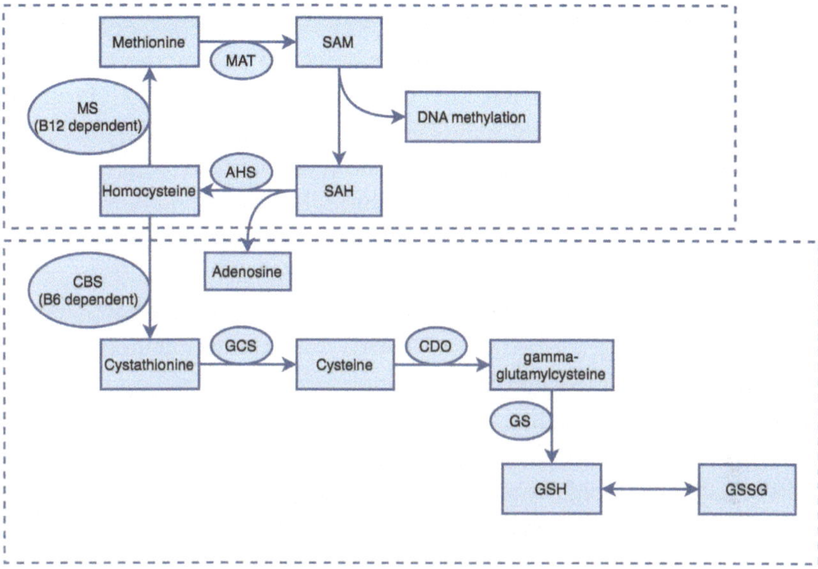

Fig. 2 In the highest rectangle is shown the methylation cycle. Adenosyl is added to methionine, by methionine adenosyltransferase, creating S-adenosylmethionine (SAM) which, through methyl donation is involved in various methylation processes, including DNA methylation. Through this donation S-adenosylhomocysteine (SAH) is created. Adenosylhomocysteine synthase converts SAH to homocysteine, which by a vitamin B12 dependent reaction by methionine synthase (MS) is converted back to methionine. Under some circumstances homocysteine can be converted to cystathionine through a vitamin B6 dependent reaction mediated by cystathionine-beta-synthase (CBS), which is further converted to cysteine by glutamate cysteine synthase. Cysteine dioxygenase (CDO) further converts cysteine to gamma-glutamylcysteine, which lastly is converted to reduced glutathione by glutathione synthase (GS).

The ratio between SAM and SAH, can be used as a measure of the patients methylation capacity and this ratio is lower in autistic children when compared to non-ASD siblings and also when compared to healthy controls.[56,64,65] Likewise, lower levels of DNA methylation as measured by percentage 5-methylcytosine, and methionine have also been shown.[31,43,49,56,64–67] It should be noted that these findings are from already diagnosed children, and as such, we do not know if these abnormal levels of methylation capacity are responsible for the development of ASD or whether these abnormalities are just a later product in the pathophysiology of the disorder. Furthermore, treatment of these imbalance seem able to rectify them, however, so far it is uncertain whether a clinical effect is seen from supplemental treatment, aimed toward normalizing the methylation metabolites.[65] A study has shown a correlation between clinical severity score and methionine and SAM/SAH ratio, respectively.[68]

3.2 The transsulfuration pathway

In the above cycle, homocysteine was remethylated into methionine, however, another possibility exists, the transsulfuration pathway. Through a reaction mediated by cystathionine beta synthase, homocysteine is removed from the above cycle, and is converted into cystathionine, which is further split into α-ketobutyrate and cysteine. Through a couple more vitamin B6 dependent steps cysteine is used to produce GSH, which, as mentioned before, is the reduced form of glutathione (see Fig. 2).

Glutathione is key in the mitochondria, as it is involved in the neutralization of hydrogen peroxide, formed by superoxide dismutase as mentioned earlier. The neutralization is mediated by glutathione peroxidase. Furthermore, glutathione is involved in cell signaling, renewal and differentiation and in regulation of enzyme and immune cell activity.[50,69,70] Studies have shown lower plasma levels of GSH and higher plasma levels of GSSG along with a reduced GSH/GSSG ratio in patients with ASD compared to controls.[49,56,64,66,71] The change in plasma GSSG and the GSH/GSSG ratio is mainly caused by an outflux of GSSG into the plasma.[48] Intracellular GSSG is kept low as to have a high GSH/GSSG ratio, thereby optimizing the antioxidative capacity.[48] This is done by either reducing GSSG back into GSH by glutathione reductase, using NADPH.[48] As such the abnormalities seen in plasma regarding GSH and GSH/GSSH, can be seen as a consequence of exposure to high levels of oxidative stress.

3.3 Abnormalities in other nutrients

Investigations into nutrients with antioxidative effects (vitamins A, B6, C and E) and nutrients involved in the above metabolic cycles and pathway (vitamin B12 and folate), have not shown any significant abnormal levels when investigated in a meta-analysis in patients with ASD, although a couple of studies investigating vitamin E have shown lower levels in patients with ASD.[31,35,43,56,66,72–76]

4. Mitochondrial abnormalities and ASD

The mitochondria is responsible for phosphorylation of adenosine diphosphate (ADP) to form adenosine triphosphate (ATP), the energy source of the body and the electron transport chain (ETC), also called the respiratory chain, is a vital part of making this possible.[77] The ETC consists of several complexes, shuttling electrons, to create a proton gradient, making ATP production possible (see Fig. 1).

The genetic basis of the respiratory chain is dual, as it is made up of both mitochondrial and nuclear DNA, consistent with the heterogeneity of the disorder and the multifactorial understanding of the etiology.[78]

Biomarkers relating to the electron transport chain in mitochondria have also been found to be abnormal in patients with ASD, where complex I and III is the most commonly found to be abnormal, but not exclusively.[79] Post-mortem brain studies from children with ASD have revealed lower levels in the cerebellum of complexes III and V, in the temporal cortex of complexes II, III and V and in the frontal cortex of only complex I.[80] As mentioned earlier, complexes I and III are the main complexes producing ROS. This study further found higher levels of oxidative stress in the cerebellum and the temporal cortex of the children and a normalization of ETC complexes levels in adulthood[80] Curiously, a reduced level of complex I activity has been associated with increased ROS production.[81]

Clinically, mitochondrial dysfunction (MD) can give rise to a wide variety of symptoms including, but not limited to, muscle weakness, rhabdomyolysis, developmental delay, seizures, loss of skills and various multisystem diseases.[82] Some of the symptoms of mitochondrial dysfunction is similar to the symptom representation seen in autism, including loss of previously learned skills, developmental delay and seizures.[83] Patients with autism, also have a higher prevalence of MD.[84] A population based study found a prevalence of MD in patients with ASD to be 7.2% in making MD one of the most common medical conditions associated with ASD.[3]

When patients with both ASD and MD have been compared to patients with only MD, higher rates of lethargy, fatigue, ataxia and gastrointestinal (GI) dysfunction have been found in the autistic patients.[85] ASD/MD patients, when compared to ASD patients without MD, have a higher rate of symptoms such as gastrointestinal (GI) dysfunction, gross motor delay, seizures and regression.[85] In addition, mitochondrial dysfunction is thought to cause altered or even impaired energy metabolism in the brain, and abnormal brain development, neurotransmission and neuronal connectivity.[78] The severity of autistic symptoms is associated with biomarkers of mitochondrial dysfunction.[86] The alterations in brain development and connectivity seen, together with a normalization in adulthood of levels of ECT complexes seen later in life, could be an explanation of the lessening of symptom severity seen in some, not all, patients as they get older and their ECT complex levels normalize. However, it should be noted that, at this point, is it not known whether MD is found in a subset of ASD patients, or if the ASD patient group has a spectrum of MD of differing severity.[85] Some patients have been observed to develop a clinical picture ASD and MD, following a fever, suggesting a vulnerability to environmental factors among patients with unidentified MD.[85]

Various non-specific biomarkers, used as an indicator for MD, including lactate, pyruvate, lactate/pyruvate ratio, creatine kinase and several others, have been found to be abnormal in patients with ASD and MD, but have not been used diagnostically, given their unspecific nature.[83,87,88]

5. A unified framework

It may be evident from reading so far, that these "systems" cannot be seen as separate actors, but rather have a complicated interplay with each other. Oxidative stress is produced in the mitochondria, ROS produced in the mitochondria contribute to the inflammatory cytokine production, and vice versa.[26,89] Pro-oxidative enzymes can furthermore stimulate the ROS production in the mitochondria.[26] Oxidative stress inhibits methionine synthase, to increase the production of glutathione.[90] This reduction, however, creates hypomethylation, which can further disturb the gene expression.[62]

Depending on the level of mitochondrial ROS different mechanisms have been proposed, including an involvement in the inflammatory process at moderate levels by effect of pyroptosis and induction of apoptosis at high levels.[26] Pyroptosis, in contrast to apoptosis, is a form of programmed cell death, which causes local inflammation, mediated by a release of several inflammatory cytokines through an increasingly porous plasma membrane.[91,92]

Signs of oxidative stress, immune dysregulation and mitochondrial dysfunction has been found widespread in the autistic brain.[93] In the frontal lobe, involved in executive functions among other functions, disturbance in the methylation cycle, lower ETC complex activities and elevations in proinflammatory cytokines such as TNF-alpha and IL-6 have been found.[80,94,95] In the temporal lobe, including the hippocampus, Brodmann area 21 and Brodmann area 22 (Wernicke's area), which is involved in long term memory and processing of sound, language and understanding, studies have found higher levels of oxidative stress, oxidative damage to DNA and mitochondrial proteins, lower GSH/GSSG ratio and lower ETC complexes activities.[96–100] In the cerebellum, responsible for motor coordination, postural stability, and modulation of cognition, neuroinflammation, signs of oxidative damage and decreased expression of genes relating to mitochondrial ETC complexes, and reduced levels of enzymes involved in glutathione synthesis have been significantly different in the ASD brain.[96,101–103] In the limbic system, involved in emotional awareness, pain and error detection, neuroinflammation and decreased expression of mitochondrial genes have been found.[102,104–106] As such, the damage caused by these interconnecting and interacting abnormalities is not seen in a single brain area but almost universally through the CNS. This is fully in accordance with the heterogeneity seen in patients with the disorder, with symptomatology and severity relating to which areas are affected, how much of the brain is affected and how severely the damage done is. When looking at the functions of the various brain areas affected, there is also quite an overlap with autistic symptoms, and even with some of the comorbid disturbances and disorders seen in patients with ASD.

6. Comorbidity

Patients diagnosed with ASD, have an increased risk of being diagnosed with a second psychiatric condition.[107] In adulthood, patients with ASD have a higher risk compared to the general population of every major psychiatric disorders including schizophrenia, bipolar disorder, depression, psychoses, suicide attempt, attention deficit and anxiety.[108] Shared genetic risk factors have been found in ASD, ADHD depression, bipolar disorder and schizophrenia, which could indicate similar pathophysiological processes.[109] Similarly, some of the metabolic abnormalities, focused on in earlier section also show linkage to conditions comorbid to ASD. Oxidative stress have been linked to hypercholesterolemia, hypertension, stroke, Parkinson's disease and renal disorders, all of which have a higher prevalence among patients with ASD.[21,108,110–113] Oxidative stress has been discussed in relation to and associated with other neurological and psychiatric disorders including Alzheimer's disease, schizophrenia, bipolar disorder and ADHD.[114–118]

Mitochondrial dysfunction has been linked to major depressive disorder, bipolar disorder and schizophrenia.[78,119,120] Antipsychotics have even been found to reduce ETC complex I activity, with an interaction between the dopamine and respiratory chain systems, as an essential part of the pathophysiology, having been proposed.[78] From rat studies glutathione have been hypothesized to have a role in the development of bipolar disorder and schizophrenia because of its redox regulatory function.[121] This is further supported by both glutamate and dopamine being redox reactive, and creating ROS during neurotransmission.[121]

Epilepsy, a neurological condition, is highly associated with autism spectrum disorder.[108] In several cases of epilepsy, having a genetic cause with a known gene mutation, signs of increased ROS production, oxidative damage, a deficiency of ECT complex I and decreased energy production from mitochondria have also been found.[122,123] Impaired response to oxidative stress is also associated with intellectual disability, with severity of intellectual disability correlating with level of ROS.[124–126] Interestingly, diminishing intelligence quotient correlates with deviation from normal head circumference, suggesting that the increased head size could reflect abnormal development, also hypothesized in regards to development of ASD.[19]

7. Future treatment options

ASD is currently viewed as a lifelong disorder, with no medical treatment. The hope is that by understanding the pathophysiology of the disorder, the disorder can either be prevented or treated. If, as many studies point toward, the disorder is caused my abnormal methylation, oxidative stress or other metabolic abnormalities there might be a possibility for treatment, as these are reversible processes. This of course necessitates that the damage already caused by these processes can also be reversed. A complex dietary supplement targeted toward oxidative stress, inflammation and mitochondrial dysfunction has showed improvement in the brain mitochondrial function of complex III and IV in rodents, resulting in reduced release of free radicals, which were furthermore associated with cognitive function.[127] The substance methylene blue, which has been suggested as a treatment for Alzheimer's disease, can function as an electron shuttler in the mitochondria, bypassing complexes I–III, thereby limiting the consequences of reduced activity in those complexes.[128,129] Furthermore it has an antioxidative effect, by decreasing the production of superoxide and the hydroxyl radical.[130,131] No studies on treatment with methylene blue in ASD have been done, but it has shown effect on development of depression in bipolar disorder in a double-blind study.[132]

N-acetylcysteine (NAC) is the major donator of cysteine to the production of glutathione, and its antioxidative effect has for many years been used in treatment of paracetamol overdose.[133] Apart from replenishing GSH levels it also have a more direct antioxidative effect, by neutralizing hydroxyl radicals and to a lesser degree hydrogen peroxide.[134] Direct administration of glutathione is not very efficient at replenishing GSH peripherally and poorly penetrates the blood-brain barrier, however, NAC can readily penetrate the blood-brain barrier and raise the brain levels of GSH.[135–138] The potential uses of NAC in psychiatry are many including schizophrenia, depression and bipolar disorder.[139] Few studies have been done one NAC treatment in ASD, but so far the treatment has shown effect on irritability, and level of GSH, while a single study found no effect of the treatment.[140–142] There is some evidence supporting treatment with vitamin and nutrient supplements in ASD. Selective treatment with vitamin B12, have shown effect on clinical symptoms, at least on some scales, which have correlated with methionine and SAM/SAH ratio.[68] Furthermore, a study investigating treatment with folinic acid showed improvements in the verbal communication, while a more comprehensive approach have also shown improvements in nonverbal IQ and autism severity.[143,144] Vitamin C have also shown some effect on especially autistic symptoms in a double-blind study.[145]

Furthermore, knowledge of the pathophysiology can possibly help prevent development of the disorder, through preventive care, given to the mother by supplementation or treatment to normalize these processes. Lastly, if neither prevention or treatment is possible, knowledge of these systems can coerce the finding of biomarkers or candidate genes to help in diagnosing the disorder, minimizing the diagnostic delay.[146]

Conclusively, neither the etiology or the pathophysiology of autism spectrum disorders are yet fully understood. However, considerable progress have been made, such that multiple risk factors and causal genes have been identified, emphasizing the multifactorial etiology of the disorder. Furthermore, potential treatment targets have been identified, and there is at least some evidence supporting treatment with alleviation of autism symptoms. There are still many research areas where studies are warranted, to further expand our understanding of the etiology and pathophysiology of this disorder, which in the most severe cases are devastating to the patient and their family.

References

1. World Health Organization. *The ICD-10 Classification of Mental and Behavioural Disorders. Clinical Descriptions and Diagnostic Guidelines*. Geneva: WHO; 1992.
2. Baxter AJ, Brugha TS, Erskine HE, Scheurer RW, Vos T, Scott JG. The epidemiology and global burden of autism spectrum disorders. *Psychol Med*. 2015;45(3):601–613. https://doi.org/10.1017/S003329171400172X.
3. Oliveira G, Diogo L, Grazina M, et al. Mitochondrial dysfunction in autism spectrum disorders: a population-based study. *Dev Med Child Neurol*. 2005;47(3):185–189. https://doi.org/10.1017/s0012162205000332.

4. Lord C, Rutter M, Goode S, et al. Autism diagnostic observation schedule: a standardized observation of communicative and social behavior. *J Autism Dev Disord.* 1989;19(2):185–212.
5. Looney JM, Childs HM. The lactic acid and glutathione content of the blood of schizophrenic patients. *J Clin Invest.* 1934;13(6):963–968. https://doi.org/10.1172/JCI100639.
6. Hallmayer J, Cleveland S, Torres A, et al. Genetic heritability and shared environmental factors among twin pairs with autism. *Arch Gen Psychiatry.* 2011;68(11):1095–1102. https://doi.org/10.1001/archgenpsychiatry.2011.76.
7. Betancur C. Etiological heterogeneity in autism spectrum disorders: more than 100 genetic and genomic disorders and still counting. *Brain Res.* 2011;1380:42–77. https://doi.org/10.1016/j.brainres.2010.11.078.
8. Ramaswami G, Geschwind DH. Genetics of autism spectrum disorder. *Handb Clin Neurol.* 2018;147:321–329. https://doi.org/10.1016/B978-0-444-63233-3.00021-X.
9. Schaefer GB. Clinical genetic aspects of ASD spectrum disorders. *Int J Mol Sci.* 2016;17(2):180. https://doi.org/10.3390/ijms17020180.
10. Muhle R, Trentacoste SV, Rapin I. The genetics of autism. *Pediatrics.* 2004;113(5):e472–e486.
11. Spence SJ. The genetics of autism. *Semin Pediatr Neurol.* 2004;11(3):196–204. https://doi.org/10.1016/j.spen.2004.07.003.
12. Schaefer GB, Mendelsohn NJ. Genetics evaluation for the etiologic diagnosis of autism spectrum disorders. *Genet Med.* 2008;10(1):4–12. https://doi.org/10.1097/GIM.0b013e31815efdd7.
13. Huguet G, Ey E, Bourgeron T. The genetic landscapes of autism Spectrum disorders. *Annu Rev Genomics Hum Genet.* 2013;14(1):191–213. https://doi.org/10.1146/annurev-genom-091212-153431.
14. Ilieva M, Svenningsen ÅF, Thorsen M, Michel TM. Psychiatry in a dish: stem cells and brain organoids modeling autism spectrum disorders (Asd). *Biol Psychiatry.* 2017;83(3):558–568. https://doi.org/10.1016/j.biopsych.2017.11.011.
15. Redcay E, Courchesne E. When is the brain enlarged in autism? A meta-analysis of all brain size reports. *Biol Psychiatry.* 2005;58(1):1–9. https://doi.org/10.1016/j.biopsych.2005.03.026.
16. Kaushik G, Zarbalis KS. Prenatal neurogenesis in autism spectrum disorders. *Front Chem.* 2016;4:12. https://doi.org/10.3389/fchem.2016.00012.
17. Hazlett HC, Gu H, Munsell BC, et al. Early brain development in infants at high risk for autism spectrum disorder. *Nature.* 2017;542(7641):348–351. https://doi.org/10.1038/nature21369.
18. Courchesne E, Karns CM, Davis HR, et al. Unusual brain growth patterns in early life in patients with autistic disorder: an MRI study. *Neurology.* 2001;57(2):245–254. https://doi.org/10.1212/WNL.57.2.245.
19. Chaste P, Klei L, Sanders SJ, et al. Adjusting head circumference for covariates in autism: clinical correlates of a highly heritable continuous trait. *Biol Psychiatry.* 2013;74(8):576–584. https://doi.org/10.1016/j.biopsych.2013.04.018.
20. Rossignol DA, Frye RE. A review of research trends in physiological abnormalities in autism spectrum disorders: immune dysregulation, inflammation, oxidative stress, mitochondrial dysfunction and environmental toxicant exposures. *Mol Psychiatry.* 2012;17(4):389–401. https://doi.org/10.1038/mp.2011.165.
21. Pizzino G, Irrera N, Cucinotta M, et al. Oxidative stress: harms and benefits for human health. *Oxid Med Cell Longev.* 2017;2017:8416763. https://doi.org/10.1155/2017/8416763.
22. Fridovich I. The biology of oxygen radicals. *Science.* 1978;201(4359):875–880.

23. Kern JK, Geier DA, Sykes LK, Geier MR. Evidence of neurodegeneration in autism spectrum disorder. *Transl Neurodegener*. 2013;2(1):17. https://doi.org/10.1186/2047-9158-2-17.
24. Michel TM, Pulschen D, Thome J. The role of oxidative stress in depressive disorders. *Curr Pharm Des*. 2012;18(36):5890–5899.
25. Ghanizadeh A, Berk M, Farrashbandi H, Alavi Shoushtari A, Villagonzalo KA. Targeting the mitochondrial electron transport chain in autism, a systematic review and synthesis of a novel therapeutic approach. *Mitochondrion*. 2013;13(5):515–519. https://doi.org/10.1016/j.mito.2012.10.001.
26. Li X, Fang P, Mai J, Choi ET, Wang H, Yang X. Targeting mitochondrial reactive oxygen species as novel therapy for inflammatory diseases and cancers. *J Hematol Oncol*. 2013;6:19. https://doi.org/10.1186/1756-8722-6-19.
27. Jastroch M, Divakaruni AS, Mookerjee S, Treberg JR, Brand MD. Mitochondrial proton and electron leaks. *Essays Biochem*. 2010;47:53–67. https://doi.org/10.1042/bse0470053.
28. Chen Q, Vazquez EJ, Moghaddas S, Hoppel CL, Lesnefsky EJ. Production of reactive oxygen species by mitochondria: central role of complex III. *J Biol Chem*. 2003;278(38):36027–36031. https://doi.org/10.1074/jbc.M304854200.
29. Jones DP, Go Y-M. Redox compartmentalization and cellular stress. *Diabetes Obes Metab*. 2010;12(suppl. 2):116–125. https://doi.org/10.1111/j.1463-1326.2010.01266.x.
30. Glasauer A, Chandel NS. Targeting antioxidants for cancer therapy. *Biochem Pharmacol*. 2014;92(1):90–101. https://doi.org/10.1016/j.bcp.2014.07.017.
31. Frustaci A, Neri M, Cesario A, et al. Oxidative stress-related biomarkers in autism: systematic review and meta-analyses. *Free Radic Biol Med*. 2012;52(10):2128–2141. https://doi.org/10.1016/j.freeradbiomed.2012.03.011.
32. Okado-Matsumoto A, Fridovich I. Subcellular distribution of superoxide dismutases (SOD) in rat liver: Cu, Zn-SOD in mitochondria. *J Biol Chem*. 2001;276(42):38388–38393. https://doi.org/10.1074/jbc.M105395200.
33. Lindenau J, Noack H, Possel H, Asayama K, Wolf G. Cellular distribution of superoxide dismutases in the rat CNS. *Glia*. 2000;29(1):25–34.
34. Vergani L, Lanza C, Rivaro P, Abelmoschi ML, Genti S. Metals, metallothioneins and oxidative stress in blood of autistic children. *Res Autism Spectr Disord*. 2011;5:286–293. https://doi.org/10.1016/j.rasd.2010.04.010.
35. Al-Gadani Y, El-Ansary A, Attas O, Al-Ayadhi L. Metabolic biomarkers related to oxidative stress and antioxidant status in Saudi autistic children. *Clin Biochem*. 2009;42(10 – 11):1032–1040. https://doi.org/10.1016/j.clinbiochem.2009.03.011.
36. Zoroglu SS, Armutcu F, Ozen S, et al. Increased oxidative stress and altered activities of erythrocyte free radical scavenging enzymes in autism. *Eur Arch Psychiatry Clin Neurosci*. 2004;254(3):143–147. https://doi.org/10.1007/s00406-004-0456-7.
37. Golse B, Debray-Ritzen P, Durosay P, Puget K, Michelson AM. Alterations in two enzymes: superoxide dismutase and glutathion peroxidase in developmental infantile psychosis (infantile autism) (author's transl). *Rev Neurol*. 1978;134(11):699–705.
38. Kovač J, Macedoni Lukšič M, Trebušak Podkrajšek K, Klančar G, Battelino T. Rare single nucleotide polymorphisms in the regulatory regions of the superoxide dismutase genes in autism Spectrum disorder. *Autism Res*. 2014;7(1):138–144. https://doi.org/10.1002/aur.1345.
39. Lismont C, Nordgren M, Brees C, Knoops B, Van Veldhoven PP, Fransen M. Peroxisomes as modulators of cellular protein thiol oxidation: a new model system. *Antioxid Redox Signal*. 2019;30(1):22–39. https://doi.org/10.1089/ars.2017.6997.
40. Radiq R, Turrensll JF, Changll LY, Bush KM, Crapoll JD, Freeman BA. Detection of catalase in rat heart mitochondria. *J Biol Chem*. 1991;266(32):22028–22034.

41. Margis R, Dunand C, Teixeira FK, Margis-Pinheiro M. Glutathione peroxidase family—an evolutionary overview. *FEBS J*. 2008;275(15):3959–3970. https://doi.org/10.1111/j.1742-4658.2008.06542.x.
42. Meguid NA, Dardir AA, Abdel-Raouf ER, Hashish A. Evaluation of oxidative stress in autism: defective antioxidant enzymes and increased lipid peroxidation. *Biol Trace Elem Res*. 2011;143(1):58–65. https://doi.org/10.1007/s12011-010-8840-9.
43. Pasca SP, Dronca E, Kaucsar T, et al. One carbon metabolism disturbances and the C677T MTHFR gene polymorphism in children with autism spectrum disorders. *J Cell Mol Med*. 2009;13(10):4229–4238. https://doi.org/10.1111/j.1582-4934.2008.00463.x.
44. Yorbik O, Sayal A, Akay C, Akbiyik DI, Sohmen T. Investigation of antioxidant enzymes in children with autistic disorder. *Prostaglandins Leukot Essent Fatty Acids*. 2002;67(5):341–343. https://doi.org/10.1054/plef.2002.0439.
45. Mostafa GA, El-Hadidi ES, Hewedi DH, Abdou MM. Oxidative stress in Egyptian children with autism: relation to autoimmunity. *J Neuroimmunol*. 2010;219(1–2): 114–118. https://doi.org/10.1016/j.jneuroim.2009.12.003.
46. Söğüt S, Zoroğlu SS, Özyurt H, et al. Changes in nitric oxide levels and antioxidant enzyme activities may have a role in the pathophysiological mechanisms involved in autism. *Clin Chim Acta*. 2003;331(1–2):111–117. https://doi.org/10.1016/s0009-8981(03)00119-0.
47. Launay JM, Ferrari P, Haimart M, et al. Serotonin metabolism and other biochemical parameters in infantile autism. *Neuropsychobiology*. 1988;20(1):1–11. https://doi.org/10.1159/000118465.
48. Eklöw L, Thor H, Orrenius S. Formation and efflux of glutathione disulfide studied in isolated rat hepatocytes. *FEBS Lett*. 1981;127(1):125–128. https://doi.org/10.1016/0014-5793(81)80357-2.
49. James SJ, Cutler P, Melnyk S, et al. Metabolic biomarkers of increased oxidative stress and impaired methylation capacity in children with autism. *Am J Clin Nutr*. 2004;80(6):1611–1617.
50. Noble M, Smith J, Power J, Mayer-Pröschel M. Redox state as a central modulator of precursor cell function. *Ann N Y Acad Sci*. 2006;991(1):251–271. https://doi.org/10.1111/j.1749-6632.2003.tb07481.x.
51. Rose S, Melnyk S, Pavliv O, et al. Evidence of oxidative damage and inflammation associated with low glutathione redox status in the autism brain. *Transl Psychiatry*. 2012;2:e134https://doi.org/10.1038/tp.2012.61.
52. Edmonson C, Ziats MN, Rennert OM. Altered glial marker expression in autistic post-mortem prefrontal cortex and cerebellum. *Mol Autism*. 2014;5(1):3. https://doi.org/10.1186/2040-2392-5-3.
53. Andoh-Noda T, Akamatsu W, Miyake K, et al. Differentiation of multipotent neural stem cells derived from Rett syndrome patients is biased toward the astrocytic lineage. *Mol Brain*. 2015;8(1):31. https://doi.org/10.1186/s13041-015-0121-2.
54. Williams EC, Zhong X, Mohamed A, et al. Mutant astrocytes differentiated from Rett syndrome patients-specific iPSCs have adverse effects on wild-type neurons. *Hum Mol Genet*. 2014;23(11):2968–2980. https://doi.org/10.1093/hmg/ddu008.
55. Fiorentino M, Sapone A, Senger S, et al. Blood–brain barrier and intestinal epithelial barrier alterations in autism spectrum disorders. *Mol Autism*. 2016;7(1):49. https://doi.org/10.1186/s13229-016-0110-z.
56. Melnyk S, Fuchs GJ, Schulz E, et al. Metabolic imbalance associated with methylation dysregulation and oxidative damage in children with autism. *J Autism Dev Disord*. 2012;42(3):367–377. https://doi.org/10.1007/s10803-011-1260-7.
57. Abdallah MW, Mortensen EL, Greaves-Lord K, et al. Neonatal levels of neurotrophic factors and risk of autism spectrum disorders. *Acta Psychiatr Scand*. 2013;128(1):61–69. https://doi.org/10.1111/acps.12020.

58. Nickl-Jockschat T, Michel TM. The role of neurotrophic factors in autism. *Mol Psychiatry*. 2011;16(5):478–490. https://doi.org/10.1038/mp.2010.103.
59. Abdallah MW, Larsen N, Mortensen EL, et al. Neonatal levels of cytokines and risk of autism spectrum disorders: an exploratory register-based historic birth cohort study utilizing the Danish newborn screening biobank. *J Neuroimmunol*. 2012;252(1–2):75–82. https://doi.org/10.1016/j.jneuroim.2012.07.013.
60. Walker CK, Krakowiak P, Baker A, Hansen RL, Ozonoff S, Hertz-Picciotto I. Preeclampsia, placental insufficiency, and autism spectrum disorder or developmental delay. *JAMA Pediatr*. 2015;169(2):154–162. https://doi.org/10.1001/jamapediatrics.2014.2645.
61. Roberts JM, Hubel CA. The two stage model of preeclampsia: variations on the theme. *Placenta*. 2009;30(suppl. A):S32–S37. https://doi.org/10.1016/j.placenta.2008.11.009.
62. Reik W, Dean W. DNA methylation and mammalian epigenetics. *Electrophoresis*. 2001;22(14):2838–2843. https://doi.org/10.1002/1522-2683(200108)22:14<2838::AID-ELPS2838>3.0.CO;2-M.
63. Rizwana R, Hahn PJ. CpG methylation reduces genomic instability. *J Cell Sci*. 1999;112(pt. 2):4513–4519.
64. James SJ, Melnyk S, Jernigan S, et al. Metabolic endophenotype and related genotypes are associated with oxidative stress in children with autism. *Am J Med Genet B Neuropsychiatr Genet*. 2006;141B(8):947–956. https://doi.org/10.1002/ajmg.b.30366.
65. James SJ, Melnyk S, Fuchs G, et al. Efficacy of methylcobalamin and folinic acid treatment on glutathione redox status in children with autism. *Am J Clin Nutr*. 2009;89(1):425–430. https://doi.org/10.3945/ajcn.2008.26615.
66. Adams JB, Audhya T, McDonough-Means S, et al. Nutritional and metabolic status of children with autism vs. neurotypical children, and the association with autism severity. *Nutr Metab (Lond)*. 2011;8(1):34. https://doi.org/10.1186/1743-7075-8-34.
67. D'Eufemia P, Finocchiaro R, Celli M, Viozzi L, Monteleone D, Giardini O. Low serum tryptophan to large neutral amino acids ratio in idiopathic infantile autism. *Biomed Pharmacother*. 1995;49(6):288–292. https://doi.org/10.1016/0753-3322(96)82645-X.
68. Hendren RL, James SJ, Widjaja F, Lawton B, Rosenblatt A, Bent S. Randomized, placebo-controlled trial of methyl B12 for children with autism. *J Child Adolesc Psychopharmacol*. 2016;26(9):774–783. https://doi.org/10.1089/cap.2015.0159.
69. Biswas S, Chida AS, Rahman I. Redox modifications of protein–thiols: emerging roles in cell signaling. *Biochem Pharmacol*. 2006;71(5):551–564. https://doi.org/10.1016/j.bcp.2005.10.044.
70. Yan Z, Banerjee R. Redox remodeling as an immunoregulatory strategy. *Biochemistry*. 2010;49(6):1059–1066. https://doi.org/10.1021/bi902022n.
71. Geier DA, Kern JK, Garver CR, Adams JB, Audhya T, Geier MR. A prospective study of transsulfuration biomarkers in autistic disorders. *Neurochem Res*. 2009;34(2):386–393. https://doi.org/10.1007/s11064-008-9782-x.
72. Krajcovicova-Kudlackova M, Valachovicova M, Mislanova C, Hudecova Z, Sustrova M, Ostatnikova D. Plasma concentrations of selected antioxidants in autistic children and adolescents. *Bratisl Lek Listy*. 2009;110(4):247–250.
73. Adams JB, George F, Audhya T. Abnormally high plasma levels of vitamin B6 in children with autism not taking supplements compared to controls not taking supplements. *J Altern Complement Med*. 2006;12(1):59–63. https://doi.org/10.1089/acm.2006.12.59.
74. Adams M, Lucock M, Stuart J, Fardell S, Baker K, Ng X. Preliminary evidence for involvement of the folate gene polymorphism 19bp deletion-DHFR in occurrence of autism. *Neurosci Lett*. 2007;422(1):24–29. https://doi.org/10.1016/j.neulet.2007.05.025.

75. Eto I, Bandy MD, Butterworth CE. Plasma and urinary levels of biopterin, neopterin, and related pterins and plasma levels of folate in infantile autism. *J Autism Dev Disord.* 1992;22(2):295–308. https://doi.org/10.1007/BF01058157.
76. Meguid NA, Hashish AF, Anwar M, Sidhom G. Reduced serum levels of 25-hydroxy and 1,25-dihydroxy vitamin D in Egyptian children with autism. *J Altern Complement Med.* 2010;16(6):641–645. https://doi.org/10.1089/acm.2009.0349.
77. Turrens JF. Mitochondrial formation of reactive oxygen species. *J Physiol.* 2003;552(pt. 2):335–344. https://doi.org/10.1113/jphysiol.2003.049478.
78. Ben-Shachar D. Mitochondrial dysfunction in schizophrenia: a possible linkage to dopamine. *J Neurochem.* 2002;83(6):1241–1251. https://doi.org/10.1046/j.1471-4159.2002.01263.x.
79. Weissman JR, Kelley RI, Bauman ML, et al. Mitochondrial disease in autism spectrum disorder patients: a cohort analysis. Schiffmann R, ed. *PLoS One.* 2008;3(11):e3815. https://doi.org/10.1371/journal.pone.0003815.
80. Chauhan A, Gu F, Essa MM, et al. Brain region-specific deficit in mitochondrial electron transport chain complexes in children with autism. *J Neurochem.* 2011;117(2): 209–220. https://doi.org/10.1111/j.1471-4159.2011.07189.x.
81. Blanchet L, Buydens MC, Smeitink JAM, Willems PHGM, Koopman WJH. Isolated mitochondrial complex I deficiency: explorative data analysis of patient cell parameters. *Curr Pharm Des.* 2011;17(36):4023–4033. https://doi.org/10.2174/138161211798764870.
82. Morava E, van den Heuvel L, Hol F, et al. Mitochondrial disease criteria: diagnostic applications in children. *Neurology.* 2006;67(10):1823–1826. https://doi.org/10.1212/01.wnl.0000244435.27645.54.
83. Rose S, Niyazov DM, Rossignol DA, Goldenthal M, Kahler SG, Frye RE. Clinical and molecular characteristics of mitochondrial dysfunction in autism spectrum disorder. *Mol Diagn Ther.* 2018;22(5):571–593. https://doi.org/10.1007/s40291-018-0352-x.
84. Haas RH. Autism and mitochondrial disease. *Dev Disabil Res Rev.* 2010;16(2):144–153. https://doi.org/10.1002/ddrr.112.
85. Rossignol DA, Frye RE. Mitochondrial dysfunction in autism spectrum disorders: a systematic review and meta-analysis. *Mol Psychiatry.* 2012;17(3):290–314. https://doi.org/10.1038/mp.2010.136.
86. Minshew NJ, Goldstein G, Dombrowski SM, Panchalingam K, Pettegrew JW. A preliminary 31P MRS study of autism: evidence for undersynthesis and increased degradation of brain membranes. *Biol Psychiatry.* 1993;33(11 − 12):762–773. https://doi.org/10.1016/0006-3223(93)90017-8.
87. Khemakhem AM, Frye RE, El-Ansary A, Al-Ayadhi L, Ben BA. Novel biomarkers of metabolic dysfunction is autism spectrum disorder: potential for biological diagnostic markers. *Metab Brain Dis.* 2017;32(6):1983–1997. https://doi.org/10.1007/s11011-017-0085-2.
88. Essa MM, Guillemin GJ, Waly MI, et al. Increased markers of oxidative stress in autistic children of the Sultanate of Oman. *Biol Trace Elem Res.* 2012;147(1–3):25–27. https://doi.org/10.1007/s12011-011-9280-x.
89. Corda S, Laplace C, Vicaut E, Duranteau J. Rapid reactive oxygen species production by mitochondria in endothelial cells exposed to tumor necrosis factor-alpha is mediated by ceramide. *Am J Respir Cell Mol Biol.* 2001;24(6):762–768. https://doi.org/10.1165/ajrcmb.24.6.4228.
90. Deth R, Muratore C, Benzecry J, Power-Charnitsky VA, Waly M. How environmental and genetic factors combine to cause autism: a redox/methylation hypothesis. *Neurotoxicology.* 2008;29(1):190–201. https://doi.org/10.1016/j.neuro.2007.09.010.
91. Zitvogel L, Kepp O, Galluzzi L, Kroemer G. Inflammasomes in carcinogenesis and anticancer immune responses. *Nat Immunol.* 2012;13(4):343–351. https://doi.org/10.1038/ni.2224.

92. Kepp O, Galluzzi L, Zitvogel L, Kroemer G. Pyroptosis—a cell death modality of its kind? *Eur J Immunol.* 2010;40(3):627–630. https://doi.org/10.1002/eji.200940160.
93. Rossignol DA, Frye RE. Evidence linking oxidative stress, mitochondrial dysfunction, and inflammation in the brain of individuals with autism. *Front Physiol.* 2014;5:150. https://doi.org/10.3389/fphys.2014.00150.
94. Muratore CR, Hodgson NW, Trivedi MS, et al. Age-dependent decrease and alternative splicing of methionine synthase mRNA in human cerebral cortex and an accelerated decrease in autism. Phillips W, ed. *PLoS One.* 2013;8(2):e56927. https://doi.org/10.1371/journal.pone.0056927.
95. Li X, Chauhan A, Sheikh AM, et al. Elevated immune response in the brain of autistic patients. *J Neuroimmunol.* 2009;207(1–2):111–116. https://doi.org/10.1016/j.jneuroim.2008.12.002.
96. Evans TA, Siedlak SL, Lu L, et al. The autistic phenotype exhibits a remarkably localized modification of brain protein by products of free radical-induced lipid oxidation. *Am J Biochem Biotechnol.* 2008;4(2):61–72. https://doi.org/10.3844/ajbbsp.2008.61.72.
97. Chauhan A, Audhya T, Chauhan V. Brain region-specific glutathione redox imbalance in autism. *Neurochem Res.* 2012;37(8):1681–1689. https://doi.org/10.1007/s11064-012-0775-4.
98. Tang G, Gutierrez Rios P, Kuo S-H, et al. Mitochondrial abnormalities in temporal lobe of autistic brain. *Neurobiol Dis.* 2013;54:349–361. https://doi.org/10.1016/j.nbd.2013.01.006.
99. Palmieri L, Papaleo V, Porcelli V, et al. Altered calcium homeostasis in autism-spectrum disorders: evidence from biochemical and genetic studies of the mitochondrial aspartate/glutamate carrier AGC1. *Mol Psychiatry.* 2010;15(1):38–52. https://doi.org/10.1038/mp.2008.63.
100. Sajdel-Sulkowska EM, Xu M, McGinnis W, Koibuchi N. Brain region-specific changes in oxidative stress and neurotrophin levels in autism spectrum disorders (ASD). *Cerebellum.* 2011;10(1):43–48. https://doi.org/10.1007/s12311-010-0223-4.
101. Ginsberg MR, Rubin RA, Falcone T, Ting AH, Natowicz MR. Brain transcriptional and epigenetic associations with autism. Landsberger N, ed. *PLoS One.* 2012;7(9): e44736https://doi.org/10.1371/journal.pone.0044736.
102. Vargas DL, Nascimbene C, Krishnan C, Zimmerman AW, Pardo CA. Neuroglial activation and neuroinflammation in the brain of patients with autism. *Ann Neurol.* 2005;57(1):67–81. https://doi.org/10.1002/ana.20315.
103. Gu F, Chauhan V, Chauhan A. Impaired synthesis and antioxidant defense of glutathione in the cerebellum of autistic subjects: alterations in the activities and protein expression of glutathione-related enzymes. *Free Radic Biol Med.* 2013;65:488–496. https://doi.org/10.1016/j.freeradbiomed.2013.07.021.
104. Anitha A, Nakamura K, Thanseem I, et al. Brain region-specific altered expression and association of mitochondria-related genes in autism. *Mol Autism.* 2012;3(1):12. https://doi.org/10.1186/2040-2392-3-12.
105. Anitha A, Nakamura K, Thanseem I, et al. Downregulation of the expression of mitochondrial electron transport complex genes in autism brains. *Brain Pathol.* 2013;23(3):294–302. https://doi.org/10.1111/bpa.12002.
106. Suzuki K, Sugihara G, Ouchi Y, et al. Microglial activation in young adults with autism spectrum disorder. *JAMA Psychiat.* 2013;70(1):49. https://doi.org/10.1001/jamapsychiatry.2013.272.
107. Hossain MM, Khan N, Sultana A, et al. Prevalence of comorbid psychiatric disorders among people with autism spectrum disorder: an umbrella review of systematic reviews and meta-analyses. *Psychiatry Res.* 2020;287:112922. https://doi.org/10.1016/j.psychres.2020.112922.
108. Croen LA, Zerbo O, Qian Y, et al. The health status of adults on the autism spectrum. *Autism.* 2015;19(7):814–823. https://doi.org/10.1177/1362361315577517.

109. Smoller JW, Kendler K, Craddock N, et al. Identification of risk loci with shared effects on five major psychiatric disorders: a genome-wide analysis. *Lancet.* 2013;381(9875):1371–1379. https://doi.org/10.1016/S0140-6736(12)62129-1.
110. Chatterjee M, Saluja R, Kanneganti S, Chinta S, Dikshit M. Biochemical and molecular evaluation of neutrophil NOS in spontaneously hypertensive rats. *Cell Mol Biol (Noisy-le-Grand).* 2007;53(1):84–93.
111. Halliwell B. Role of free radicals in the neurodegenerative diseases. *Drugs Aging.* 2001;18(9):685–716. https://doi.org/10.2165/00002512-200118090-00004.
112. Galle J. Oxidative stress in chronic renal failure. *Nephrol Dial Transplant.* 2001;16(11):2135–2137. https://doi.org/10.1093/ndt/16.11.2135.
113. Rodrigo R, Fernandez-Gajardo R, Gutierrez R, et al. Oxidative stress and pathophysiology of ischemic stroke: novel therapeutic opportunities. *CNS Neurol Disord Drug Targets.* 2013;12(5):698–714. https://doi.org/10.2174/1871527311312050015.
114. Joseph N, Zhang-James Y, Perl A, Faraone SV. Oxidative stress and ADHD. *J Atten Disord.* 2015;19(11):915–924. https://doi.org/10.1177/1087054713510354.
115. Butterfield DA, Drake J, Pocernich C, Castegna A. Evidence of oxidative damage in Alzheimer's disease brain: central role for amyloid beta-peptide. *Trends Mol Med.* 2001;7(12):548–554. https://doi.org/10.1016/s1471-4914(01)02173-6.
116. Zhang M, Zhao Z, He L, Wan C. A meta-analysis of oxidative stress markers in schizophrenia. *Sci China Life Sci.* 2010;53(1):112–124. https://doi.org/10.1007/s11427-010-0013-8.
117. Brown NC, Andreazza AC, Young LT. An updated meta-analysis of oxidative stress markers in bipolar disorder. *Psychiatry Res.* 2014;218(1–2):61–68. https://doi.org/10.1016/j.psychres.2014.04.005.
118. Wu JQ, Kosten TR, Zhang XY. Free radicals, antioxidant defense systems, and schizophrenia. *Prog Neuropsychopharmacol Biol Psychiatry.* 2013;46:200–206. https://doi.org/10.1016/j.pnpbp.2013.02.015.
119. Scaini G, Rezin GT, Carvalho AF, Streck EL, Berk M, Quevedo J. Mitochondrial dysfunction in bipolar disorder: evidence, pathophysiology and translational implications. *Neurosci Biobehav Rev.* 2016;68:694–713. https://doi.org/10.1016/j.neubiorev.2016.06.040.
120. Gardner A, Pagani M, Wibom R, Nennesmo I, Jacobsson H, Hällström T. Alterations of rCBF and mitochondrial dysfunction in major depressive disorder: a case report. *Acta Psychiatr Scand.* 2003;107(3):233–239. https://doi.org/10.1034/j.1600-0447.2003.02188.x.
121. Dean OM, van den Buuse M, Bush AI, et al. A role for glutathione in the pathophysiology of bipolar disorder and schizophrenia? Animal models and relevance to clinical practice. *Curr Med Chem.* 2009;16(23):2965–2976. https://doi.org/10.2174/.092986709788803060.
122. Pearson-Smith J, Patel M. Metabolic dysfunction and oxidative stress in epilepsy. *Int J Mol Sci.* 2017;18(11):2365. https://doi.org/10.3390/ijms18112365.
123. Kunz WS, Kudin AP, Vielhaber S, et al. Mitochondrial complex I deficiency in the epileptic focus of patients with temporal lobe epilepsy. *Ann Neurol.* 2000;48(5):766–773.
124. Sengupta A, Das U, Manna K, et al. An association study of severity of intellectual disability with peripheral biomarkers of disabled children in a rehabilitation home, Kolkata, India. *Sci Rep.* 2019;9(1):13652. https://doi.org/10.1038/s41598-019-49728-3.
125. Bosshard M, Aprigliano R, Gattiker C, et al. Impaired oxidative stress response characterizes HUWE1-promoted X-linked intellectual disability. *Sci Rep.* 2017;7(1):15050. https://doi.org/10.1038/s41598-017-15380-y.

126. Carmeli E, Bachar A, Rom O, Aizenbud D. Oxidative stress and nitric oxide in sedentary older adults with intellectual and developmental disabilities. *Adv Exp Med Biol.* 2015;884:21–27. https://doi.org/10.1007/5584_2015_166.
127. Aksenov V, Long J, Liu J, et al. A complex dietary supplement augments spatial learning, brain mass, and mitochondrial electron transport chain activity in aging mice. *Age (Dordr).* 2013;35(1):23–33. https://doi.org/10.1007/s11357-011-9325-2.
128. Oz M, Lorke DE, Petroianu GA. Methylene blue and Alzheimer's disease. *Biochem Pharmacol.* 2009;78(8):927–932. https://doi.org/10.1016/j.bcp.2009.04.034.
129. Wen Y, Li W, Poteet EC, et al. Alternative mitochondrial electron transfer as a novel strategy for neuroprotection. *J Biol Chem.* 2011;286(18):16504–16515. https://doi.org/10.1074/jbc.M110.208447.
130. Salaris SC, Babbs CF, Voorhees WD. Methylene blue as an inhibitor of superoxide generation by xanthine oxidase. A potential new drug for the attenuation of ischemia/reperfusion injury. *Biochem Pharmacol.* 1991;42(3):499–506. https://doi.org/10.1016/0006-2952(91)90311-r.
131. Kelner MJ, Bagnell R, Hale B, Alexander NM. Potential of methylene blue to block oxygen radical generation in reperfusion injury. *Basic Life Sci.* 1988;49:895–898. https://doi.org/10.1007/978-1-4684-5568-7_146.
132. Naylor GJ, Martin B, Hopwood SE, Watson Y. A two-year double-blind crossover trial of the prophylactic effect of methylene blue in manic-depressive psychosis. *Biol Psychiatry.* 1986;21(10):915–920. https://doi.org/10.1016/0006-3223(86)90265-9.
133. Scalley RD, Conner CS. Acetaminophen poisoning: a case report of the use of acetylcysteine. *Am J Hosp Pharm.* 1978;35(8):964–967.
134. Aruoma OI, Halliwell B, Hoey BM, Butler J. The antioxidant action of N-acetylcysteine: its reaction with hydrogen peroxide, hydroxyl radical, superoxide, and hypochlorous acid. *Free Radic Biol Med.* 1989;6(6):593–597. https://doi.org/10.1016/0891-5849(89)90066-x.
135. Neuwelt EA, Pagel MA, Hasler BP, Deloughery TG, Muldoon LL. Therapeutic efficacy of aortic administration of N-acetylcysteine as a chemoprotectant against bone marrow toxicity after intracarotid administration of alkylators, with or without glutathione depletion in a rat model. *Cancer Res.* 2001;61(21):7868–7874.
136. Farr SA, Poon HF, Dogrukol-Ak D, et al. The antioxidants alpha-lipoic acid and N-acetylcysteine reverse memory impairment and brain oxidative stress in aged SAMP8 mice. *J Neurochem.* 2003;84(5):1173–1183. https://doi.org/10.1046/j.1471-4159.2003.01580.x.
137. Witschi A, Reddy S, Stofer B, Lauterburg BH. The systemic availability of oral glutathione. *Eur J Clin Pharmacol.* 1992;43(6):667–669. https://doi.org/10.1007/BF02284971.
138. Borgström L, Kågedal B. Dose dependent pharmacokinetics of N-acetylcysteine after oral dosing to man. *Biopharm Drug Dispos.* 1990;11(2):131–136. https://doi.org/10.1002/bdd.2510110205.
139. Dean O, Giorlando F, Berk M. N-acetylcysteine in psychiatry: current therapeutic evidence and potential mechanisms of action. *J Psychiatry Neurosci.* 2011;36(2):78–86. https://doi.org/10.1503/jpn.100057.
140. Dean OM, Gray KM, Villagonzalo K-A, et al. A randomised, double blind, placebo-controlled trial of a fixed dose of *N*-acetyl cysteine in children with autistic disorder. *Aust N Z J Psychiatry.* 2017;51(3):241–249. https://doi.org/10.1177/0004867416652735.
141. Wink LK, Adams R, Wang Z, et al. A randomized placebo-controlled pilot study of N-acetylcysteine in youth with autism spectrum disorder. *Mol Autism.* 2016;7(1):26. https://doi.org/10.1186/s13229-016-0088-6.

142. Hardan AY, Fung LK, Libove RA, et al. A randomized controlled pilot trial of oral N-acetylcysteine in children with autism. *Biol Psychiatry*. 2012;71(11):956–961. https://doi.org/10.1016/j.biopsych.2012.01.014.
143. Frye RE, Slattery J, Delhey L, et al. Folinic acid improves verbal communication in children with autism and language impairment: a randomized double-blind placebo-controlled trial. *Mol Psychiatry*. 2018;23(2):247–256. https://doi.org/10.1038/mp.2016.168.
144. Adams JB, Audhya T, Geis E, et al. Comprehensive nutritional and dietary intervention for autism spectrum disorder—a randomized, controlled 12-month trial. *Nutrients*. 2018;10(3):369. https://doi.org/10.3390/nu10030369.
145. Dolske MC, Spollen J, McKay S, Lancashire E, Tolbert L. A preliminary trial of ascorbic acid as supplemental therapy for autism. *Prog Neuropsychopharmacol Biol Psychiatry*. 1993;17(5):765–774. https://doi.org/10.1016/0278-5846(93)90058-Z.
146. Constantino JN, Charman T. Diagnosis of autism spectrum disorder: reconciling the syndrome, its diverse origins, and variation in expression. *Lancet Neurol*. 2016;15(3):279–291. https://doi.org/10.1016/s1474-4422(15)00151-9.

CHAPTER THIRTEEN

In vitro models for ASD-patient-derived iPSCs and cerebral organoids

Sonja Simone Hohmann[a,*], Mirolyuba Ilieva[a,b,c], Tanja Maria Michel[a,b,c]

[a]Psychiatry in the Region of Southern Denmark, Odense University Hospital, Odense, Denmark
[b]Department of Psychiatry, Department of Clinical Research, University of Southern Denmark, Odense, Denmark
[c]BRIDGE—Brain Research—Inter-Disciplinary Guided Excellence, Department of Clinical Research, University of Southern Denmark, Odense, Denmark
*Corresponding author: e-mail address: hohmann.sonja@googlemail.com

Contents

Abstract

Autism spectrum disorder (ASD) is a set of pervasive neurodevelopmental disorders. The causation is multigenic in most cases, which makes it difficult to model the condition *in vitro*. Advances in pluripotent stem cell technology has made it possible to generate *in vitro* models of human brain development. Induced pluripotent stem cells (iPSCs) can be generated from somatic cells and have the ability to differentiate to all of the body's cells. This chapter aims to give an overview of the iPSC technology for generating neural cells and cerebral organoids as models for neurodevelopment and how these models are utilized in the study of ASD. The combination of iPSC technology and the genetic modification tool CRISPR/Cas9 is described, and current limitations and future perspectives of iPSC technology is discussed.

Progress in Molecular Biology and Translational Science, Volume 173
ISSN 1877-1173
https://doi.org/10.1016/bs.pmbts.2020.04.019

1. Modeling human neurodevelopment *in vitro*

Autism or Autism spectrum disorder (ASD) is a genetic highly heterogenous group of disorders characterized by an impairment in social interaction and communication as well as restricted or repetitive behavior and interests.[1] ASD is known to have a genetic predisposition. It is highly heritable, the concordance rate in monogenetic twins is between 60% and 90%.[2] However, idiopathic ASD with unknown causes makes up the majority of cases.[1] The multigenic origin of ASD makes it notoriously difficult to model. The generation of relevant and disease-specific models could lead to novel discoveries of biomarkers, molecular and cellular pathways, and therapeutic targets.[3] To accomplish this, models recapitulating human neurodevelopment are needed.

1.1 Pluripotent stem cells

Human *in vitro* culture systems have traditionally been based on immortalized cell lines with genetic and epigenetic aberrations or patient-derived primary cells.[4] However, the genetic heterogeneity of autism makes it difficult to produce relevant cellular models, and the inaccessibility of the human brain *in vivo* eliminates the possibility of obtaining primary cells. In addition, the fact that the underlying causes of ASD originate in the fetal neurodevelopment, *post-mortem* brain tissue can only provide an incomplete picture of the pathogenesis of ASD.[3,5] Moreover, ASD is an umbrella term for a multitude of neurodevelopmental disorders, and the genetic architecture of these disorders is highly complex and heterogeneous.[6–8] The core symptoms of ASD vary in severity and are often accompanied by other comorbid conditions and diseases, including anxiety, sleep disorders, aggression, intellectual disability.[9] In Timothy syndrome (TS), a neurodevelopmental disorder associated with cognitive impairment, cardiac defects, immune deficiency, and 60% of patients fulfill the criteria for ASD.[10,11] Rett syndrome is a neurodevelopmental disorder associated with regression in development in language in females, while males suffer from severe congenital encephalopathy leading to an early death.[11] Tuberous sclerosis complex (TSC) is another neurodevelopmental disorder caused by mutations in TSC1 and TSC2, the genes coding for hamartin and tuberin, respectively. The clinical features of TSC include intellectual disability and autism.[11]

The pathophysiology of ASD remains elusive, and research has identified more than one thousand genes to be involved in ASD.[12] Also, the underlying genetic architecture of most ASD cases is unknown. This highly complex group of disorders is therefore difficult to model with traditional cell culture technology.[6]

Stem cell models are a promising solution for these challenges, as they can model aspects of brain development *in vitro* and have thus evolved to be a valuable experimental system to understanding disease mechanisms in ASD.[13]

The stem cell repertoire is made up of embryonic stem cells (ESCs) and adult stem cells. Adult stem cells are multipotent, i.e., they have the ability to only differentiate into specific cell types referred to by their tissue origin.[14] ESCs are undifferentiated cells derived from inner cell mass of the blastocyst that possess the stem cell characteristics of self-renewal and pluripotency. Pluripotent stem cells (PSCs), such as ESCs, can differentiate into all three germ layers while maintaining pluripotency, which makes them a valuable tool for *in vitro* models of neurodevelopment.[14] However, the use of human embryos for research is ethically problematic, which brought forth the production of induced pluripotent stem cells (iPSCs). Takahashi and Yamanaka developed a method to reprogram differentiated cells to an embryonic-like state, by introducing the four reprogramming factors Oct3/4, Sox2, c-Myc, and Klf4 into mouse embryonic or adult fibroblasts they successfully induced a pluripotent stem cell state. These factors have been identified among 24 candidate genes as being able to induce and maintain pluripotency in mouse fibroblasts in ES cell culture conditions.[15] The use of iPSCs offers a ground-breaking platform for studying neurodevelopmental disorders, as they can be used to generate neural cells that recapitulate the development of the brain with patient-specific genetic abnormalities.[16] The reprogramming of patient-derived somatic cells (fibroblasts, blood and epithelial cells) to iPSCs allows the *in vitro* study of the abnormal neurodevelopment in disorders such as ASD.

1.2 Generation of PSC-derived neural cells

The iPSC technology has led to insights into relevant cell types involved in disease mechanisms of ASD. An overview of the different iPSC-based methodologies is shown in Table 1. Fig. 1 provides a schematic overview of the utilization of the iPSC technology. With iPSC technology, it is possible to generate neural progenitor cells (NPCs) that give rise to glial and neuronal cells. In ASD research iPSC technology has led to the identification of cell-specific mechanisms underlying ASD.[3,36] Moreover, cell-specific models are

Table 1 Methods for the generation of neural cells and organoids from pluripotent stem cells.

Reference	Procedure	Cell/tissue types	Results
17	No growth factors or inductive signals, no serum, no co-culture	Mouse ESCs to NPCs	Optimization of neuronal and glial cell production from mouse ESCs
18	SFEB method; serum-free; ROCK inhibitor Y27632	hESCs to NPCs (Bf1 + cortical and basal telencephalic progenitors)	ROCK inhibition diminishes apoptosis and increases cloning efficiency
19	SFEBq method for 3D aggregates	Mouse and human ESCs to self-organized apico-basally polarized cortical tissues	Recapitulation of embryonic corticogenesis
20	EB, stromal feeder co-culture; no growth factors	hESCs to NECs	NECs can be maintained in neural induction medium with bFGF for several months until further use
21	No EB, no growth factors; dual SMAD inhibition with Noggin (inhibitor of BMP) and SB431542 (inhibitor of Lefty/Activin/TGFβ)	hESCs to midbrain dopamine and spinal motoneurons	Rapid uniform neural conversion under adherent culture conditions; shorter differentiation periods (19 days) compared to stromal feeder-mediated induction protocols
22	Dual SMAD inhibition; Noggin replaced by dorsomorphin; retinoic acid	hESCs and hiPSCs to cortical projection neurons	The differentiation process recapitulates cortical development; hESCs differentiate to NSCs and NPCs, then to cortical projection neurons that actively fire action potentials, undergo synaptogenesis, and form neural circuits *in vitro*
[23]	Cortical-like 3D culture based on SFEBq; small molecules, growth factors (FGF2, inhibitors of BMP, Wnt/β-catenin, and TGFβ/activin), nodal pathway to induce forebrain fate	hiPSCs to 3D structures containing polarized radial glia, intermediate progenitors, and layer-specific cortical neurons	Enriched transcriptome in transcription factors controlling the specification, growth, and patterning of the dorsal telencephalon, in correlation with human cerebral cortical wall at 8–10 weeks after conception

24	Transdifferentiation with forced induction of Sox2	Mouse and human fibroblasts to iNCs	iNCs express NSC markers and resemble NSCs in morphology, self-renewal, ability to form neurospheres, and gene expression profile
25	Differentiation with forced induction of Ngn2 or NeuroD1	hESCs and hiPSCs to iNCs	Nearly 100% yield and purity in less than 2 weeks; neurons form mature pre- and postsynaptic specializations, and integrate into existing synaptic networks when transplanted into mouse brain
26	Nodal/activin/TGFβ inhibitor, BMP inhibitor	hESCs to neocortex	Self-organized cortical tissue; multi-layered structure: subplate, cortical plate, and Cajal-Retzius cell zones; three progenitor zones: ventricular, subventricular, and intermediate
27	Cerebral organoids; self-organizing capacity of iPSCs, EB formation, embedding in Matrigel; no growth factors	Mouse and human ESCs and iPSCs to cerebral organoids	Discrete brain regions: cerebral cortex with progenitor populations producing mature cortical neuron subtypes; model for microcephaly
28	Differentiation and transdifferentiation with forced induction of ASCL1	Mouse and human fibroblasts and ESCs to iNCs	Neurons express mature neuronal markers, exhibit typical passive and active intrinsic membrane properties, and form functional pre- and postsynaptic structures; primarily excitatory iNCs
29	3D culture; no extracellular matrix; SMAD inhibitor, FGF2, EGF, BDNF, and NT3	hiPSCs to cerebral cortex-like structure (human cortical spheroids (hCSs))	hCSs contain neurons from cortical layers; at 2.5 months transciptionally correlates with mid-fetal prenatal brain (19–24 post-conception weeks); electrophysiologically mature neurons surrounded by astrocytes, functional synapses; only excitatory neurons of the dorsal telencephalon

Continued

Table 1 Methods for the generation of neural cells and organoids from pluripotent stem cells.—cont'd

Reference	Procedure	Cell/tissue types	Results
30	Co-culture of different cell types to model cellular interaction within the developing brain; self-assembled into 3D constructs	hESCs to NPCs, endothelial cells, mesenchymal cells, microglia/macrophage precursors	Transcriptionally express neurogenesis, vasculature development, and microglia lineages
31	Miniaturized spinning bioreactor (SpinΩ); dual SMAD inhibitors (dorsomorphin and A-83); EB embedded in Matrigel; GSK-3β inhibitor (CHIR99021), recombinant WNT3A protein, SMAD inhibitors SB431542 and LDN193189, SHH agonists (recombinant SHH and Purmorphamine), FGF-8	hiPSCs to forebrain-specific organoids	Forebrain organoids oSVZ-like region containing NPCs; all 6 cortical layers; GABAergic neuronal subtypes; midbrain organoids: TH+ cells, $FOXA2^+$, NURR1, and PITX3; hypothalamic lineage
32	dNS and CTraS methods; dual SMAD inhibition with dorsomorphin and SB431542; Wnt agonist CHIR99021	hPSCs to NPCs	CTraS gives rise to cells at the transitional differentiation state that are committed to all three germ layers
33	CRISPR/Cas9 induced deletion of RB1; generation of RB1-KO cerebral organoids with EB formation, ROCK inhibition, Matrigel embedding, orbital shaker	RB1 mutant hESCs to cerebral organoids	RB loss caused cell cycle and pro-apoptotic effects in Sox2+ NSCs and NPCs, and in DCX+ and Tuj1+ neurons; increased S phase entry in DCX+ cells; increased CCNA2 and BAX gene expression; Tuj1+ neuronal migration; upregulated VLDLR gene expression

34	CRISPR/Cas9 induced deletion of PTEN; PTEN mutant cerebral organoid generation by EB formation, Matrigel embedding; no growth factors	PTEN mutant hESCs to cerebral organoids	PTEN mutant cerebral organoids develop a markedly expanded VZ/SVZ, increased overall size, and a larger surface area organized into continuous cortical folds
35	CRISPR/Cas9 induced deletion of NOTCH2NL; cerebral organoids generated by EB formation, small molecules and recombinant proteins: NOGGIN, DKK1, SB431542 (directed diff. based on Ref. 19)	NOTCH2NL mutant hESCs to cortical organoids	Transcriptomic analyses of organoids with NOTCH2NL deletion showed enrichment for genes involved in neuron differentiation

The table summarizes procedures that have been developed for the generation of neural cells and cerebral organoids utilizing mouse and human embryonic stem cells (ESCs) or induced pluripotent stem cells (iPSCs). Cell types used for differentiation, and differentiated cell and tissue types are shown, as well as the major findings and results based on the specific procedure. ASCL1, Achaete-Scute family BHLH transcription factor 1; BAX, BCL2 associated X; BDNF, brain-derived neurotrophic factor; bFGF, basic fibroblast growth factor; BMP, bone morphogenic protein; CCNA2, cyclin A2; CTras, chemically transitional EB-like state; DCX, doublecortin; DKK1, Dickkopf-related protein 1; dNS, direct neurospheres; EB, embryoid body; EGF, epidermal growth factor; FGF-8, fibroblast growth factor 8; FOXA2, forkhead box A2; GSK-3β, glycogen synthase kinase 3 beta; iNC, induced neuronal cell; NEC, neuroepithelial cell; NeuroD1, neuronal differentiation 1; Ngn2, neurogenin 2; NOTCH2NL, Notch 2 N-terminal like; (h)NPC, (human) neural progenitor cell; NT3, neurotrophin 3; NURR1, nuclear receptor related 1; oSVZ, outer subventricular zone; PITX3, pituitary homebox 3; PTEN, phosphatase and tensin homolog; RB1-KO, retinoblastoma 1 knockout; ROCK, Rho associated coiled-coil containing protein kinase; SFEB(q), serum-free culture of EB-like aggregates (quick); SHH, sonic hedgehog; SMAD, mothers against decapentaplegic; Sox2, SRY-box transcription factor 2; TGFβ, transforming growth factor beta; TH, tyrosine hydroxylase; Tuj1, neuron-specific class III β-tubulin; VLDLR, very low density lipoprotein receptor; VZ, ventricular zone; Wnt, wingless-type MMTV integration site; WNT3A, Wnt family member 3A.

a useful tool to identify therapeutic targets and test pharmacological agents and other therapeutics interventions in relevant cell types.[3,13] Early differentiation protocols are based on embryoid body (EB) formation, stromal feeder co-culture or the presence or absence of specific growth factors, and consist of three major steps; aggregation of iPSCs ("embryoid body" formation), differentiation of multipotent primitive neuroepithelial cells (NECs), and generation of region-specific NECs.[20,21] However, the ability of PSCs to spontaneously acquire NPC identity has also been demonstrated by a protocol using no growth factors or inductive signals, and in the absence of serum.[17,37] Another serum-free approach to differentiating mouse EB derived from PSCs to a neural fate has been demonstrated and termed serum-free culture of embryoid body-like aggregates (SFEB). This method has since also been used for human PSCs and was later optimized with a quick-aggregation procedure (SFEBq-modified quick-aggregation SFEB procedure), which has been shown to be more effective for inducing differentiation.[18,19]

In an attempt to avoid using the EB method due to EBs heterogeneous nature, a new approach emerged in 2009. Based on research demonstrating a crucial role for SMAD signaling during neural induction, Chambers et al. created a differentiation method for PSCs to neural cells based on the inhibition of the SMAD signaling pathway.[21]

Protocols using SMAD inhibition have since been optimized and combined with other small molecules and factors that have shown to reduce variation in the differentiation efficiency.[22,32] SMAD inhibition was then combined with inhibition of Wnt signaling to enhance the differentiation and to change the state of PSCs to a chemically transitional EB-like state (CTraS).[32]

In 2013, a different approach has been developed where mouse and human fibroblasts were directly transdifferentiated to induced neuronal cells (iNCs) by the forced induction of different transcription factors.[24,25,28]

A dramatic time difference for generating neural cells has been observed between mouse and human PSCs. Neurons generated from human PSCs are visible approximately 20 days after differentiation, whereas mouse PSCs generate neurons after only 5 days.[17,38,39] However, when studying human features of diseases, it is often desirable to use human models, such as patient-derived iPSCs.[37] The time difference is thought to be an intrinsic feature of the cells, as transplanting human neurons into mice did not accelerate the process of maturation.[40–42] In addition, the timing of iPSC-derived neural cell differentiation is comparable to that seen in the development of the human fetal brain.[37] This observation points out the fact that human cell models are preferable to animal cells in the research of human neurodevelopment. iPSC technology has proven to be a powerful tool for modeling human neurodevelopment.[4] In addition, time differences in differentiation methods and maturation of cells have been reported and are a continuous subject of improvement.[43]

1.3 Utilizing iPSC technology in the study of ASD

1.3.1 Neural progenitor cells

Neural progenitor cells (NPCs) are multipotent cells that are fate restricted. NPCs don't exhibit self-renewal but can give rise to neurons and glial cells.[12,44–46] NPCs play a central role in neurogenesis and are therefore central in the research area of developmental neurobiology.[47] NPCs have been shown to play a role in the pathogenesis of macrocephaly in non-syndromic ASD,[3] as well as the abnormal interplay between Reelin-DAB1 and mTORC1 networks.[48] In addition, NPCs generated from ASD patient-derived iPSCs have been shown to display an altered proliferation.[3] In 2011, patient-specific iPSC-derived cortical NPCs and neurons have led to the discovery of defected calcium signaling in TS. In addition, the study identified TS patient-derived neurons to show abnormal expression of tyrosine hydroxylase and increased production of norepinephrine and dopamine.[10] Recently, NPCs derived from TSC patients were reported to exhibit enlarged cell size and altered proliferation and neurite outgrowth. In this study, a heterozygous iPSC line was generated from a TSC patient harboring a heterozygous mutation of the TSC1 gene. To investigate the role of TSC1 in neurodevelopment, an isogenic wild type (WT) and an isogenic TSC1 homozygous iPSC line were generated using the CRISPR/Cas9 genetic manipulation technology. Differentiated heterozygous and homozygous TSC1 NPCs showed a genotype-dependent increase in cell size, proliferation and neurite outgrowth compared to WT NPCs. Transcriptomic analyses showed differentially expressed genes related to ASD, where up- and downregulation was further increased or decreased in homozygous NPCs compared to heterozygous NPCs.[49]

1.3.2 Neurons

Neurons are generated from NSCs during brain development.[46] A transcriptomic study showed a module of genes related to the synapses to be upregulated in iPSC-derived neurons from patients and unaffected individuals, while it is downregulated in post-mortem brain tissue of patients with ASD. In this study, neurons have been generated by the EB method and the addition of retinoic acid.[50] *Post-mortem* brain tissue studies showed neural networks to be abnormally formed in patients with ASD, and the morphology of neurons to be altered.[12,51–53] Neurons derived from Rett syndrome (RTT) patient iPSCs have shown to have a reduced number of dendritic spines and synapses, and abnormal electrophysiology caused by abnormalities in calcium spikes.[54] Another study combined the EB method with dual SMAD inhibition to generate neurons from iPSCs derived from ASD patients carrying deletions in the PTCHD1-AS gene to show the physiological impacts of these deletions on neurons. In this study, these deletions were shown to have an impact on the neurophysiology of excitatory synapses.[55] SH3- and multiple ankyrin repeats protein (SHANK) mutations are recurrently found in ASD patients and are thus among the subjects of interest in iPSC technology-based ASD research. SHANK proteins are synaptic scaffolding proteins present in postsynaptic densities (PSD) of excitatory neurons.[56,57] iPSC-derived neurons from ASD patients displaying SHANK3 haploinsufficiency showed a significant reduction in dendritic spine densities, and the volumes of whole spines and spine heads.[56] Hyperconnectivity and an altered transcriptome in neurons from ASD patients with SHANK2 mutations have been identified in iPSC-derived neurons. SHANK2-knockout (KO) iPSC line and a gene correction of mutated iPSC lines were generated using CRISPR/Cas9 technology.[58]

1.3.3 Glial cells

Post-mortem and positron-emission tomography (PET)-imaging studies have shown activated microglia in the brains of ASD patients and given their role in neuroinflammation, they are thought to be involved in the pathogenesis of ASD. Moreover, human *post-mortem* ASD brain tissue exhibits an altered synaptic density, which is believed to be caused by dysfunctional synaptic pruning, a process that changes and reduces the number of neurons, synapses and axons, and microglia.[59,60]

Astrocytes are a type of glial cells with ectodermal, neuroepithelial origin and are involved in homeostasis and defense mechanisms in the central nervous system (CNS). They play essential regulatory roles in neurotransmitter recycling, synaptogenesis, and a supportive role for neurons and the formation of neural networks.[12,61]

Reactive astrocytes have been identified to be the main source of cytokines in *post-mortem* brain tissue of autistic patients.[59] Astrocytes derived from RTT patient iPSCs caused a reduced size of soma and number of terminal ends, and shorter neurite length in co-cultured neurons.[62] More recently, a large computational study of *post-mortem* brain tissue from individuals with ASD showed astrocytic debris to be present in the cerebrospinal fluid (CSF)-brain barrier, suggesting astrocytes to be involved in ASD pathology.[63] Astrocytes generated by the EB method and dual SMAD inhibition from iPSCs derived patients with ASD have shown to have an increased level of the pro-inflammatory cytokine interleukin-6 (IL-6).[1,64] Oligodendrocytes derived from TSC patient iPSCs have been shown to enhance the increased network activity and increased dendritic branching of neurons in co-culture. Moreover, TSC oligodendrocytes showed a decreased maturation, but increased proliferation.[65]

1.4 The progression from 2D to 3D models

While cell type-specific models have granted insight into disease mechanisms of ASD, 2D culture models of the developing brain are limited in their ability to recapitulate the interaction of different cell types in the developing brain, because 2D cultures can mainly be used to generate homogenous cell populations.[66] However, in order to be able to study the full aspect of ASD, the model system must be able to provide insight into cell-cell interactions of different cell types. The emergence of 3D cell cultures holds this potential, as they are able to more accurately reflect the cellular "ecosystems" of complex tissues.[67]

Three-dimensional (3D) cell cultures emerged in 2011 with the first self-organizing optic cups from human PSCs in floating culture.[68] The 3D self-organizing properties were found to be provided by transferring cells to Matrigel droplets, a supportive extracellular matrix.[27] Since then, iPSCs have been shown to self-organize into a variety of brain regions, such as hindbrain, midbrain, forebrain, and retinal tissues.[27,37,66,69,70] Due to this variety, these tissues have been termed cerebral organoids (from Latin "of or relating to the brain").

Initially, organoid technology was based on the generation of embryoid bodies (EB) from ESC. EBs are aggregates of pluripotent stem cells and are generated by detaching ESC colonies from the culture plate to form aggregates that float in the medium. In the presence of neural induction medium, the EBs form a structure that is called neural tube-like rosettes due to the morphological resemblance to the neural tube. At this stage, the cells express neuroectodermal markers.[20] The Matrigel approach to generating cerebral organoids is based on the aforementioned protocol, where the neuroectodermal marker-expressing neural tube-like rosettes are embedded in Matrigel droplets to provide a growth matrix and placed in a spinning bioreactor to increase oxygen and nutrient absorption. This method leads to rapid formation of cerebral organoids. Cerebral organoids have since been successfully generated from both ESCs and iPSCs.[27]

In 2015, human cerebral spheroids (hCS) have been generated by the combination of the Matrigel approach and dual SMAD inhibition.[21,29] Cerebral organoids have been shown to express several markers specific to brain regions. Immunohistochemical staining of whole organoids revealed the presence of forebrain, midbrain and hindbrain regions as well as regional sub specification.[27] In addition, reverse transcriptase PCR (RT-PCR) of whole organoids revealed pluripotency markers to decrease while forebrain markers remained highly expressed during organoid development, suggesting a recapitulation of human fetal brain development.[27] Interestingly, the regionalization of cerebral organoids is entirely based on self-organization. Differentiation protocols for the induction of cerebral organoids that are patterned to specific brain regions and tissue identities allow for the generation of, e.g., midbrain or hypothalamic organoids.[71,72] These patterned organoids can be co-cultured to generate combined cerebral organoids, as has been shown in a study fusing dorsal and ventral forebrain organoids.[73] However, the generation of patterned organoids requires the addition of growth factors, whereas self-organizing whole-brain organoids rely on intrinsic mechanisms.[5,27]

Immunohistochemical staining of cerebral organoids have shown the presence of cells with glial identities that increase with time, and cells with astrocytic and oligodendrocytic morphology have been identified.[74] The cellular diversity has also been confirmed by single cell RNA sequencing that confirmed the presence of NPCs, neurons, and glial cells in cerebral organoids.[75]

1.5 Utilizing cerebral organoid models in the study of ASD

Although ASD research utilizing the cerebral organoid technology is still in its infancy, it is developing quickly and has already brought forth some findings related to ASD. In 2015, Mariani et al. generated cerebral organoids derived from four families that included a family member affected by ASD and an increased head circumference. One to three healthy first-degree family members served as controls. Transcriptomic and pathway analyses of ASD-derived cerebral organoids revealed a differential gene expression of genes involved in GABAergic inhibitory neuron development, cell proliferation, neuronal differentiation, and synaptic assembly compared to control organoids.[76]

The chromodomain helicase DNA-binding protein 8 (CHD8) has been identified in multiple studies to be a candidate gene for ASD.[77] In a transcriptomic study, CRISPR/Cas9 technology was used to generate an $CHD8^{+}/^{-}$ iPSC line that was subsequently used to generate cerebral organoids. RNA sequencing of cerebral organoids showed that CHD8 regulates the expression of genes implicated in ASD. Remarkably, a significant overlap of differentially expressed genes identified in this study shows overlap with the DEGs found in the Mariani study.[76,77]

In a recent study RAB39b-associated macrocephaly/autism, cerebral organoids were generated and found to have a dysregulated PI3K-AKT-mTOR signaling that leads to a dysregulation of cortical neurogenesis, which contributes to macrocephaly and autistic-like behaviors.[78]

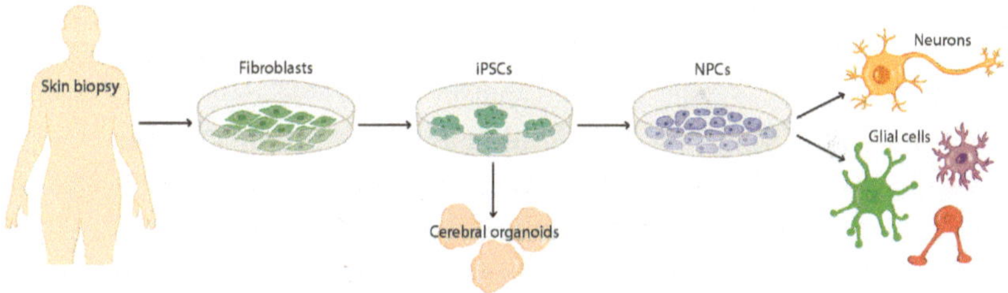

Fig. 1 iPSC technology for the generation of patient-derived neural cells and cerebral organoids. Schematic overview of the procedure. A skin biopsy is taken from patients and unaffected individuals. Fibroblasts are cultured and reprogrammed to induced pluripotent stem cells (iPSCs) by introducing the four reprogramming factors Oct3/4, Sox2, c-Myc, and Klf4. iPSCs form colonies in culture form aggregates (embryoid bodies (EBs)). EBs can be differentiated to neural progenitor cells (NPCs) and subsequently to neural cells by the addition of growth factors, small molecules, introduction of transcription factors, or the absence of serum. The inhibition of the SMAD signaling pathway makes it possible to bypass the EB-state. Transdifferentiation of fibroblasts to neural cells through the forced induction of different transcription factors is possible. Cerebral organoids can be generated from iPSCs. Neural induction medium induces EBs to form neural-like rosettes that can be embedded in droplets of Matrigel, a supporting extracellular matrix. In a spinning bioreactor, cerebral organoids are rapidly formed by self-organization. *Courtesy of Eva Frederikke Høy Helms.*

2. CRISPR/Cas9 and iPSC technologies

Genome editing tools, such as CRISPR/Cas9, are increasingly used in combination with iPSC technology, and it has been useful for generating iPSC lines where the only genetic difference between lines is the disease-causing mutation.[4,37,79,80] CRISPR/Cas9 is a nuclease-based system that is used to induce targeted mutations and gene insertions. Cas9 is a nuclease that is targeted to a specific genomic locus by a guide RNA and then causes double-strand breaks to activate endogenous DNA repair processes.[81] In the field of iPSC and cerebral organoids technology, CRISPR/Cas9 has been mostly utilized for the genetic modification of cerebral organoid founder cells.[81] In a study utilizing CRISPR/Cas9 technology, a human retinoblastoma1 knock out (RB1-KO) ESC line was generated to investigate the role of RB1 in human neurodevelopment.[82] One study investigating human cortical growth and folding utilized CRISPR/Cas9 technology to generate an hESC line that was used to generate genetically modified cerebral organoids.[34] Another CRISPR/Cas9 modified hESC-derived cerebral organoid model was generated to investigate the role of human-specific NOTCH2NL in the evolution of a larger human neocortex. NOTCH2NL duplications have been shown to be involved in autism and macrocephaly, and the study found NOTCH2NL deletion to increase neuron differentiation.[35,83] Genetically modifying cerebral organoid founder cells has the advantage of generating cerebral organoids with all cells containing the modification. However, CRISPR/Cas9 technology can also be utilized in a later stage of cerebral organoid generation.[81] By co-injecting Ras and TP53 plasmids close to the surface of 4 months-old cerebral organoids and subsequent electroporation-mediated transfection, a cerebral organoid-based glioblastoma model was generated. Both plasmids contained fluorescent markers that showed take-over of genetically modified cells of the organoid beginning 2 weeks post-electroporation.[84] In a concept paper, CRISPR/Cas9 genome editing has recently been proposed to be used to tag iPSC-derived neural cells by inserting a fluorescent reporter cassette to trace the fate of specific neural cells during development.[85] CRISPR/Cas9-mediated gene editing can be utilized to generate cell lines with patient-specific mutations or to "repair" patient-specific mutations to generate an isogenic control cell line.[81] In the field of ASD research, this is a major advantage.

3. Limitations and future directions

Despite the benefits of the use of iPSC technology as a model system of neurodevelopment, it also comes with some limitations. One of the main limitations of iPSC-based studies is the limited amount of patient-derived cell lines that limit the possibility of quantitative analyses.[5] Future improvements of the cerebral organoid system should address scalability. One way of improving the usability of the iPSC technology on a larger scale is to generate biobanks of cells derived from patients and unaffected individuals to overcome the challenge of statistically insignificant numbers of cell lines. Russo et al. generated a biobank with about 300 stem cell lines obtained from the tooth pulp of ASD patients.[12] Another biobank with more than 200 fibroblast, iPSC, NSC, and glial cell lines was created from ASD patients and normal unaffected volunteers as controls.[86]

Another major limitation of the cerebral organoid technology is the organoid-to-organoid variability. Self-organizing cerebral organoids display a difference in cellular composition between organoids and batches, reflecting the intrinsic heterogeneity within cerebral organoid models.[66,75,87] This sample-to-sample variability limits quantitative analyses and is thought to be caused in part by the variation in the number of neural rosettes that are formed during the differentiation process.[88] This challenge has been addressed by developing protocols for generating brain area-specific organoids.[72] Moreover, 3D bioprinting has been proposed to revolutionize the field by increasing the throughput and reproducibility of cerebral organoid generation.[89] 3D bioprinting uses cells and biocompatible materials as an ink (bioink) to build 3D tissues, and it is possible to use more than one cell type, which allows for accurate replication of the complexity of brain tissue.[89,90] This method holds the potential to reduce organoid-to-organoid variability by being able to precisely control the structure of cerebral organoids.[89] As another source of variability, the bioreactor-based batch effects are mentioned.[66,72] The bioreactor approach of generating cerebral organoids is based on NASA-developed technology and simulates microgravity that has the potential to increase cell-cell interactions by reducing shear stress as well as induction of differentiation.[91,92] In addition, it increases oxygen diffusion and thereby reduces apoptosis. Miniaturized bioreactors have been produced to decrease the volume of media, which makes the method more applicable and facilitates scaling.[66]

Limitations of iPSC-based *in vitro* models for ASD include the lack of vascularization for better transportation of oxygen and nutrients to avoid inner necrosis and the lack of nutrient exchange.[5,16] To overcome this challenge, cerebral organoids have been successfully transplanted into mouse cortex, where they exhibited enhanced survival and robust vascularization.[93] Another approach has shown endothelial vascular growth factor (VEGF) supplementation to successfully promote blood vessel formation in cerebral organoids *in vitro*.[94] Incorporating circulating blood in current model system will enhance the use of iPSC technology in modeling ASD.[16]

Despite the current limitations of cerebral organoid systems, they display a higher complexity and homogeneity compared to 2D models and thus offer a wider and more complex picture of heterogeneous disorders such as ASD.[5] Technologies such as bioprinting and genetic modification with CRISPR/Cas9 hold the potential to combat the current limitations of cerebral organoid reproducibility and variability.[90,95]

References

1. Russo FB, Freitas BC, Pignatari GC, et al. Modeling the interplay between neurons and astrocytes in autism using human induced pluripotent stem cells. *Biol Psychiatry*. 2018;83(7):569–578.
2. Jamain S, Radyushkin K, Hammerschmidt K, et al. Reduced social interaction and ultrasonic communication in a mouse model of monogenic heritable autism. *Proc Natl Acad Sci U S A*. 2008;105(5):1710–1715.
3. Marchetto MC, Belinson H, Tian Y, et al. Altered proliferation and networks in neural cells derived from idiopathic autistic individuals. *Mol Psychiatry*. 2017;22(6):820–835.
4. Beneto N, Cozar M, Castilla-Vallmanya L, et al. Neuronal and astrocytic differentiation from Sanfilippo C syndrome iPSCs for disease modeling and drug development. *J Clin Med*. 2020;9(3):644.
5. Ilieva M, Fex Svenningsen A, Thorsen M, Michel TM. Psychiatry in a dish: stem cells and brain organoids modeling autism spectrum disorders. *Biol Psychiatry*. 2018;83(7):558–568.
6. Aigner S, Heckel T, Zhang JD, Andreae LC, Jagasia R. Human pluripotent stem cell models of autism spectrum disorder: emerging frontiers, opportunities, and challenges towards neuronal networks in a dish. *Psychopharmacology (Berl)*. 2014;231(6):1089–1104.
7. Masi A, DeMayo MM, Glozier N, Guastella AJ. An overview of autism spectrum disorder, heterogeneity and treatment options. *Neurosci Bull*. 2017;33(2):183–193.
8. Sharma SR, Gonda X, Tarazi FI. Autism spectrum disorder: classification, diagnosis and therapy. *Pharmacol Ther*. 2018;190:91–104.
9. Liu X, Campanac E, Cheung HH, et al. Idiopathic autism: cellular and molecular phenotypes in pluripotent stem cell-derived neurons. *Mol Neurobiol*. 2017;54(6):4507–4523.
10. Pasca SP, Portmann T, Voineagu I, et al. Using iPSC-derived neurons to uncover cellular phenotypes associated with Timothy syndrome. *Nat Med*. 2011;17(12):1657–1662.
11. Adams JW, Cugola FR, Muotri AR. Brain organoids as tools for modeling human neurodevelopmental disorders. *Physiology (Bethesda)*. 2019;34(5):365–375.

12. Russo FB, Brito A, de Freitas AM, Castanha A, de Freitas BC, Beltrao-Braga PCB. The use of iPSC technology for modeling autism spectrum disorders. *Neurobiol Dis.* 2019;130:104483.
13. Young-Pearse TL, Morrow EM. Modeling developmental neuropsychiatric disorders with iPSC technology: challenges and opportunities. *Curr Opin Neurobiol.* 2016;36:66–73.
14. Das AK, Pal R. Induced pluripotent stem cells (iPSCs): the emergence of a new champion in stem cell technology-driven biomedical applications. *J Tissue Eng Regen Med.* 2010;4(6):413–421.
15. Takahashi K, Yamanaka S. Induction of pluripotent stem cells from mouse embryonic and adult fibroblast cultures by defined factors. *Cell.* 2006;126(4):663–676.
16. Freitas BC, Mei A, Mendes APD, Beltrao-Braga PCB, Marchetto MC. Modeling inflammation in autism spectrum disorders using stem cells. *Front Pediatr.* 2018;6:394.
17. Ying QL, Stavridis M, Griffiths D, Li M, Smith A. Conversion of embryonic stem cells into neuroectodermal precursors in adherent monoculture. *Nat Biotechnol.* 2003;21(2):183–186.
18. Watanabe K, Ueno M, Kamiya D, et al. A ROCK inhibitor permits survival of dissociated human embryonic stem cells. *Nat Biotechnol.* 2007;25(6):681–686.
19. Eiraku M, Watanabe K, Matsuo-Takasaki M, et al. Self-organized formation of polarized cortical tissues from ESCs and its active manipulation by extrinsic signals. *Cell Stem Cell.* 2008;3(5):519–532.
20. Xia X, Zhang SC. Differentiation of neuroepithelia from human embryonic stem cells. *Methods Mol Biol.* 2009;549:51–58.
21. Chambers SM, Fasano CA, Papapetrou EP, Tomishima M, Sadelain M, Studer L. Highly efficient neural conversion of human ES and iPS cells by dual inhibition of SMAD signaling. *Nat Biotechnol.* 2009;27(3):275–280.
22. Shi Y, Kirwan P, Livesey FJ. Directed differentiation of human pluripotent stem cells to cerebral cortex neurons and neural networks. *Nat Protoc.* 2012;7(10):1836–1846.
23. Mariani J, Simonini MV, Palejev D. Modeling human cortical development in vitro using induced pluripotent stem cells. *Proc Natl Acad Sci USA.* 2012;109(31): 12770–12775. https://doi.org/10.1073/pnas.1202944109.
24. Ring KL, Tong LM, Balestra ME, et al. Direct reprogramming of mouse and human fibroblasts into multipotent neural stem cells with a single factor. *Cell Stem Cell.* 2012;11(1):100–109.
25. Zhang M, Scholer HR, Greber B. Rapid and efficient generation of neurons from human pluripotent stem cells in a multititre plate format. *J Vis Exp.* 2013;(73):e4335.
26. Kadoshima T, Sakaguchi H, Nakano T. Self-organization of axial polarity, inside-out layer pattern, and species-specific progenitor dynamics in human ES cell-derived neocortex. *Proc Natl Acad Sci USA.* 2013;110(50):20284–20289. https://doi.org/10.1073/pnas.1315710110.
27. Lancaster MA, Renner M, Martin CA, et al. Cerebral organoids model human brain development and microcephaly. *Nature.* 2013;501(7467):373–379.
28. Chanda S, Ang CE, Davila J, et al. Generation of induced neuronal cells by the single reprogramming factor ASCL1. *Stem Cell Rep.* 2014;3(2):282–296.
29. Pasca AM, Sloan SA, Clarke LE, et al. Functional cortical neurons and astrocytes from human pluripotent stem cells in 3D culture. *Nat Methods.* 2015;12(7):671–678.
30. Schwartz MP, Hou Z, Propson NE. Human pluripotent stem cell-derived neural constructs for predicting neural toxicity. *Proc Natl Acad Sci USA.* 2015;112(40): 12516–12521. https://doi.org/10.1073/pnas.1516645112.
31. Qian X, Nguyen HN, Song MM. Brain-region-specific organoids using mini-bioreactors for modeling ZIKV exposure. *Cell.* 2016;165(5):1238–1254. https://doi.org/10.1016/j.cell.2016.04.032.

32. Fujimori K, Matsumoto T, Kisa F, Hattori N, Okano H, Akamatsu W. Escape from pluripotency via inhibition of TGF-beta/BMP and activation of Wnt signaling accelerates differentiation and aging in hPSC progeny cells. *Stem Cell Rep*. 2017;9(5): 1675–1691.
33. Matsui T, Nieto-Estevez V, Kyrychenko S, Schneider JW, Hsieh J. Retinoblastoma protein controls growth, survival and neuronal migration in human cerebral organoids. *Development*. 2017;144(6):1025–1034. https://doi.org/10.1242/dev.143636.
34. Li Y, Muffat J, Omer A, et al. Induction of expansion and folding in human cerebral organoids. *Cell Stem Cell*. 2017;20(3):385–396.e383.
35. Fiddes IT, Lodewijk GA, Mooring M, et al. Human-specific NOTCH2NL genes affect Notch signaling and cortical neurogenesis. *Cell*. 2018;173(6):1356–1369.e1322.
36. Martinez-Cerdeno V, Noctor SC. Neural progenitor cell terminology. *Front Neuroanat*. 2018;12:104.
37. Kelava I, Lancaster MA. Stem cell models of human brain development. *Cell Stem Cell*. 2016;18(6):736–748.
38. Watanabe K, Sasai Y. In vitro differentiation of telencephalic precursors from ES cells. *Tanpakushitsu Kakusan Koso*. 2005;50(6 Suppl):711–716.
39. Lancaster MA, Knoblich JA. Generation of cerebral organoids from human pluripotent stem cells. *Nat Protoc*. 2014;9(10):2329–2340.
40. Espuny-Camacho I, Michelsen KA, Gall D, et al. Pyramidal neurons derived from human pluripotent stem cells integrate efficiently into mouse brain circuits in vivo. *Neuron*. 2013;77(3):440–456.
41. Maroof AM, Keros S, Tyson JA, et al. Directed differentiation and functional maturation of cortical interneurons from human embryonic stem cells. *Cell Stem Cell*. 2013;12(5): 559–572.
42. Nicholas CR, Chen J, Tang Y, et al. Functional maturation of hPSC-derived forebrain interneurons requires an extended timeline and mimics human neural development. *Cell Stem Cell*. 2013;12(5):573–586.
43. Deneault E, White SH, Rodrigues DC, et al. Complete disruption of autism-susceptibility genes by gene editing predominantly reduces functional connectivity of isogenic human neurons. *Stem Cell Rep*. 2019;12(2):427–429.
44. Hsieh J, Aimone JB, Kaspar BK, Kuwabara T, Nakashima K, Gage FH. IGF-I instructs multipotent adult neural progenitor cells to become oligodendrocytes. *J Cell Biol*. 2004;164(1):111–122.
45. Stiles J, Jernigan TL. The basics of brain development. *Neuropsychol Rev*. 2010;20(4):327–348.
46. Arai Y, Taverna E. Neural progenitor cell polarity and cortical development. *Front Cell Neurosci*. 2017;11:384.
47. Gage FH, Temple S. Neural stem cells: generating and regenerating the brain. *Neuron*. 2013;80(3):588–601.
48. Sanchez-Sanchez SM, Magdalon J, Griesi-Oliveira K, et al. Rare RELN variants affect Reelin-DAB1 signal transduction in autism spectrum disorder. *Hum Mutat*. 2018;39(10):1372–1383.
49. Martin P, Wagh V, Reis SA, et al. TSC patient-derived isogenic neural progenitor cells reveal altered early neurodevelopmental phenotypes and rapamycin-induced MNK-eIF4E signaling. *Mol Autism*. 2020;11:2.
50. Griesi-Oliveira K, Fogo MS, BGG P, et al. Transcriptome of iPSC-derived neuronal cells reveals a module of co-expressed genes consistently associated with autism spectrum disorder. *Mol Psychiatry*. 2020;.
51. Belichenko PV, Hagberg B, Dahlstrom A. Morphological study of neocortical areas in Rett syndrome. *Acta Neuropathol*. 1997;93(1):50–61.

52. Armstrong DD, Dunn K, Antalffy B. Decreased dendritic branching in frontal, motor and limbic cortex in Rett syndrome compared with trisomy 21. *J Neuropathol Exp Neurol.* 1998;57(11):1013–1017.
53. Sajdel-Sulkowska EM, Xu M, McGinnis W, Koibuchi N. Brain region-specific changes in oxidative stress and neurotrophin levels in autism spectrum disorders (ASD). *Cerebellum.* 2011;10(1):43–48.
54. Marchetto MC, Carromeu C, Acab A, et al. A model for neural development and treatment of Rett syndrome using human induced pluripotent stem cells. *Cell.* 2010;143(4):527–539.
55. Ross PJ, Zhang WB, Mok RSF, et al. Synaptic dysfunction in human neurons with autism-associated deletions in PTCHD1-AS. *Biol Psychiatry.* 2020;87(2):139–149.
56. Gouder L, Vitrac A, Goubran-Botros H, et al. Altered spinogenesis in iPSC-derived cortical neurons from patients with autism carrying de novo SHANK3 mutations. *Sci Rep.* 2019;9(1):94.
57. Jiang YH, Ehlers MD. Modeling autism by SHANK gene mutations in mice. *Neuron.* 2013;78(1):8–27.
58. Zaslavsky K, Zhang WB, McCready FP, et al. SHANK2 mutations associated with autism spectrum disorder cause hyperconnectivity of human neurons. *Nat Neurosci.* 2019;22(4):556–564.
59. Vargas DL, Nascimbene C, Krishnan C, Zimmerman AW, Pardo CA. Neuroglial activation and neuroinflammation in the brain of patients with autism. *Ann Neurol.* 2005;57(1):67–81.
60. Hutsler JJ, Zhang H. Increased dendritic spine densities on cortical projection neurons in autism spectrum disorders. *Brain Res.* 2010;1309:83–94.
61. Ben Haim L, Rowitch DH. Functional diversity of astrocytes in neural circuit regulation. *Nat Rev Neurosci.* 2017;18(1):31–41.
62. Williams EC, Zhong X, Mohamed A, et al. Mutant astrocytes differentiated from Rett syndrome patients-specific iPSCs have adverse effects on wild-type neurons. *Hum Mol Genet.* 2014;23(11):2968–2980.
63. DiStasio MM, Nagakura I, Nadler MJ, Anderson MP. T lymphocytes and cytotoxic astrocyte blebs correlate across autism brains. *Ann Neurol.* 2019;86(6):885–898.
64. Lee CH, Ingrole RSJ, Gill HS. Generation of induced pluripotent stem cells using elastin like polypeptides as a non-viral gene delivery system. *Biochim Biophys Acta Mol Basis Dis.* 2020;1866(4):165405.
65. Nadadhur AG, Alsaqati M, Gasparotto L, et al. Neuron-glia interactions increase neuronal phenotypes in tuberous sclerosis complex patient iPSC-derived models. *Stem Cell Rep.* 2019;12(1):42–56.
66. Quadrato G, Brown J, Arlotta P. The promises and challenges of human brain organoids as models of neuropsychiatric disease. *Nat Med.* 2016;22(11):1220–1228.
67. Rowe RG, Daley GQ. Induced pluripotent stem cells in disease modelling and drug discovery. *Nat Rev Genet.* 2019;20(7):377–388.
68. Danjo T, Eiraku M, Muguruma K, et al. Subregional specification of embryonic stem cell-derived ventral telencephalic tissues by timed and combinatory treatment with extrinsic signals. *J Neurosci.* 2011;31(5):1919–1933.
69. Iefremova V, Manikakis G, Krefft O, et al. An organoid-based model of cortical development identifies non-cell-autonomous defects in Wnt signaling contributing to miller-Dieker syndrome. *Cell Rep.* 2017;19(1):50–59.
70. Bershteyn M, Nowakowski TJ, Pollen AA, et al. Human iPSC-derived cerebral organoids model cellular features of lissencephaly and reveal prolonged mitosis of outer radial glia. *Cell Stem Cell.* 2017;20(4):435–449.e434.
71. Monzel AS, Smits LM, Hemmer K, et al. Derivation of human midbrain-specific organoids from neuroepithelial stem cells. *Stem Cell Rep.* 2017;8(5):1144–1154.

72. Qian X, Nguyen HN, Jacob F, Song H, Ming GL. Using brain organoids to understand Zika virus-induced microcephaly. *Development*. 2017;144(6):952–957.
73. Bagley JA, Reumann D, Bian S, Levi-Strauss J, Knoblich JA. Fused cerebral organoids model interactions between brain regions. *Nat Methods*. 2017;14(7):743–751.
74. Renner M, Lancaster MA, Bian S, et al. Self-organized developmental patterning and differentiation in cerebral organoids. *EMBO J*. 2017;36(10):1316–1329.
75. Dang J, Tiwari SK, Agrawal K, Hui H, Qin Y, Rana TM. Glial cell diversity and methamphetamine-induced neuroinflammation in human cerebral organoids. *Mol Psychiatry*. 2020;.
76. Mariani J, Coppola G, Zhang P, et al. FOXG1-dependent dysregulation of GABA/glutamate neuron differentiation in autism spectrum disorders. *Cell*. 2015;162(2):375–390.
77. Wang P, Mokhtari R, Pedrosa E, et al. CRISPR/Cas9-mediated heterozygous knockout of the autism gene CHD8 and characterization of its transcriptional networks in cerebral organoids derived from iPS cells. *Mol Autism*. 2017;8:11.
78. Zhang W, Ma L, Yang M, et al. Cerebral organoid and mouse models reveal a RAB39b-PI3K-mTOR pathway-dependent dysregulation of cortical development leading to macrocephaly/autism phenotypes. *Genes Dev*. 2020;34:580–597.
79. Doudna JA, Charpentier E. Genome editing. The new frontier of genome engineering with CRISPR-Cas9. *Science*. 2014;346(6213):1258096.
80. Hockemeyer D, Jaenisch R. Induced pluripotent stem cells meet genome editing. *Cell Stem Cell*. 2016;18(5):573–586.
81. Fischer J, Heide M, Huttner WB. Genetic modification of brain organoids. *Front Cell Neurosci*. 2019;13:558.
82. Matsui T, Nieto-Estevez V, Kyrychenko S, Schneider JW, Hsieh J. Retinoblastoma protein controls growth, survival and neuronal migration in human cerebral organoids. *Development*. 2017;144(6):1025–1034.
83. Bernier R, Steinman KJ, Reilly B, et al. Clinical phenotype of the recurrent 1q21.1 copy-number variant. *Genet Med*. 2016;18(4):341–349.
84. Ogawa J, Pao GM, Shokhirev MN, Verma IM. Glioblastoma model using human cerebral organoids. *Cell Rep*. 2018;23(4):1220–1229.
85. Filice F, Schwaller B, Michel TM, Grunblatt E. Profiling parvalbumin interneurons using iPSC: challenges and perspectives for autism spectrum disorder (ASD). *Mol Autism*. 2020;11(1):10.
86. Brick DJ, Nethercott HE, Montesano S, et al. The autism spectrum disorders stem cell resource at Children's Hospital of Orange County: implications for disease modeling and drug discovery. *Stem Cells Transl Med*. 2014;3(11):1275–1286.
87. Lancaster MA, Corsini NS, Wolfinger S, et al. Guided self-organization and cortical plate formation in human brain organoids. *Nat Biotechnol*. 2017;35(7):659–666.
88. Yang G, Shcheglovitov A. Probing disrupted neurodevelopment in autism using human stem cell-derived neurons and organoids: an outlook into future diagnostics and drug development. *Dev Dyn*. 2020;249(1):6–33.
89. de la Vega L, Lee C, Sharma R, Amereh M, Willerth SM. 3D bioprinting models of neural tissues: the current state of the field and future directions. *Brain Res Bull*. 2019;150:240–249.
90. Salaris F, Rosa A. Construction of 3D in vitro models by bioprinting human pluripotent stem cells: challenges and opportunities. *Brain Res*. 2019;1723:146393.
91. Goodwin TJ, Schroeder WF, Wolf DA, Moyer MP. Rotating-wall vessel coculture of small intestine as a prelude to tissue modeling: aspects of simulated microgravity. *Proc Soc Exp Biol Med*. 1993;202(2):181–192.

92. Goodwin TJ, Prewett TL, Wolf DA, Spaulding GF. Reduced shear stress: a major component in the ability of mammalian tissues to form three-dimensional assemblies in simulated microgravity. *J Cell Biochem.* 1993;51(3):301–311.
93. Daviaud N, Friedel RH, Zou H. Vascularization and engraftment of transplanted human cerebral organoids in mouse cortex. *eNeuro.* 2018;5(6):ENEURO.0219-18.2018.
94. Ham O, Jin YB, Kim J, Lee MO. Blood vessel formation in cerebral organoids formed from human embryonic stem cells. *Biochem Biophys Res Commun.* 2020;521(1):84–90.
95. Salaris F, Colosi C, Brighi C, et al. 3D bioprinted human cortical neural constructs derived from induced pluripotent stem cells. *J Clin Med.* 2019;8(10):1595.

Index

Note: Page numbers followed by "*f*" indicate figures and "*t*" indicate tables.

E

F

Q

R

S

FSC
www.fsc.org
MIX
Papier | Fördert
gute Waldnutzung
FSC® C083411

Kolibri 360 GmbH
Ferdinand-Jühlke-Straße 7
99095 Erfurt, Deutschland
produktsicherheit@kolibri360.de